Gerhard Schlemmer, Jan Schlemmer

Instrumental Analysis

Also of interest

Elemental Analysis.
An Introduction to Modern Spectrometric Techniques
Schlemmer G, Balcaen, Todolí, Hinds, 2019
ISBN 978-3-11-050107-0, e-ISBN 978-3-11-050108-7

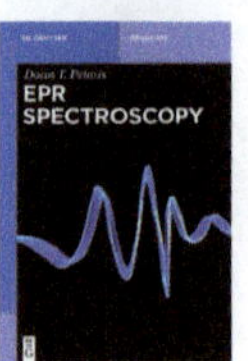

EPR Spectroscopy
Petasis, Hendrich, 2022
ISBN 978-3-11-041753-1, e-ISBN 978-3-11-041756-2

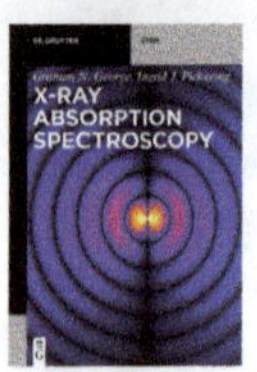

X-ray Absorption Spectroscopy
George, Pickering, 2023
ISBN 978-3-11-057037-3, e-ISBN 978-3-11-057044-1

Forensic Chemistry.
Fundamentals
Grossman, 2021
ISBN 978-3-11-071878-2, e-ISBN 978-3-11-071881-2

Bioanalytical Chemistry.
From Biomolecular Recognition to Nanobiosensing
Ugo, Marafini, Meneghello, 2021
ISBN 978-3-11-058909-2, e-ISBN 978-3-11-058916-0

Gerhard Schlemmer,
Jan Schlemmer

Instrumental Analysis

Chemical IT

DE GRUYTER

Authors
Dr. Gerhard Schlemmer
99423 Weimar
Germany

Dr. Jan Schlemmer
88250 Weingarten
Germany

ISBN 978-3-11-068964-8
e-ISBN (PDF) 978-3-11-068966-2
e-ISBN (EPUB) 978-3-11-068972-3

Library of Congress Control Number: 2022933836

Bibliographic information published by the Deutsche Nationalbibliothek
The Deutsche Nationalbibliothek lists this publication in the Deutsche Nationalbibliografie; detailed bibliographic data are available on the Internet at http://dnb.dnb.de.

Cover image: Jan Schlemmer, Gerhard Schlemmer
Typesetting: Integra Software Services Pvt. Ltd.
Printing and binding: CPI books GmbH, Leck

www.degruyter.com

Contents

1 Analysis of matter: IT for life sciences, natural sciences, technological progress, and quality assurance

Karl R. Popper was a philosophers of science and one of the most prominent philosophers of the twentieth century. He defined a criterion to distinguish between science and pseudoscience. A theory is called scientific if it makes statements and/or prognoses which can be principally verified or falsified or where it is possible to show under which boundary conditions the theory does no longer hold. A theory which cannot be falsified by operating experience or research is pseudoscience. Verification and falsification are fundamental elements of analytical sciences

1.1 The question

The treasurer in the royal bank wants to know how close the gold content in the bullions comes to 100%. The medical doctor needs to understand how the gold content in the blood of the patient treated with anti-rheumatism medicine decays with time. The director of the metallurgical company needs to be sure that the chromate content in the wastewater is below the level set by decree, and the customs officer must be sure that 100,000 tons of oil in the supertanker do not contain too much sulfur. Millions of samples need to be tested every day, millions of questions are asked to the analytical laboratory. Connected with the answer are issues about immense monetary values, purposeful medical treatment, litigable evidence, or a solid base for legislation concerning the environment or consumer protection. In many cases the correct analytical information is of utmost importance, in other cases it may be completely worthless. The responsibility of the analytical laboratory is outrageous; however, the responsibility to ask the question correctly is of utmost importance as well.

1.1.1 Define what is to be measured

Setting up an analytical method is an important process. What molecules or elements need to be determined? Is there a product specification or legislation that defines what species are required to be determined? If not, then meet with stakeholders to find out what is the exact analytical question and why the analysis is run. The why is important. If specifications are wrong, the result will be unsatisfactory for the client. The analyst in many cases can support the specification process substantially. He ought to know the specifics of techniques and methods.

https://doi.org/10.1515/9783110689662-001

1.1.2 How important is this analysis?

Most analyses have some importance. Some of the analyses, however, are more important than others (monetary value, time sensitive medical decisions, final product quality, etc.). The resources spent (time, labor, money) are generally aligned with the importance of the results from the analysis. It may also dictate how fast the samples have to get to the lab for analysis and how fast the analysis must be done.

1.1.3 What is the sample, how is it sampled, and how does it get to the lab?

Knowledge of the sample is essential. What is the sample composition (solid, liquid, or gas)? This will have a bearing on sample preparation. Who takes the sample and what methods and devices are used is of vital concern. This can be the single biggest source of error in the whole analysis. Although sampling is outside the scope of this book, it is recommended to consult relevant literature about it [1, 2]. The other question that needs to be answered is: how does the sample get to the laboratory and who has the responsibility for this activity? Defining this is important to attaining a smooth and timely flowing operation.

1.1.4 Analytical diversity

Analytical information will consist of measured values. These may look similar, but their reliability might be different. Depending on the method used to obtain them they will be more or less suitable (timewise, economically, with respect to complexity) to provide the relevant information. The analyst should be able to judge on the best method or select between the ones he has at hand. However, in addition to the requirements of the stakeholder discussed above, an expectation concerning analytical quality is a must. The latter specification is called "analytical figures of merit, FOM [3, 4]." Figures of merit are quality criteria. They are defined by standards and are an essential part of the answer.

> *Alchemists turned into chemists when they stopped keeping secrets.*
>
> (Eric S. Raymond, author, software developer)

1.2 The answer

What is the most important aspect of analytical data? The immediate answer will be that they must be accurate. Stakeholders as well as analytical chemists want to give the correct reply to the inquiry. However, it needs to be exactly defined what is meant by the term “accurate.” Even when the term is clearly defined, a real proof of accuracy is often very difficult. First and foremost, the integrity of the sample coming to the lab may be questionable, second the suitable sample preparation for the selected technique will play an important role for the result obtained by the analytical instrument. Third, the analytical instrument may be biased by the composition of the sample. The techniques described in this book are to a greater or lesser extent sensitive to these so-called “matrix effects.” The term “accuracy” includes both the trueness of the result and the range where the result can be expected. It must be made clear that the demand on accuracy depends very strongly on the question. For the control of most results from a soil analysis it may be enough to state, that the value determined is clearly and far below a threshold level. If the result is very close to the limit, the demand on accuracy becomes high. If the concentration of gold is determined in a bullion, the demand on accuracy will always be extremely high.

1.2.1 Measurand

The original information obtained from an analytical instrument – the measurand – is a figure, read out from an electronic device. It is usually an electrical quantity (e.g., a digitalized current or voltage) which is entirely dependent on the design of the instrument and algorithms used during the measurement process. In many cases, there is not even a physical equation behind the process which allows to use theory for calibration of the apparatus. In a later paragraph it will be shown, how many parameters and processes can influence this single number. Most of the techniques described in this book are based on relative measurement systems. This means, that the sample, during, before or after measurement is compared to standards of known mass or concentration. The relation between measurand and mass or concentration is called calibration function. Few of the techniques come close to the demand of an absolute measurement device, where a read-out value can be clearly related to an analyte mass or concentration without applying standards. To give a simple example of an absolute measurement device: a balance can be gauged with a single standard to be accurate between a minimum and a maximum value with a certain standard deviation. Once calibrated, the indicated mass is the accurate result for a specified amount of time. From everyday life we are used to an important property of such absolute instruments: we can use the instrument to measure the quantity of interest, e.g., the mass of a chicken or the mass of grain with the same procedure. However, this is generally no longer true for relative instruments, where we therefore need to discuss the notion of trueness.

1.2.2 Trueness

Provided the calibration solution is representative for the chemical species to be analyzed, the measurand is connected to a concentration or mass of the analyte. It may be expressed in mg/L, ng, % of the sample, percent of agreement to an expectation etc. However, the sample composition, may bias the value in the sample compared to the reference solution. This question is extremely important and one of the major complications in analytical work. In fact, the composition of the sample will often have an influence on the sensitivity (the term will be explained below) or cause a systematic bias to the measurand. Samples are often unknown in composition, and they may have a wide span in concentration. Human serum as an example is close in composition while urine can vary substantially. Wastewater may be simple or extremely complex while tap waters will be much more similar. It is therefore difficult to compose standards which come close to a variety of samples to be determined routinely. Methods to compensate for the bias are numerous in analytical chemistry. They will be mentioned in the following paragraphs, and an overview will be given in the sections presenting the individual analytical techniques.

1.2.3 Repeatability and precision

We know that measured values of the same target quantity are slightly different. We quantitate that using the statistical quantity "standard deviation." Which sources we include may differ from case to case, however. The standard deviation of results obtained from the same sample by different techniques, different types of sample preparation, different operators, will usually be significantly higher than that obtained from one instrument within a very short time obtained from one sample prepared by the same operator. We are using the terms precision and repeatability to distinguish. It is important to mention that each technique is capable of providing a characteristic "best" repeatability under optimal, conditions. Techniques which are able to do the same type of analysis, e.g., ICP-OES and ICP-MS may provide different repeatability. Repeatability may therefore be a selection criterion for the analytical technique. It must be emphasized, that repeatability, of course, depends very strongly on the concentration of the sample for measurement relative to the detection capability of the selected technique. The function relating repeatability and concentration is different for different techniques. Understanding the relevant sources of error at, e.g., low and high concentrations is very important for method development. Tuning for best repeatability will cost time and money as the measurement conditions will be usually more sophisticated, and lengthy. Statistically trivial is the fact that averaging over a bigger number of measurements will provide better repeatability. Optimizing for best precision will be more challenging as we must consider sample preparation steps and the human error factor as well. If we combine the information of an unbiased result (trueness) with the repeatability, taking the number of repetitions into account one will be able to make a statement on measurement accuracy.

1.2.4 Sensitivity

Sensitivity is an output response to an input number of atoms or molecules to be measured. This response may be blown up (electronically or physicochemically) or it may be attenuated. Sensitivity becomes meaningful only if we compare it with the baseline noise (the standard deviation) at "zero" mass or concentration. Only then it will show how significant the result is with respect to repeatability. The sensitivity of the techniques under discussion are orders of magnitude different. Some of them use arbitrary numbers (intensities), or, in the case of absorption methods, a physically defined magnitude (absorbance). Sensitivity is often used as a quality criterion for an analytical technique, but it should never be over-emphasized. Not seldom the analytical conditions become much more stable if method development is optimized towards ruggedness (see below) rather than towards sensitivity. If I stands for the quantity indicated by the instrument when measuring a known concentration c, the sensitivity S at this very concentration is expressed by eq. (1.1).

Equation 1.1: Sensitivity of an analytical measurement:

$$S = I/c \tag{1.1}$$

S depends on instrument settings. S and c are directly proportional only for a certain mass or concentration range (see below).

1.2.5 Limit of detection

Are we able to detect the analyte qualitatively or quantitatively? How many individual results do we need to decide? How is our detection limit connected with precision and hence with accuracy? The limit of detection is clearly defined by international standards [5]. If our zero measurement is zero and our sample does not distort sensitivity and precision, the determination of the detection limit or the limit of quantitation is an easy task. We need to know the repeatability of the blank, and the reading of a standard close to the limit of quantitation. If the conditions defined above are fulfilled, the easiest equation to define the detection limit is eq. (1.2).

Equation 1.2: Simplified calculation of the detection limit of an analytical method:

$$C_{\mathrm{dl}} = 3\sigma_{\mathrm{bl}}/S \tag{1.2}$$

C_{dl} is the concentration at the detection limit, σ_{bl} is the standard deviation of a blank reading with net intensity 0, and S is the sensitivity determined with a standard close to the detection limit. The determination of the detection limit becomes more reliable if the confidence bands of a 10-point calibration curve, spanning a range from the limit of quantitation (LOQ) to 10 times LOQ are used. The detection limits of the techniques published by the instrument manufacturers and those attainable in the laboratory may differ substantially. Limit of detection is certainly an important criterion to select an instrumental technique. But one should always keep in mind that the technique with the lowest detection limit may not be the best one for the application!

1.2.6 Calibration

The techniques discussed in this textbook are relative (1.2.1). Analytical measurements must be calibrated.

The intensity of the measurand I is therefore related with the help of reference solutions to concentration. The corresponding terms are sensitivity S and slope of a calibration curve.

Favored algorithm in international standards is a linear calibration function (eq. (1.3)).

Equation 1.3: Linear calibration function of an analytical measurement:

$$I = b_0 + b_1 c \tag{1.3}$$

I is the intensity reading, b_0 is the theoretical intersection of the calibration curve at zero concentration. Ideally it matches the absorbance of a reference solution with zero addition of the analyte under consideration (blank solution); b_1 is the slope of the line; c is the concentration added to the reference solutions. Figure 1.1 features a typical linear calibration curve. The example is taken from an elemental determination using Atomic Absorption Spectrometry.

The basis for the linear calibration curve is the simple linear regression. A number n of reference solutions with added concentrations c_1, c_2,. . . are related to the instrument output I_1, I_2, . . ., n related pairs (c_n-I_n) are fit in a way that a linear curve describes best their position in the two-dimensional plane.

Prerequisite for a linear regression is obviously that b_1 is strictly independent of the concentration. This is usually the case within a certain concentration range which depends strongly on the physical principle behind the technique. Emission techniques usually have a wider linear span compared to absorption techniques. Linear calibration curves usually allow best accuracy if the criterion of linearity is followed strictly. If b_1 is changing (usually it becomes smaller) with increasing c, systematic errors by the regression algorithm will be introduced. In this case nonlinear algorithms are in many cases the much better solution for analytical quality (see Section 1.2.7).

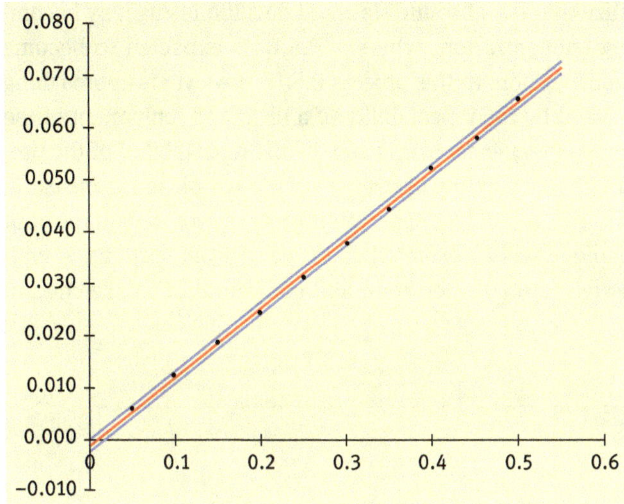

Figure 1.1: Linear calibration curve. The measurand (absorbance) is plotted against analyte concentration.

It has been shown that in many of the techniques described in this text, the frequency of random measurements usually follows a Gaussian distribution. Although the number of measurements is usually not big enough to fully justify this statistical approach, the Gaussian shaped distribution is assumed to be a suitable approach for data evaluation, and it is the base of most analytical standards. The mean value of a measurement with n replicates under this assumption is defined by eq. (1.4). The standard deviation is calculated by eq. (1.5).

Equation 1.4: Mean value of a measurement with n replicates:

$$\bar{X} = \frac{1}{n}\sum_{i=1}^{n} x_i \tag{1.4}$$

Equation 1.5: Standard deviation of a measured value X under the assumption of a Gaussian distribution:

$$s = \sqrt{\frac{\sum_{i=1}^{n}(x_i - \bar{x})^2}{n-1}} \tag{1.5}$$

Relative standard deviation (R.S.D.) is the standard deviation related to the mean value of the measurement. R.S.D. is often indicated in % of the mean value (eq. (1.6)). Equation 1.6: Relative standard deviation of a measurement:

$$\text{R.S.D.} = s/\bar{x} \tag{1.6}$$

It is very important to realize that the absolute standard deviation is usually becoming larger with increasing concentration, whereas R.S.D. is expected to become smaller with increasing concentration. In this process R.S.D., however is approaching a minimum. This is determined by the repeatability of a technical component of the system, which is *concentration independent*. This may be the repeatability of the process of analyte generation, the repeatability of the feed of the sample to the measurement cell, the repeatability of the analyte separation process, etc. This important characteristic of each technique will be exemplified in the relevant sections. A typical plot which relates relative standard deviation and concentration is sketched in Figure 1.2.

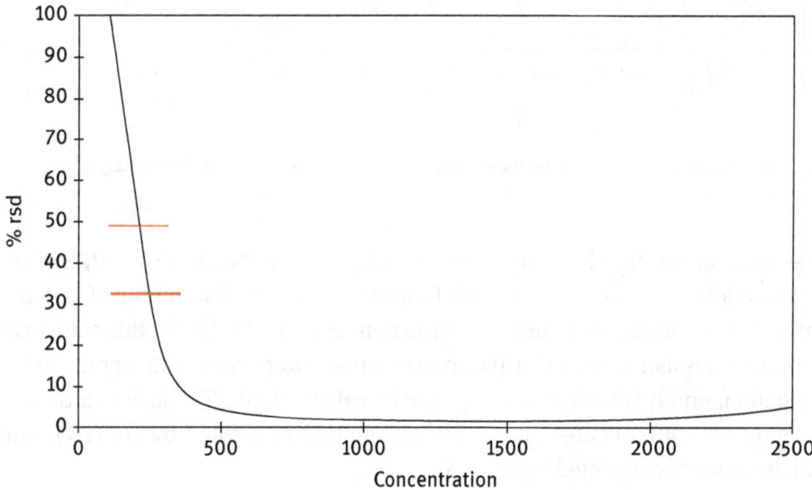

Figure 1.2: Relative standard deviation in % as a function of the concentration exemplified. The red bars indicate the limits of detection using the 2σ and the 3σ criterion.

Standard deviation combined with sensitivity yields the respective repeatability depending on concentration or mass of analyte. This value indicates at which concentration the measurement is capable of distinguishing between presence or absence of the analyte under investigation, at which level a quantitative determination becomes possible, and at which level a measurement can be run with defined repeatability of, e.g., 10% r.s.d. or better. Important terms used *are limit of detection (l.o.d.)* and *limit of quantitation (l.o.q.)*

1.2.7 Working range

Many technical and chemical effects may influence the sensitivity at higher concentration. These may include spectroscopic effects such as the spectral overlap of absorbance and emission profiles of the source. Saturation effects of certain species may become visible. Carry-over in flow parts of the equipment, or chemical effects during formation or decay of species may play a role. Examples will be discussed in the respective sections of the individual techniques. The first indication of such effects is non-linearity of the calibration curve. Usually, it will bend toward the concentration axis; the sensitivity drops with increasing concentration. Sometimes the curve may be bent toward the ordinate (intensity axis) as well, or it may become s-shaped. When approaching this region, the linear regression model must no longer be used. This is of utmost importance if the method of standard additions is used (see section 1.2.8). In case of non-linearity, a curve with two coefficients is the much more suitable model.

Equation 1.7: 2-Coefficient equation for the description of non-linear calibration curves.

$$I = b_0 + b_1 c + b_2 c^2 \tag{1.7}$$

This type of curve is often available in the instrument software. In Figure 1.3 a non-linear calibration curve is plotted.

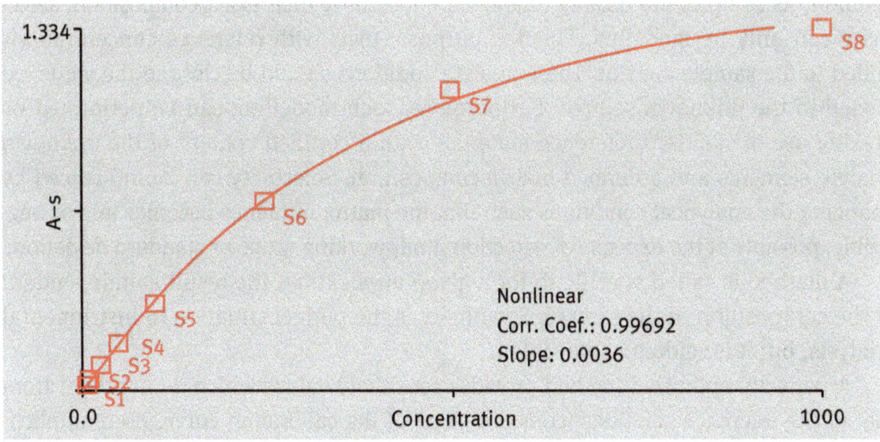

Figure 1.3: Non-linear calibration in atomic absorption spectroscopy. Absorbance plotted against concentration.

We have defined the lower part of the working range using important terms such as "limit of detection," "limit of quantitation." The upper part of the working range, however, is not defined by figures of merit. At very high concentrations the slope of the calibration curve often becomes very small. A reasonable suggestion for defining the upper range of the working curve may therefore be based on sensitivity or on reciprocal normalized sensitivity (see Section 2.6). If the sensitivity is, e.g., 1.5 times lower than in the linear portion of the curve, even calibration curves with two coefficients should no longer be used. Another criterion might be a significant increase in R.S.D. at high concentrations.

1.2.8 Selectivity, specificity, interferences

The analyte species are often a trace or ultra-trace component of the composition of the sample. Many effects, generated by one or several major constituents of the sample (matrix), may therefore influence the response of an analytical measurement. The influence may be chemical, physical, or spectroscopic. Thus, it may affect the entire sample preparation and measurement process. The result may be a constant positive or negative effect on the intensity (additive effect) or it may bias the slope of the calibration curve by a constant factor (multiplicative effect). The effect is called *interference* if it cannot be calibrated properly. It will then distort the trueness of the result. The analytical process of method development is almost always centered around compensation of matrix effects by using suitable chemical and technical means and/or calibration methods. A method is called selective, if sensitivity and/on standard deviation of the measurement of the unknown sample does not differ from the data of the calibration solution. As samples are usually unknown, concerning their exact composition, selectivity can only be modelled. For this purpose, tests with reference concentrations added to the sample are run. These analyte additions should be close to the value expected in the unknown sample. Furthermore, such modelling can be performed by making use of standard reference materials with a certified content of the unknown analyte elements and a defined matrix composition. Selectivity can be influenced by changing the analytical conditions such that the matrix influence becomes less or negligible, possibly at the expense of detection limit, working range or standard deviation.

A method is called specific if, for a given application, the result is independent of the composition of the sample. Specificity is the perfect situation in instrumental analysis, but it is seldom achievable.

A perfectly optimized method provides sensitivity values which do not differ from that of the reference solutions (constant slope of the calibration curve, no multiplicative effects) and do not result in a positive or negative bias parallel to the measurand axis (additive effect). In this case the result of the measurement is expected to be true; the certified concentration of a standard reference material will be found within the statistical limits. Trueness is the lack of systematic errors of a measurement. Superimposed to possible systematic errors are the random errors discussed above. These random errors may of course be influenced by the matrix as well. Trueness and precision will result in a possible total bias relative to the "true value." This is understood by the term "accuracy" which includes systematic errors and random errors.

Instead of calibrating with simple calibration solutions, the analyte can be added to the unknown samples and the result can be calculated from the calibration curve which includes the unknown concentration in the sample. This method is called *standard addition*, and it is a recommended procedure in standard methods. It must be emphasized though, that trueness and accuracy will benefit only if

- the calibration curve is strictly linear
- the systematic error is multiplicative and not additive
- the added concentrations are very similar to the unknown concentration.

The final aim of an analytical measurement would be to use a reference which allows to compare the result from an unknown sample with this reference and verify or correct the result based on this standard. This principle is hardly possible in analytical chemistry. Reference materials with a composition as close as possible to the unknown sample are accepted as a proof for trueness and accuracy, though they only constitute a probable validation. The stated concentrations in standard reference materials are determined in extensive interlaboratory measurement projects using analytical methods based on different physicochemical principles. Traceability is based on these standards.

1.2.9 Ruggedness

A method is called "rugged" if small changes in the measurement conditions, such as temperature, gas flows, photon intensity, and exact chemical composition of the sample, will not introduce systematic or random errors. These will remain constant within defined boundaries of the measurement conditions. Ruggedness can be tested by a deliberate slight change of the measurement conditions. Monitoring of the result and the standard deviations under different conditions will show whether these changes influence repeatability, trueness, or accuracy. Long-term stability can be monitored by running a control sample repeatedly over time.

1.2.10 Time of analysis

Time of analysis consists of many factors. The workflow in the laboratory is as important as sample preparation. The pursued analytical quality (FOM) plays an important a part as the time for preparing and servicing of the analytical instruments, and the time required once the equipment is running automatically and unattended. Focusing on instruments, a few main questions need to be answered (the following list is just a selection of the most important ones):

- How many different analyte elements or molecules per sample need to be determined?
- How many samples must be run in an attended day shift or in an unattended night shift?
- What are the requirements for each of the analytes with respect to limit of quantitation, reproducibility, control of trueness?
- How much sample mass or volume is available for the total analysis?
- How long is the automated operation time before activity of an operator is required?
- Can all analytes be determined with only one instrumental technique out of one sample vessel?

Only very well-equipped laboratories will be able to distribute the samples between instruments to perfectly meet optimized analysis time. In most cases trade-offs will be necessary. It must be kept in mind that

- Instruments running without being productive are often generating cost (e.g., ICP techniques; see Chapter 2).
- Extensive automated quality control may extend time and cost of analysis significantly.
- The time for controlling the results, which were previously acquired automatically, is usually substantial as well, and must not be disregarded when planning the total analysis time.

1.2.11 Importance of the results

The importance of the results will assist in guiding:

- the number of replicates analyzed for the sample
- the number of quality controls to be analyzed
- the limits for the quality controls, and how frequently the quality controls are analyzed.

1.2.12 Post analysis details

At the end of the day, a lot of data from an analytical working day or from a night shift are printed out and/or saved on the computer. They may as well be resident on a Laboratory Information Management System (LIMS). Modern instrumentation usually includes plenty of functions which control the analytical figures of merit. Still, the critical assessment of the protocol by the responsible analyst is obligatory. The final report will probably be much shorter than the entire protocol of the analysis and should include only the necessary information for the client. The figures in the final protocol should be reported as physically reasonable values including error and confidence information.

Even then, i.e., when the results are controlled and reported in the correct way, it may be necessary to keep the samples for possible re-assessment in case of doubt. The procedure of sample storage and/or disposal is usually defined in the operating procedures and in the quality assurance handbook of the laboratory. While planning the analysis, including the technique used for the determinations, it must be secured that enough sample is kept for re-assessment.

1.3 The race between scientific idea, technical realization, application, and international standard

Does a physical effect like emission of radiation, absorption of energy, fluorescence, adsorption to a sorbent, and movement of ions in an electric field have the potential to be used in analytical chemistry? The basic idea of many modern analytical techniques has been developed at scientific institutes. The appropriate analytical instruments are developed and specified in the R&D departments of analytical instrument manufacturers. The specifications are often set by chemists or physicists and the instruments are designed by mechanical and electronic engineers. The software package, the interface to the user, is written by software engineers. The prototypes and the preproduction instrument units are tested and verified by analytical chemists and, finally, the products are sold by people with diverging expert knowledge. The user of analytical instruments is predominantly concerned with the solution of an analytical problem with high-quality, acceptable input of laboratory time, level of expert skills, financial investment, and cost of ownership. It becomes obvious that extensive communication between academia, industry, and analytical laboratories, as well as between marketing, R&D, production, and customer support inside the companies is required to end up with the optimal tools for a given set of analytical problems.

The research and development resources of instrument manufacturers for testing and developing new techniques are limited. R&D is usually used to capacity with improvement of existing equipment or updating known techniques with improved and cost-reduced equipment. The new ideas must show the potential of a significant improvement over existing techniques and methods before a project in the R&D department is started. This improvement may be a significant benefit in analytical quality such as a new type of information, much lower detection limits, much less sample requirement, reduced analysis time, lower cost of analysis, or portable instruments. One example may be the laser-induced plasma spectroscopy (LIPS) as compared to inductively coupled plasma optical emission spectroscopy (ICP-OES). Often a proof-of-principle phase is started in cooperation with a research institute which is partly funded by industry or public research projects. Not seldom small companies or start-ups will design the first product before the market leaders of analytical instrumentation start their own development project as well.

Once an idea made the cut to an industrial project, the development objects change significantly. In addition to the specified analytical quality, production cost, estimated sales price, safety, ease of use, cost of operation, serviceability, and finally appearance become important. Industrial development projects are much, much more expensive than academic research projects and the proof of principle [6]. Among the important deliverables to the future customer are extensive technical documentation such as instruction manual, service manual, and at least a few application notes. The latter is usually referred to as "cookbook." Methods must be available which allow the user to judge on the basic instrumental integrity. This procedure is called instrument performance verification (IPV) and it is a part of method validation [7].

The self-conception of science and R&D in academia and industry is and must be significantly divergent. In Table 1.1, a few of the substantial differences are listed.

Table 1.1: Selected criteria for academic and industrial R&D.

Character	Academia	Industry
Incentive	Scientific interest/publication	Market needs/unique offering
Invented device	Research tool	Product
Components	Selected hand-crafted components	Commercially available parts
Success criteria	Outstanding results from limited number of experiments	Long-term stability/ruggedness of target performance
Benchmark	Fine-tuned unique specimen	Average performing system
Appearance	Breadboard	Designed product
Economy	–	Binding cost specification

The first commercial product is usually introduced by one manufacturing company only. Some of the techniques may even never find another following producer. This will limit the application of the technique to selected fields or laboratories or analytical research. New instrumental techniques will be standardized to local or international norms or standards only, if more than one commercial supplier stands behind the product. As a consequence, the new technique will first find its way into research institutes where new fields of application can be investigated and published. Future oriented laboratories with application fields not strictly regulated, may use the new technique to offer more attractive services. This starting phase of making use of a new technique in the market is very important, though. Limitations and problems of the technique are usually found and reported only by investigation of different sample types and various analytical requirements. These results will then foster technical improvement and new developments. It usually takes more than a decade until a new technique becomes technically and application wise mature.

International standards for the analysis of species x in matrix y are a way of standardizing analytical knowledge. Standards simplify method development for laboratories and increase the confidence into the analytical results. Local and international bodies are staffed by analysts of experienced laboratories, members of regulative bodies, academic scientists, and specialists from instrument manufacturing companies. These are meeting in set time intervals, and decide on the refresh period of current standards or on projects of standardizing a new method or technique. The process is time consuming and expensive. It requires that instruments from various manufacturers are operating in laboratories which have already proven to be of high analytical standard. The application range of the method must be evaluated as well as the sample preparation requirements, the effects of the matrices to be expected, the type of calibration etc. Once the basic standard is written and the document is approved, an interlaboratory test must be organized, run, and evaluated. Only if the results of the participating laboratories are in the specified range of the analytical figures of merit, the standard can be approved locally, and presented to the next level of international standardizing bodies. Not seldom methods cannot be standardized due to a lack of consistent interlaboratory test data or a lack of participating laboratories. Development and approval of a new standard usually requires at least half a decade. Once in use it will not be touched again in the next 1½ decades unless significant quality problems should be reported. International standards are extremely important in many fields where the result is essential for human health, consumer protection, environmental protection, legislation, international trade, etc. The standards help to compare data from various laboratories and increase the confidence in the information content. On the other hand, the method is behind methods published by scientific institutes or instrument developments by about a decade. It must be kept in mind that analytical tests are a potential source of contamination and chemical waste as well. Old methods often require much more chemicals, solvents, and other resources than their modern equivalent and are therefore non-economic. Sometimes they may even block developments at the manufacturers' sites.

1.4 Appearance and development of analytical methods: a historic overview

The base of analytical methods is often very old. The light of the stars has been observed by our ancestors already in the stone ages, and complex separation of substances date back to the first high cultures where, as an example, perfumes were distilled in India already 3,000 years ago. However, the quantitative classification of mixtures of substances was based on chemical reactions followed by gravimetry till the twentieth century.

Natural philosophy and physics were the fundaments of analytical information starting from the eighteenth century and intensified in the nineteenth century. The works of Galvani and Volta started understanding of and extensive research in electrochemistry. Faraday, Hittorf, Nernst, and others contributed to understanding of mass transport and movement of ions in the electric field.

Although mankind observed rainbows from time immemorial, splitting of light and the explanation of the phenomenon date back to about 1700. Newton used a glass prism to split the light in 1666 and explained the effect with a particle model of light, whereas Huygens described light separation with its wave character. Looking at the stars with separated light at the beginning of the nineteenth century revealed that the rainbow spectrum was not continuous but showed pronounced missing lines (Wollaston, Fraunhofer see figure 1.4). Systematic mapping of the lines allowed Bunsen and Kirchhoff to relate spectral lines to chemical elements at about the middle of the nineteenth century. Qualitative elemental spectroscopy was born. Hitherto unknown elements were found by identifying the lines of the sun spectrum. Lockyer detected the unknown element Helium this way. Spectra generated by heating various salts in a hot burner led to similar findings. In this case emission lines rather than absorption lines were the base for the detection of

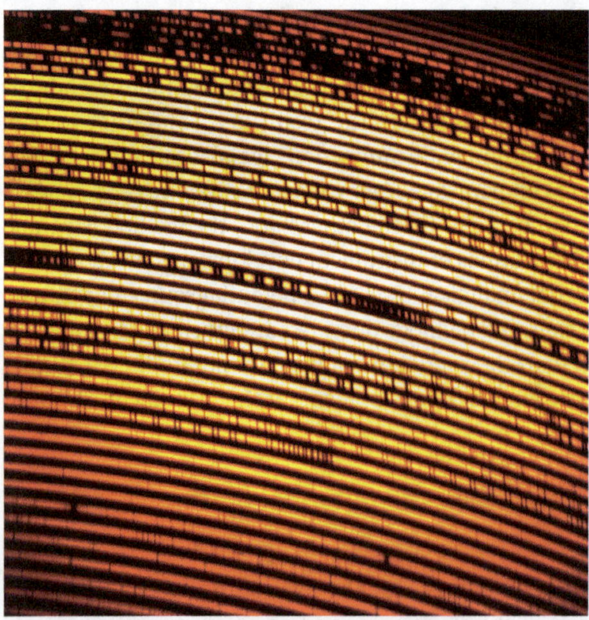

Figure 1.4: Sun spectrum obtained with high a high resolution Echelle spectrometer. Dark regions indicate missing lines. Courtesy of ISAS, Institute of Applied Spectroscopy, Berlin, Germany.

rubidium and cesium by Bunsen and Kirchhoff. All these findings were qualitative. It took almost another century to develop quantitative analytical instruments.

Historically, spectroscopy was first used in the visible wavelength range between 380 and 740 nm. It seemed likely that radiation should be existing with shorter and longer wavelength than those in the visible range. Herschel detected the part of the spectrum toward longer wavelengths (infrared spectrum) and Ritter toward shorter wavelengths (ultraviolet spectrum) at the turn to the nineteenth century. The elemental optical spectroscopy was extended to the UV range (down to 160 nm) and to the infrared range (up to 900 nm) in the twentieth century. In 1895, Röntgen detected a radiation in the very short wavelength range below 1 nm. This radiation was quickly used for medical purposes. Its analytical use was first investigated at about 1930; its commercial use started 20 years later.

At the beginning of the twentieth century the character of light was well known. Different types of spectra were observed starting from discrete lines to band type structures. It was obvious that these phenomena could be used to obtain chemical information about elements and molecules. The first quantitative determination of molecules was reported in 1920. However, though research revealed new principles, such as the Raman technique in 1928, the first commercial spectrometers appeared only in the 1940s. After World War II, the commercial research and development obtained pronounced momentum, and consecutively the techniques started to be used in the commercial laboratory.

The analytical technique most widely used, to obtain chemical information is the separation of mixtures of substances. The technique is originally closely related to the filtration process, which has been practically used by mankind back in early civilizations. Attempts to extract salt from seawater with the help of selected soil were already reported in Greece at Aristoteles' time. The scientific development of chemical compounds started in the nineteenth century with the description of paper chromatographic effects (Runge 1855). A more detailed description of adsorption process and the different behavior of compounds along adsorptive material was given by Michael Tswett shortly after the turn of the twentieth century. He was the father of liquid chromatography. He used the term "chromatography" as he separated colored natural mixtures with the help of the sorbent calcium carbonate. He compared the separation process with the process of separating light with optical means. However, it took till the middle of the twentieth century until columns were used to separate amino acids (Martin and Synge). At about that time the separation of compounds from gas mixtures were reported as well.

Although processes which could be related to chemical information technology were used from early civilizations to the age of enlightenment, real research started at about 1700 and culminated in industrial and commercial use of these principles in the form of commercial analytical instrumentation in the second half of the twentieth century.

The ancient Greece and the Greek city states were the nucleus of an intense observation of nature and fundamental consideration of the encircling world. Aristoteles around 350 BC engaged himself deeply with very "modern questions" such as root cause, transformation, structure, and dynamic changes in the world in substance, elements and atomism, and dynamics of space. Fundamental methods of natural science were defined and used at that time. However, the apparatus was missing to consolidate the considerations. With the rebirth of the classics started today's natural science. Around 1700 and in the eighteenth century logic and scientific imagination reached an amazing flush. Many fundamental scientific insights stem from this time.

Elemental spectroscopy was of paramount interest because of its closeness to astronomy. When Bunsen and Kirchhoff related sharp spectral lines to elements in the middle of the nineteenth century, astronomy had a new and powerful tool for research. Astrophysics was born. There was a lack of sources to generate free atoms or even excited atoms for using spectroscopy for the purpose of chemical information at that time right away. Bunsen's burner was strong enough to generate and excite a few of the alkaline and earth alkaline elements but the bulk of the elements could not be singularized this way. The analytical applications suffered from the lack of reliable source of free atoms or excited atoms and ions. A reliable source for generation of atoms and ions was the spark technology which was developed for radio transmission purposes but could be applied to spectroscopy as well [8]. The second fundamental problem to use spectroscopy for quantitative analysis was quantitation of line intensities. Chemical photo plates were available, but the analysis process required generation of a stable line emission, processing of the photo plate and densitometric evaluation of the lines on the plate. The possibility to quantitate photons and transfer them to electrons with the help of the photoelectric effect [9] was the breakthrough for quantitative emission spectroscopy. A second potential atom source available at that time were ethine-oxygen burners which were used for welding since the beginning of the twentieth century. Burners with a broad firing zone combined ethine with air or nitrous oxide and are used for atomic absorption spectrometry since about 1955.

Electrical discharges in gases are known since about 1700 (the early eighteenth century again!). Hauksbee, an assistant of Newton, generated a glow discharge of mercury vapor at reduced pressure in a glass bulb in a potential gradient. At about 1930 plasma was described by Peter Debye as a "spheric medium with a high concentration of electrons." Glass discharge lamps were commercially available at that time. It was to be expected that discharges would be used in spectroscopy as well. It took until about 1970 before it became a niche technology in optical emission spectrometry. Pure high-energy plasmas at ambient pressure were described since about 1950. Powered by high-energy, high-frequency electromagnetic fields the application was first focused on recasting of metals, technical coating, welding, etc. This type of plasma was called the inductively coupled plasma (ICP). It was obvious that the potential for optical spectroscopy would sooner or later be recognized. Experiments took place in several countries since about 1960–65. Researchers such as Greenfield [10], and Fassel and coworkers [11] laid the fundamentals for a rapidly developing analytical technique, the ICP-OES, the optical emission spectroscopy with atomization, and predominantly ionization, of the sample in an ICP. Not much later it was recognized that the high density of charged particles generated in an ICP could be used in elemental mass spectrometry as well. Researchers mainly in the United States [12] developed prototypes for the first commercial spectrometers at about 1980. Since then, it became the number 1 instrument for high-performance elemental analysis next to ICP-OES. Other plasma sources, such as laser light or microwave radiation, are nowadays gaining importance and may become a simpler alternative to the ICP in the next decade.

Parallel to OES, atomic absorption spectrometry was developed between 1950 and 1960. Basis were the Fraunhofer lines from the sun spectrum. Elements could obviously absorb discrete energy from a continuous spectrum. If this energy would be provided by a source, delivering exactly this element-specific line, a technique would be possible which would allow extremely specific determinations. This was the idea of Walsh [13] in Australia and Alkemade [14] in the Netherlands when they invented atomic absorption spectrometry as a quantitative analytical tool. First experiments were performed with flames with the ethine burners described above. Later a graphite furnace [15] became a powerful alternative source of atoms since 1965. For about 30 years, AAS remained the number one tool for quantitative elemental analysis. Slightly later, atomic fluorescence spectroscopy (AFS) was proposed by Winefordner and coworkers [16] as an alternative to AAS. The advantages, such as improved signal to noise ratio, and larger linear dynamic range, could be proven. Complex possible interferences, however, caused a fast decline of interest into this variety of atomic spectroscopy. If matrix can be excluded completely, e.g., in the ultra-trace detection of mercury, AFS established itself as the "gold standard."

A second, completely different way for the quantitative determination of elements in extremely small concentrations is to keep them in a stable environment in solution and to use the electrochemical characteristic of their ions for quantitation. Electroanalytical methods were introduced into the analytical laboratory in parallel to spectroscopy in the twentieth century but about two decades earlier. Electrogravimetry, the deposition of elements at the cathode of an electrolysis arrangement, was already used shortly before the turn to the twentieth century (Luckow, Gibbs, Classen). Methods which are based on the direct reading of electric magnitudes like voltage and current were commercially used only few years later. The development of the potentiometric methods [17] is based on specific electrodes which are selective for specific ions. Conductometry, based on the resistance of an electrolyte solution method, became popular as an extremely inexpensive analytical tool already in the beginning of the twentieth century [18]. A special kind of potentiometry, the polarography, is based on mercury drops [19]. This powerful method for a lot of elements was developed in parallel to the methods described above. Measurement of electric parameters, mainly in solution, requires sophisticated electrode materials as well as advanced electronic circuitry. Both mechanical and electronic setups underwent a rapid development and improvement during the second half of the twentieth century. Electroanalytical methods, for some time, were the most powerful techniques for ultra-trace elemental determinations [20]. With the development of graphite furnace AAS and ICP-MS their importance became less pronounced. Electroanalytical methods require a lot of chemical knowledge on interfering reactions. They are still an important tool for chemical information today including specifically the bioanalytical field.

Molecular spectroscopy methods developed in a similar pace as those of atomic spectroscopy, somewhat earlier, however. The fundamental knowledges on the separation of light and the interaction of wavelength selected light with molecules were known at the beginning of the twentieth century. Optical systems were available to provide enough optical resolution, both for discrete lines and for bands. The ultraviolet range of the spectrum was known as well as the long wavelength infrared part. It was known that an interferogram could be used instead of a dispersive spectrum with potential advantages, specifically in the longer wavelength range of the spectrum (*Michelson*, 1890). A Fourier Transformation would be necessary to calculate the spectrum from the interferogram (*Rayleigh*, Nobel Prize 1905). The missing link in molecular spectroscopy was the quantitative detection of photons and its representation as a numerical value with the help of an electric circuitry. The first real routinely usable infrared spectrometer was built in 1937 by E. Lehrer [21]. Interesting enough, Lehrer was researcher at a big chemical factory in Germany rather than coworker in a company focused on analytical instruments. Real rapid commercial development of analytical instruments started in the second half of the twentieth century, just like in elemental spectroscopy. In addition to the technical requirements mentioned above, the commercial incentive and the founding of instrument manufacturers were the driving forces. In infrared spectroscopy, representative for all other spectroscopic techniques, the question of simultaneous versus sequential recording of the spectrum was discussed early in instrument development. Multiplexing of a complete spectrum versus sequential selective evaluation was and is still today a field of discussion. However, for IR, manufacturers and the users clearly decided for simultaneous recording using interferometers and the Fourier transform technology. Although the basic knowledge of the possible advantages of the Fourier Transform process to spectroscopy was already known in the nineteenth century, computing power was required to develop an automated analytical instrument. The first commercial instrument appeared in 1969. Since about 1975 new dispersive instruments gradually disappeared from the market. IR and FTIR technologies are extended to the near infrared range today. The long wavelength molecular information plays an important role in all fields of analytical application.

A molecular technology developed in parallel to the infrared spectroscopy is the Raman spectroscopy (C. Raman, 1928). Unlike IR, Raman scattering is an inelastic scattering process [22]. It is much less sensitive than IR and signal to noise is hampered by fluorescence effects. The big advantage, however, is that the matrix effect of water is close to nil. This makes the technique very attractive to bioanalytical applications. Raman developed rapidly in the 1940s because of simple instrument requirements, lost almost all market against the much more powerful IR spectrometers in the 1950s and 1960s, and regained importance as a completely independent technique when strong photon sources (LASER) became commercially available and when bioanalytical questions started to dominate the request for novel analytical methods.

UV spectroscopy developed due to a clear request for information from the military: vitamins in the food preserves. This fostered the production of commercial instruments in 1940. After World War II, the development of commercial instruments was rapid. Referencing the measurement beam with a second beam which was not passing through the sample was an original development in UV [23]. Miniaturization, simultaneous detection of spectra, a moderate cost detector for other techniques, such as chromatography, became domains for UV spectroscopy. Bioanalytical requirements fostered the development of instruments which can quantify extremely small sample volumes. The technique will be found in most commercial laboratories today.

Other than in atomic spectroscopy, where fluorescence became routinely implemented for few elements only, it is widely used for molecular detection, in particular in the bioanalytical field. First observations date back to 1845 already. Fluorescence was first implemented in microscopy at the beginning of the twentieth century. Although most applications are connected with microscopy, molecular fluorescence has become a powerful and cost-efficient analytical tool for numerous molecular analytical questions since the late twentieth century [24].

The commercial development in chromatography started with liquids which were separated in columns carrying the sorbent. In 1941 Martin and Synge [25] promoted the development of commercial equipment by separating amino acids on a column filled with silica gel carrying water as the stationary phase. The development of gas-liquid phase separation followed 12 years later [26]. Studies of exchange and adsorption phenomena between liquids and solids, between liquids and liquids, and between gas and liquids were investigated by many researchers before commercial instruments appeared on the market. As described earlier, electronics and detector technology had to be developed to evaluate the substances coming from the separation process. It was the thermal conductivity detector (TCD), developed in 1954, which allowed to detect gases such as CO_2, SO_2, N_2, and noble gases by measuring their thermal conductivity in a heated cell between two temperature niveaux. A simple electrical bridge arrangement determines the heat in the sample cell compared to a reference cell with a pure carrier gas flow. A leap in sensitivity, specifically for volatile hydrocarbons, uses the ionization potential of the sample coming from the column. Compounds in the gas flow are ionized in a hydrogen flame and the generated electrons are captured and quantitated. The flame ionization detector (FID) was developed in 1958. The electron capture detector (ECD), developed in parallel, enabled the sensitive chromatography of halogenated hydrocarbons and many other organic compounds. It is based on the extinction of electrons from a primary ionization process, in this case stimulated by a radioactive ß-emitter. In earlier instruments, the detectors described above are often coupled in tandem arrangements. The most powerful leap forward both in separation and in detection capability was coupling of chromatography with mass spectroscopy. First experiments by Gohlke and McLafferty started in parallel with the development of GC. The first instrument combinations – interesting enough, a time-of-flight MS was used – worked in the late fifties where the successful coupling was first published [27].

In parallel, the chromatographic process was improved in speed and was miniaturized. Packing columns with smaller particle sorbent and increasing the pressure with high-performance pumps (high-performance liquid chromatography, HPLC). This became commercially available about 1970; pioneering developments were published three years earlier [28].

An important modification of gas chromatography is the head space technique, developed in parallel with gas chromatography. The idea is to detect mainly compounds dissolved in liquids (e.g., CH_4 in water) by gas chromatography of a closed cell where liquid and gas phase are in a thermodynamic equilibrium. It took almost 20 years from the first ideas to the first commercial instrument in 1967. An important driving force for this technique, among others, was the need for alcohol detection in blood. An excellent overview is found in literature [29].

Thin layer chromatography (TLC) was first reported in in the 1940s [30]. The simple and cheap instrumentation requirements the high-performing yield of separation and the fast separation process made it an attractive alternative to columns. It advanced rapidly in the 1950s. Beyond separation in one direction, the process can be continued in a second direction sequentially (two-dimensional DC) or by simultaneous application of centrifugal forces to a circular chromatographic plate. Detectors are usually densitometers which work in the visible and ultraviolet range. Nowadays thin layer chromatography can be coupled to mass spectrometry which greatly advances its specificity.

In the middle of the 1960s a system related to liquid chromatography was described by Giddings and coworkers [31]. Columns with stationary phase are replaced by flow channels. Different fields between two orthogonal flows are used as separating source, such as gravity, heat, centrifugal forces, etc. The technique is mainly used to separate polymers, biopolymers, and proteins. Modern developments concern mainly bioanalytical questions. Special columns are, e.g., doped with special ligands which can bind to specific proteins with high selectivity (affinity chromatography).

It becomes obvious that all the techniques described above were commercialized in the second half of the twentieth century. Although the scientific base was already available more than two decades earlier, the demonstrator equipment was not fit for routine. It required the availability of electronic equipment as well as suitable detectors. It required the capability of reproducible production with very tight tolerances, technical documentation, etc. With other words, it required the upcoming of companies interested in this type of business. Today, analytical instrumentation is a business with annual revenues bigger than 50 billion dollars.

> *Simplicity is a great virtue, but it requires hard work to achieve it and education to appreciate it. And to make matters worse: complexity sells better.*
>
> (Edsger W. Dijkstra; information scientist, 1984)

1.5 Physics or chemistry: the periodic change of the yellow jersey

In the movie *Star Trek*, Dr. McCoy uses a simple device, the “Tricorder,” to detect and assess diseases. A Tricorder type of equipment based solely on wave-type information is the dream of each chemical laboratory. All the necessary steps to bring the sample into a form suited for evaluation in an analytical instrument (sampling, storage, digestion, matrix separation, pre-concentration) are huge sources of potential contamination, analyte loss, non-representative sampling spot, pollution of laboratory and environment, time loss and source of immense cost of analysis. It seems to be obvious that academic and industrial scientists are desperately searching for techniques which provide the required information by just pointing with a probe at the sample under investigation. Specific interaction of light with electrons seems to be a viable way toward this goal. The exact specification of the type of information required is essential to judge whether a method will be suited to provide this information. Stored steel samples of different alloy composition can very well be safely recognized and selected with a handheld X-ray fluorescence device pointed toward the specimen. If the exact composition of an ore is required to decide on a profitable mining, a complex chemical sample pretreatment process followed by a thorough multi-element elemental analysis by ICP-OES will be the method of choice. If the total element concentration pattern in surface water is required, a direct ICP-MS analysis is the method of choice. If the question concerns the exact element species, e.g., chromate or methylmercury, sample pretreatment, coupling of methods, or UV molecular analysis may be used. Rapid determination of bacteria with a fast and yet accurate test would be highly preferential over a method which required body fluids to be incubated for several days before an accurate analytical result is available.

Once methods based on the evaluation of spectra were commercially available, the yellow jersey was worn by physics. The most important instrumental developments were related to better optical or mass resolution, higher detector sensitivity, and lower detector noise. Physical means to distinguish analyte information from background information, and sophisticated algorithms to extract analyte information from the bulk of background noise supported the specificity of the methods. Sources for sample illumination became much more powerful which made hitherto less attractive methods, such as Raman spectroscopy, very powerful and competitive again. On the side of sample handling, new and powerful methods were developed to separate samples more efficiently, to volatilize liquids with less waste, to optimize the feed of the sample into the atomizer, and to make the atomizer itself much, much more powerful. High energy system for the direct analysis of samples were developed (e.g., laser ablation and electrothermal atomizer). Direct methods such as the X-ray based techniques gained ground against methods which require dissolved samples. The main instrumental developments with their benefits and disadvantages will be described in the following sections. The introduction of each new technique was accompanied by the believe that the analytical work will come closer to "absolute," "standardless" information. Once the methods were in use in the routine laboratory for some time, it turned out that there is always a lot of cross interference from the bulk of matrix surrounding the analyte. It must be pointed out that the analyte species is often to be found in an excess of millions of other compounds which all may generate erroneous information. Sophisticated physical means correcting artifacts from matrix were invented such as the *Zeeman effect* technology in graphite furnace AAS or the *collision cell* technology in ICP-MS. Though extremely powerful, all methods based on physical means have limitations. Chemical methods to focus on specific detection of the analyte were therefore always of utmost importance to support the on-board features of the instrument. Specific gases are added to the collision cell of an ICP-MS instrument. Modifiers are added to flame and graphite furnace AAS to control unwanted chemical reactions or physical effects. Tracer elements or isotopes are added to quantitate matrix effects. Finally, various ways of analyte preconcentration or matrix separation were applied to shift the analyte to matrix ratio to a more favorable ratio. In elemental analysis and classical molecular analysis, modern developments till today concern fine tuning of detection, illumination, sample flow, means of matrix correction and sophisticated algorithms making mathematical use of the huge amount of data coming from the detector. Nevertheless, the analytical society is aware that selective chemistry will always be required, and the "Tricorder" is available for very specific purposes only.

The rapid development of bioanalytical methods brought the yellow jersey back to chemistry, more precisely, back to biochemistry. Molecular biological information is usually not directly analyzable from available samples. The traces of DNA must be amplified before they can be further evaluated. This is done with relatively simple technique, a controlled heating and cooling of the process, but with complex chemistry involved. The very small amounts of DNA are doped with buffer, nucleotides, and primer and duplicated within 5 to 10 min. Depending on the final amount of sample required an amplification by a factor 10^6 to 10^9 can be obtained in just a few hours. The amplification process and yield can be tracked by detection with the fluorescence method. In almost all types of bioanalytical analysis a lot of chemistry and biochemistry is involved. Chromatography, usually coupled to mass spectrometry, is the method of choice for specific quantitation of species. In the last few years, a lot of research capacity is devoted toward applying direct spectroscopic techniques for, e.g., the detection and quantitation of bacteria from body fluids or body tissue. Raman spectroscopy is gaining ground. It will be interesting to see whether the yellow jersey is again going to physical methods.

Literature

[1] Zhang J, Zhang C. Sampling and sampling strategies for environmental analysis. Int. J. Environ. Anal. Chem. 2012, 92(4), 466–478.

[2] Kratochvil WBD, Taylor JK. Sampling for chemical analysis. Anal. Chem. 1984, 56(5), 113–129.

[3] Bruno T. Figures of Merit, Taylor and Francis Group, LLC, New York, NY, 2014.

[4] Compendium of Chemical Terminology, 2nd Edition, IUPAC, Research Triangle Park, NC, 1993, ISO 3534-1.

[5] Chemical analysis-Decision limit, detection limit and determination limit under repeatability conditions – Terms, methods, evaluation 2008, DIN 32645-11

[6] Schlemmer G. Industrial research and development for instrumental analytics: Requirements, skills, strategic objectives. Anal. Bioanal. Chem. 2012, 403, 1195–1198.

[7] Chan CC, Lam H, Lee YC, Zhang XM. Analytical Method Validation and Instrument Performance Verification, John Wiley & Sons, Hoboken, New Jersey Wiley, 2004, ISBN: 9780471259534.

[8] Golloch A. (Ed.) Handbook of Rare Earth Elements; Chapter 7, De Gruyter, 2017, ISBN: 9783110365238.

[9] Zworykin VK, Morton GA, Malter L. Proceedings of the Institute of Radio Engineers, 1936, 24/3.

[10] Greenfield S, Jones ILI, Berry CT. High pressure plamas as spectroscopic emission sources. Analyst 1964, 89, 713–720.

[11] Dickenson GW, Fassel VA. Emission spectrometric detection of elements at nanogram per milliliter levels using induction coupled plasma excitation. Anal. Chem. 1969, 41, 1021–1024.

[12] Houk RS, Fassel VA, Flesch GD, Svec HJ, Gray AL, Taylor CE. Inductiveley coupled argon plasma as an ion source for mass spectrometric determination of trace elements. Anal. Chem. 1980, 52(14), 2283–2289.

[13] Walsh A. The application of atomic absorption spectra to chemical analysis. Spectrochim. Acta 1955, 7, 108.

[14] Alkemade CTJ, Milatz JMW. A double-beam method of spectral selection with flames Appl. Sci. Res. 1955, 4, 289.

[15] L'vov BV. The analytical use of atomic absorption spectra. Spectrochim. Acta 1961, 17, 761.

[16] Winefordner JD. Principles, methodologies, and applications of atomic fluorescence spectrometry. J. Chem. Educ. 1978, 55, 72.

[17] Die Böttgerw. Anwendung des Elektrometers als Indikator beim Titrieren von Säuren und Basen. Zeitschrift Für Physikalische Chemie 1897, 24, 253–301.

[18] Kolthoff, IM. Konduktometrische Titrationen, Verlag Theodor Steinkopff, Dresden/Leipzig. 1923.

[19] Heyrovsky J Application and Utilization of Electrochemistry in Organic Chemistry Chem. Listy 1922, 16(256–264).

[20] Scholz F. Ed. Electroanalytical Methods, 2nd Edition, 2010, Springer Heidelberg Dordrecht London New York, ISBN 978-3-642-02915-8.
[21] Die Geschichte einer Innovation: Infrarot-Spektroskopie; www.basf.com; Titelgeschichte
[22] Raman CV. A New Radiation. Indian J. Phys. 1928, 2, 397–398.
[23] Yang CC., Legallais V. Ein schnell und empfindlich aufzeichnendes Spektrophotometer für den sichtbaren und ultravioletten Bereich. I. Beschreibung und Leistung. Review of Scientific Instruments 25 , 801 (1954)
[24] Lakowicz JR. Principles of Fluorescence Spectroscopy, Springer, Boston MA, 1999.
[25] Martin JP, Synge LM. A new form of chromatogram employing two liquid phases. Biochem. J. 1941, 35, 1358–1368.
[26] James AT, Martin AJ. Gas-liquid partition chromatography: The separation and microestimation of volatile fatty acids from formic acid to dodecanoic acid. Biochem. J. 1952, 50, 679–690.
[27] Gohlke RS. Time-of-flight mass spectrometry and gas liquid partition chromatography. Anal. Chem. 1959, 31/4, 535.
[28] Horvath CG, Preiss BA, Lipsky SR. Fast liquid chromatography. Investigation of operating parameters and the separation of nucleotides on pellicular ion exchangers. Anal. Chem. 1967, 39(12), 1422–1428.
[29] Kolb B, Ettre LS. Static Headspace-Gas Chromatography, Theory and Practice, John Wiley & Sons, Hoboken, NJ, USA, 2006.
[30] Meinhard JE, Hall NF. Surface chromatography. Anal. Chem. 1949, 21(1), 185–188.
[31] Giddings JC. A new separation concept based on a coupling of concentration and flow nonuniformities. Sep. Sci. 1966, 1, 123.

2 Information based on physical methods

Mankind tends to believe his wits. It becomes quickly obvious that these are often misleading our discrimination. Is the blue sphere still blue if we look at it in a completely dark room? Although there is no color without light the sphere remains blue. The color of the sphere is a dispositional property which becomes active under certain conditions. In this chapter and in Chapter 3 we will make use of these properties for analytical purposes. Although the absorption of light in the ultraviolet spectrum does not become visible to our senses, we make use of the dispositional property of the species to detect and quantitate it. What we need is the apparatus to activate and detect a characteristic dispositional property of the species under investigation.

2.1 Optical spectroscopy

2.1.1 Electromagnetic waves

Man can see in a relatively short wavelength range between 380 (violet) and 780 nm (dark red).

Beyond, to the left side of the spectrum is the short ultraviolet wavelength range, to the right side, toward longer wavelengths is the infrared range. These sections of the spectrum, which are not directly accessible to the human eye, are again subdivided into near UV, far UV, near infrared (NIR), etc. The classification stems rather from technical peculiarities and application fields than from physical reasons. The human eye has the highest sensitivity at about 560 nm, between yellow and green during daylight and about 500 nm, dark green, in the night and about 500 nm, dark green, at night (see figure 2.1).

Like the eye, the technical components used in optical spectroscopy have an optimal range where radiation can be transmitted, shaped, transferred into electrons, and thus be used for optical spectroscopy. In this chapter we will define the spectral range from the far UV at 130 nm to about 1,000 µm in the far infrared, a wavelength range spanning about 5 orders of magnitude. Wavelength is related to frequency and energy via eq. (2.1).

Equation 2.1: Correlation between energy, frequency, and wavelength:

$$E = h \cdot \nu = h \cdot c/\lambda \tag{2.1}$$

where E is the energy, h is the Planck's constant = 6.626×10^{-34} J
h = 6.626×10^{-34} Js,
ν is the frequency, and
c is the speed of light in vacuum = 299, 792, 458 m/s.

The energy of a photon at 300 nm, as an example, is 5.9×10^{-32} J.

https://doi.org/10.1515/9783110689662-002

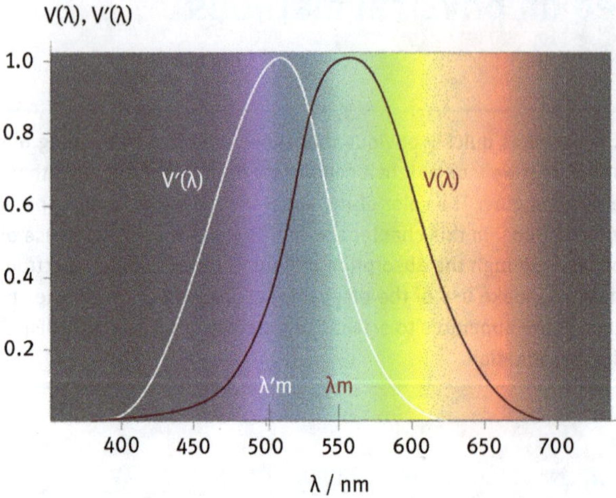

Figure 2.1: Wavelength range of the human eye: the relative intensity (ordinate) of the wavelength (abscissa) registered by the human eye is plotted for daylight (black curve) and dark environment (white curve). To the left side the ultraviolet range, to the right side the infrared range starts. Figure courtesy of Gigahertz Optik GmbH, Germany.

It is the energy of the photons which interacts with molecules or atoms in different ways, or the energy which is emitted by atoms or molecules in form of radiation after they have been thermally excited which provides the specific signal which is related to chemical information. Even if we are talking about many interacting particles (we may talk about 10^{15}) the energy flows are very small. Sophisticated equipment is required to quantitate the interactions between analyte and light. In an analytical measurement the wavelength specific flow of photons with and without the target species is compared during analysis.

The spectral range with its divisions is sketched in Figure 2.2.

2.1.2 Interaction of photons with electrons

From the energy considerations mentioned above it is comprehensible that photons in the infrared range will act in a different way than photons in the short UV range. If a high-energy impulse is given to an electron the result may be a series of events in the atom or molecule hit. For discussion of various effects in the instrumental sections later, a short discussion of electronical orbits will be helpful.

Neutral atoms contain as many electrons as protons, roughly half their atomic mass number in the periodic system. These electrons are in orbitals with different energies, symmetries, and numbers of possible population. Orbitals are mathematical functions which describe the behavior of electrons in an atom.

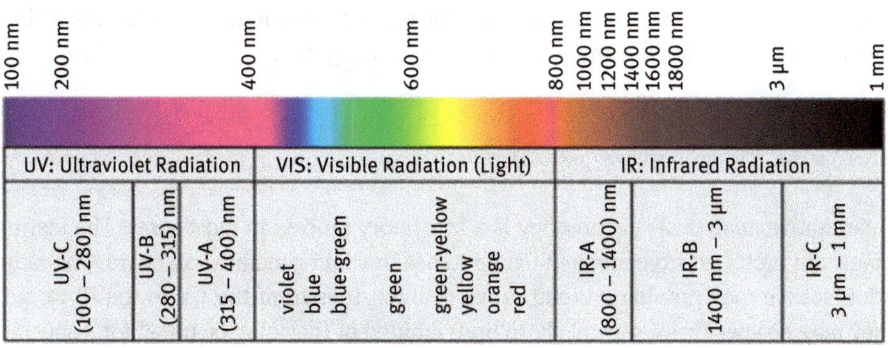

UV: Ultraviolet Radiation			VIS: Visible Radiation (Light)	IR: Infrared Radiation		
UV-C (100 – 280) nm	UV-B (280 – 315) nm	UV-A (315 – 400) nm	violet, blue, blue-green, green, green-yellow, yellow, orange, red	IR-A (800 – 1400) nm	IR-B 1400 nm – 3 µm	IR-C 3 µm – 1 mm

Figure 2.2: Electromagnetic spectral range used in optical spectroscopy. Source: Courtesy of Gigahertz Optics GmbH, Germany.

The orbitals are categorized by quantum numbers *n*, *l*, *m*, *s*. Filling of the orbitals is following quantum mechanical rules. An atom in ground state has a defined configuration of electrons. Besides the filled orbitals atoms contain a high number of available orbitals which can be occupied for a short time when thermal or electromagnetic energy is applied to the atom or molecule. Whether or not a transition is possible is defined by selection rules. If the energy applied to an electron exceeds a certain barrier, which is depending on the type of atom, the electron is leaving the atomic union leaving a positively charged ion behind. Each transition from one state to another state extracts a clearly defined energy from the applied electromagnetic wave or emits a sharp wavelength when the electrons fall back to a lower energy level.

The situation is clearly defined in atoms. It is more complicated in molecules. To simplify the situation, atomic orbitals are linearly combined to molecule orbitals [32]. The number of orbits in a molecule equals the sum of the number of orbits of the bound atoms. However, the total numbers of orbits are obviously much bigger than in atoms and so are the possible transitions. Besides orbits with binding functions orbits with anti-binding functions result from the calculations. Stable molecules require a binding orbit with lower energy than that of the atomic orbits. Non-binding orbits at a higher energy level are formed at the same time. Lower energy levels have binding functions, orbits at higher levels have anti-binding functions. These orbits follow the same filling rules as atoms. Electrons interacting with photonic energy will transfer to various energy levels, depending on the energy of the interacting photon (the wavelength of the photon). The short-term habituation of excited electrons will influence the stability of the molecule. Movements like vibrations and rotations which will be discussed in the respective sections of UV, near-infrared, infrared, and microwave spectroscopic methods can be traced back to electrons in excited energy levels. All these transitions are sharp with respect to energy. However, other than in individual atoms, different reactions are activated in the molecule which appears as an overlay of multiple sharp lines. The spectral resolution of the instruments shows the overlays as broader band phenomena which are nevertheless characteristic for the analyte molecules investigated.

2.1.3 Generation of light for optical spectroscopy

Quantitative analytical spectroscopy is a laboratory process in most cases. The instruments are protected from ambient light and use the light generated on board. The radiation source may provide a broad range of intensities from the UV to the IR range, they may be specific for several sharp lines emitted or they may be tuned for a specific portion of the spectrum. In almost all cases the source should have a high photon output, a spatially limited emission sphere, a good short and long-time constancy of emission whereby short time usually represents 20 ms to 1 s, whereas we are talking about minutes to hours when referring to "long time." As a first example we look at a high-performance continuum source (Figure 2.3). This lamp operates at 300 W continuous output power. The lamp is mainly filled with Xenon pressurized at 1.7 MPa [33]. The emission is realized in form of an extremely short electrical arc between two electrodes from special tungsten alloys. The arc is less than 1 mm and is circling around the spherically shaped cathode. The extremely high arc temperature of about 10,000 K is strongly focused on a sphere with a diameter of less than 0.2 mm, generating a high-intensity emission spectrum over the wavelength range interesting for Atomic Absorption Spectrometry or UV/VIS spectrometry from 190 to 900 nm.

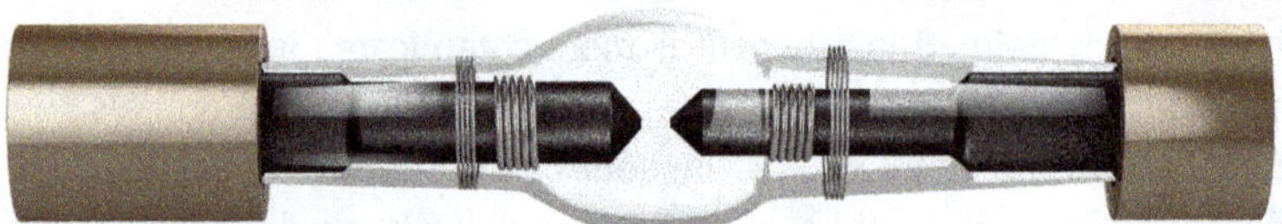

Figure 2.3: Xenon continuum source with high energy output (GLE, Berlin, Germany). Courtesy of Analytik Jena, Jena, Germany.

The emission interacting with molecules or atoms is usually focused onto or through a small cell. If absorption is the method of choice, the path of interaction between photons and analyte should be long and thin. This requires a significant part of the generated radiation to be collected and focused through the space of interaction with the analyte. A source for analytical purposes should therefore emit light in a favored direction such that it can be collected by suitable optical components. This requires a small spot where the radiation is generated like in the example shown above. The luminous intensity (I_v) is magnitude of interest. It is a measure of the wavelength-weighted power emitted by a light source in a particular direction per unit solid angle. The SI unit of luminous intensity is the candela (cd), an SI base unit. It is defined as the dihedral angle density (eq. (2.2)). The light flux ϕ_v is divided by the dihedral angle Ω.

Equation 2.2: Luminous intensity

$$I_v = \phi_v / \Omega \tag{2.2}$$

The dihedral angle Ω is the quotient of the area of a sphere A_t illuminated by the light source at a distance of r according to eq. (2.3). The standardized unit is 1, i.e., a source at a distance of 1 m generates an area of 1 m^2 on the surface of the sphere around the source.

Equation 2.3: Dihedral angle

$$\Omega = A_t / r^2 \tag{2.3}$$

The luminous intensity (cd) relates to power (W) by referring to the maximum sensitivity of the human eye at 555 nm (540 THz).

To compare the intensity of sources, the standardized dihedral angle, the steradian, is used. The luminous intensity per wavelength is obviously most important for spectroscopy. The luminous intensity for monochromatic light of a particular wavelength λ is given by eq. (2.4).

Equation 2.4: Luminous intensity for monochromatic light of wavelength λ:

$$I_v = 683\,\psi\,(\lambda) \cdot I_e \tag{2.4}$$

where I_v is the luminous intensity in candelas (cd),
683 is the conversion factor into watts,
I_e is the radiant intensity in watts per steradian (W/sr),
$\psi(\lambda)$ is the standard luminous function.

Continuum radiation sources used in spectrometry

Most of the radiation sources used in spectrometry are much less sophisticated and significantly cheaper than the high-performing xenon arc described above.

Xeneon lamps are generally capable of covering the spectral range between 190 and 1,100 nm used for UV-VIS spectrometry. The radiance at 200 nm is about 10 times lower than at 400 nm. To cover the low wavelength (UV) range, they are equipped with quartz bulbs. They are usually operated in pulsed mode with short flashes.

Halogen lamps are covering the visible range between 320 and 1,100 nm. Their emission maximum is between 800 and 1,000 nm. The radiance is strongly depending on the operating power. Their typical radiance is in the range of 100 to 1,000 $mW/cm^2 \cdot sr \cdot nm$. The lamps are inexpensive and are used in different types of analytical instruments.

D_2 lamps emit their maximum in the short wavelength range at about 200 nm. There they are a factor of 10 stronger than halogen lamps. Up to about 400 nm they decay by a factor of about 10. Deuterium lamps are low pressure arcs filled with deuterium (D_2) at a pressure of a few mbar. The cathode is preheated, and the arc is ignited at about 500 V between anode and cathode. Once the arc is active, the gain drops by a factor of about 5. The continuous arc is operated at roughly 50 W. These lamps are used, e.g., in UV, AA spectrometry, and HPLC. With specific quartz bulbs they can be operated down to 160 nm.

Long wavelength emitters

Longer wavelength radiation is increasingly sensed as warmth by man. Conversely, heated bodies are used as radiation source for infrared analysis. Bodies which are completely absorbing radiation (black body) are ideal to emit radiation (Planckian radiator). Each body which is warmer than zero degrees Kelvin (0 K) is emitting electromagnetic radiation. According to Planck and Wien, the intensity and the wavelength maximum of emitters are dependent on the temperature of the body. A black body at 10,000 K emits its maximum at about 350 nm. At 5,000 K the maximum is around 1,000 nm, however, with 20 times lower intensity. Emitters used in spectroscopy are no ideal black body emitters. Their intensity is therefore lower than predicted by theory. The heated emitters for IR radiation are usually cylindrical bodies heated to between 1,200 and 2,000 K. They are made from ceramic material such as silicon carbide (glow bar) or from mixtures from ceramics (ZrO_2, Y_2O_3). Heating comes from the ohmic resistance of the material. In case of the "Nernst-glower," the element must be preheated to reduce the resistivity before it acts as an ohmic resistor. The Nernst-glower is the most popular source for IR spectroscopy. Toward shorter wavelength the classical sources, described above can already be used.

Another powerful source for the entire wavelength range are mercury vapor lamps at various pressure inside the bulb. For the far IR range, high-pressure lamps are used. These lamps generate a plasma-type discharge.

Sources emitting a discrete line spectrum

In some cases, a source emitting a narrow line with high luminous intensity with a very low broad band emission is of utmost importance. These line emitters are mostly used in atomic absorption spectrometry (AAS) or atomic fluorescence spectrometry (AFS). The emission line of this type of lamp should be extremely narrow (in the range of 0.001 nm) and should be narrower than the absorption line of the elements to be determined (see Sections 2.1.6 and 2.1.7). This requires that the pressure in the light source is lower than that in the atomization cell. The source is therefore constructed to operate at reduced pressure and at relatively low temperature. Thus, the line source is the component which ultimately defines the spectral resolution of the absorbance measurement.

The light source, most used for AAS, is the hollow cathode lamp (Figure 2.4.)

The lamp consists of a glass cylinder with at least two electrodes. The cylinder is usually filled with neon or argon at a pressure of about 5 mbar. The cathode is a

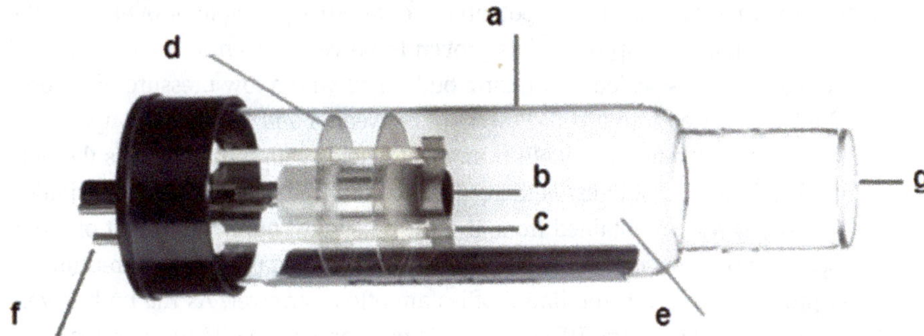

Figure 2.4: Schematic diagram of a hollow cathode lamp: a: Glass cylinder, gas tight; b: metal cylinder filled with the element excited for emission; c: metal anode; d: insulator; e: noble gas filling at reduced pressure; f: electrical connector pins; g: exit window, mostly quartz.

hollow cylinder made from or coated with the elements to be emitted. Upon the application of about 600 volts between the anode and cathode, a glow discharge is initiated. In the discharge, noble gas cations are accelerated toward the cathode and, upon impact, sputter analyte atoms from the cathode surface. These atoms are excited by collisions in the discharge and emit radiation while returning to lower energy states. The discharge process does not take place under conditions of thermal equilibrium. The gas temperature is low compared with the electron temperature of the excited atoms. This results in high emission intensity with a profile much narrower than the profile of the absorbing atoms.

The glow discharge in a hollow cathode lamp stabilizes within a few microseconds. The lamps can therefore be pulsed (modulated) at a rapid rate of up to several hundred Hertz. The performance of the hollow cathode lamps with respect to emission line width, intensity and lifetime depends on various aspects:

- The energy required to stimulate a discharge. Transitions at short wavelengths require a higher excitation energy. Additionally, the emitted radiation is more strongly absorbed by air or flame gases. Emission lines of elements with primary resonance lines at short wavelengths therefore are usually less intense than those at higher wavelengths.
- Interaction of emitted intensity and width of the emission profile. Both increase with increasing current. Beyond an optimum current which is element dependent, the intensity still increases while the line is broadened significantly due to the increased gas temperature and cathode temperature.
- Control of the operating current with respect to the lifetime of the lamp. Elements which can be easily volatilized, such as Hg, Cd, Bi, Ag, will be rapidly transported from the cathode to the walls of the glass cylinder if the cathode exceeds a certain temperature. The lifetime of the lamp will be significantly shortened under these conditions.

For elements with a high excitation potential, yet another principle of operation, the Electrodeless Discharge Lamp (EDL) has proven to be very powerful. A small mass of the analyte element(s) is sealed in a quartz bulb filled with a low pressure of a noble gas. The bulb is inserted into a coil. Powers of between 5 and 15 watts are applied to the coil to generate high-frequency electromagnetic fields within the bulb. As the bulb warms up, the element of interest is increasingly volatilized and collisional excitation takes place. The spectrum obtained from an EDL is usually "purer" than that of a hollow cathode, as no other metallic material apart from the analyte, is required for the excitation process. Moreover, the line profiles are often narrower. As the bulb is very small, and is surrounded by the HF coil, there is no pronounced difference in temperature and element transport and condensation to the walls of the bulb are limited. This type of the lamp is less suited for rapid electronic modulation. The temperature of the bulb and therefore the emission intensity and the emission profile of the atomic line stabilizes relatively slowly.

LASER and LASER diode

In the lamps described above, one or several atoms are stimulated by electrical or thermal power. Electrons gain a short time higher energy state which is lost again by emission of light. The emission is spontaneous and random. The number of excited electrons is always lower than the number of electrons in the most stable ground state. Using a third state with an energy higher than a comparatively long lived excited state one can however invert this situation. In such a situation, photons emitted from the excited state will trigger the release of further photons by excited electrons in a process called stimulated emission: LASER = Light Amplification by Stimulated Emission of Radiation. Technically, the stimulating waves are homogenized in frequency and direction by mirrors. A standing wave is fed to the element to be excited. The light emitted by stimulated emission is leaving the system through a permeable mirror. LASER sources are available since about 1965. Laser light has revolutionized innumerable processes in research, engineering, and production. Lasers are widely used in science, medicine, industry, etc. Laser emission lines are very sharp with respect to wavelength and the luminous intensity per wavelength is very high. Lasers are used in analytical instruments either as a source for illumination or as a high-energy source for volatilization of elements and excitation of atoms. Obviously, wavelength emitted, length, intensity, and repeatability of pulse and repetition rate are key criteria for the application in instrumental analytics.

Whereas the first commercial lasers were emitting at wavelength longer than the visible range, Excimer Lasers can reach wavelengths shorter than 200 nm.

Technically, Lasers differ by the medium in which spontaneous excitation is stimulated (solid state, gas, dye) and by the source which is pumping (gas lamp, flash lamp, diode, electron beam, etc.).

Most lasers are monochromatic and allow a small wavelength tuning in the range of a few nm by technical means. A wider range tuning requires optical components and/or organic lasing media (dye lasers). The use of lasers in spectroscopy usually requires sophisticated systems and makes the laser a powerful yet expensive source. The fundamental components of a laser are sketched in Figure 2.5.

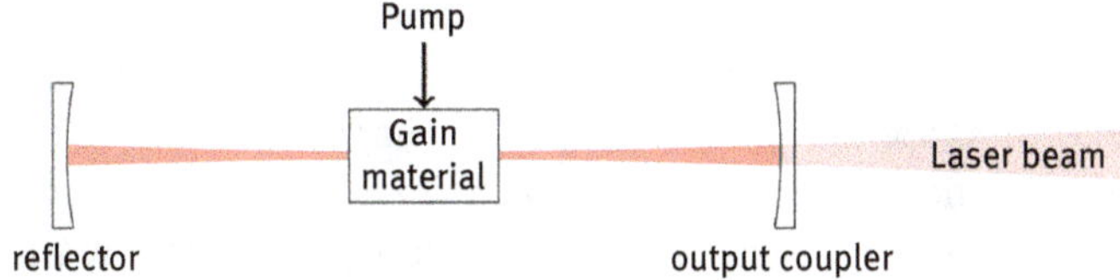

Figure 2.5: Components of a laser.

Light emitting diodes (LEDs)

LEDs are nowadays a frequently used medium for illumination. The process of photon generation is based on p-n junctions of semiconductor crystals. An electric current is flowing from the anode via the crystal to the cathode. Transition of electrons from the layer with an excess of electrons to states of lower energy in the layer with a deficit of electrons frees energy, which is emitted in the form of monochromatic light. The type of semiconductor used defines the emitted wavelength. White light can be generated by mixing the three fundamental colors red, blue, and green. White light is generated as well by illuminating an actively luminescing layer with the light of a blue diode, for example. Still the emitted wavelength of monochromatic diodes can be modulated only marginally by the electrical conditions of stimulation, and by temperature. Their application in spectroscopy is therefore limited to processes where defined small wavelength ranges are excited.

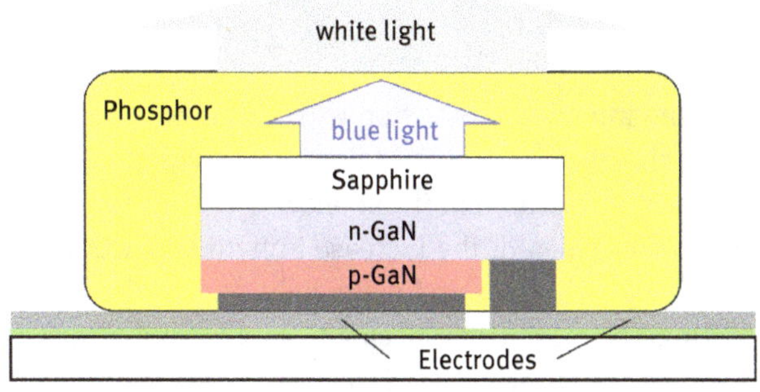

Figure 2.6: Sketch of a light emitting diode (LED).

Emitters for very short wavelength

X-ray radiation is generated by a low-pressure reservoir with high potential between anode and cathode. Electrons emitted from the (heated) cathode hit the anode with high energy. By impact on the anode electrons from inner orbits of the anode material (chromium, tungsten, molybdenum, rhodium) are knocked out. Electrons from outer orbits fill the holes and release characteristic radiation of very short wavelengths, which correspond to the quantum transition. Besides this characteristic radiation, bremsstrahlung and a lot of heat are generated. Bremsstrahlung is generated when fast electrons are changing their speed by impact with the anode material. These electrons slow down substantially. X-ray radiation of a wider spectrum is generated. Toward the short wavelength bremsstrahlung is limited by the maximum energy of electrons accelerated in the electric field. Toward longer wavelengths the emission intensity is quite homogeneous. Both, bremsstrahlung and characteristic radiation are used for X-ray spectroscopy. The radiation is leaving the tubes through windows, usually beryllium, which are permeable for the short wavelength range. The generation of X-rays is sketched in Figure 2.7.

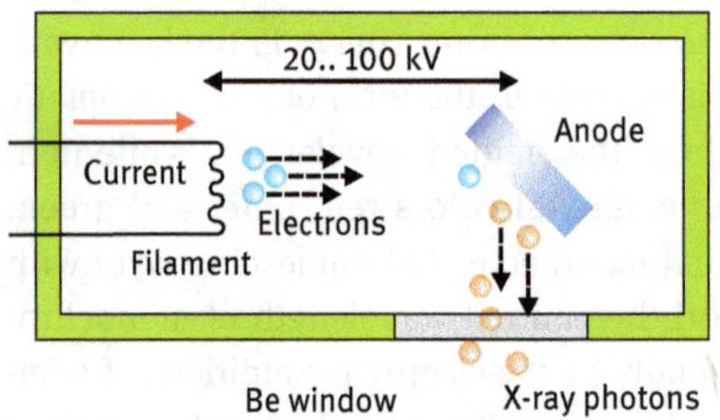

Figure 2.7: Function principle of an X-ray tube. Courtesy of Malvern Pananalytical Ltd.

2.1.4 Wavelength selection

Light must be energy specific for the intended interaction with atoms or molecules. If absorbance is to be quantitated, the difference of light flux before and after interaction is measured. The light reduction is minimal if the entire light of the source would be considered. If emission is considered, the increase in light flux would be minimal considering the number of specific lines and white light next to the line of interest. Even if a source is emitting several narrow specific lines, like in the case of a hollow cathode lamp, optical components are needed to separate the wavelength of interest from the non-wanted straylight. Just like the prism used by Newton for separation of the white sunlight into rainbow colors, the optical system of an analytical instrument provides the selection. The discriminating system may be very simple such as a filter which cuts off radiation below or above a certain wavelength and passes all the other wavelengths of the source. It may be a complex assembly of gratings and prisms which cuts out just 1 pm from the entire spectrum. In this section we will discuss the basic requirements and principles of wavelength selection.

If we consider the absorption process, the main specification for wavelength separation is to pass as much of the interactive photons and withhold as many of the non-interacting photons which we will call "straylight." Straylight may originate from the light source, the spectrometer itself, and the source (atomizer) holding or providing atoms or molecules. In case of emission methods, the sample itself is illuminating the spectrometer. The function of the spectrometer is to separate the analyte species emission from the emission of other species in the sample. As we are talking mainly about very high excitation temperatures, the source of excitation (flame, plasma, spark) contributes strongly to the emitted light itself. In fluorescence, which is basically an emission method, the exciting light must be separated from the specific emission. In all cases the intensity change of the light representing the analyte interaction (signal) compared to all other light effects (background) should be maximal.

In the following section the transfer of photons into a readable signal will be discussed. It is important to emphasize that the characteristic of the detector is extremely important for the selection of the spectrometer and vice versa.

An obviously fundamental difference is, if the spectrum is projected simultaneously on the detector, or if every wavelength or wavelength range is selected and represented such that a sequential scan of the spectrum is finally available. Another important difference is between producing a real image of the spectrum or a mathematical transform of it. An interferometer obtains the analytical information from interferences of waves. The species of interest in this case change the intensity of certain wavelengths and hence the characteristic of the interfering wave. A Fourier transformation is required to obtain the real spectrum.

The photometer part of the instrument is imaging the emission source on the atomizer, or the measurement cell or cuvette, and finally onto the detector. The emission source has a certain spherical shape with an area (F) emitting the photons under an angle (Ω). The number of photons, guided through the photometer, the "Etendue" is invariant (eq. 2.5).

Equation 2.5: Etendue of a photometer

$$\Delta F \cdot \Delta \Omega = \text{const} \tag{2.5}$$

The image of the entire emission spot must be guided through all limiting geometrical restrictions, such as windows, lenses, mirrors, atomizers, optical slits, etc. to obtain the maximum possible Etendue. This principle, and not the size of mirrors or windows, defines the performance of the photometer.

Analytical instruments such as AAS and UV, spectrometers if not limited to a single wavelength, are more demanding: the source essentially emits a spectrum of lines, mixed with lines from other elements and from components or supplies of the source. In the atomizer, many atoms, molecules and particles are generated, the analyte atoms usually being only a small fraction – less than a thousandth – of the total absorbing or emitting species. Furthermore, a lot of light is emitted in the atomizer which reaches the detector. The line used for the measurement must therefore be isolated by a monochromator or polychromator. In AAS, as an example, the emission profile of the selected resonance line is sharp, the resolution of the monochromator or polychromator for AAS can be limited to about 0.2–2 nm, 100–1,000 times wider than the absorption profile. Selecting the bandpass, depending on the spectrum in the neighborhood of the measurement line, is sufficient to reduce the stray light level for all elements to less than 1%, and still provide enough photons at the detector for good counting statistics. The situation in plasma optical emission spectrometry (see Section 2.10) is completely different. Often, thousands of lines are populating a wavelength range of a few nanometers only. The resolution must be in the range of few picometers only, at the prize of intensity limitation of the emitted light.

In a typical monochromator or polychromator, the radiation source, imaged on the entrance slit, is collimated onto a grating or a prism and separated into wavelengths (dispersed). These wavelengths, separated images of the entrance slit, are either focused on an active pixel sensor and read out quasi simultaneously, or on an exit slit which separates the wavelengths. The exit slit may iluminate a single element sensingtotal photon flux or a detector providing additional spatial resolution. The wider the exit slit of a given monochromator, the broader the wavelength window reaching the detector. At the same time, the dimensions of the entrance and exit slits determine the amount of light reaching the detector. If the spectrum is not swayed relative to the detector and is read out quasi simultaneously, we talk about a spectrograph, if the dispersive element is moving to select the wavelength region of interest, we refer to it as monochromator.

2.1.4.1 Basic considerations of wavelength separation by a grating

In the case of separation of wavelengths by prisms, the concepts of "light-rays" from geometrical optics provide a good explanation [34]. Understanding the function of a grating requires the description of electromagnetic radiation by waves. Water waves may provide some intuition about the behavior of the more abstract electromagnetic waves.

If we image a place on the water, where there is a rapid rise in the water level – e.g., a bump resulting from a stone thrown into a pond as displayed in the left of Figure 2.8 – this bump will give rise to a wave moving outward, which will quickly reduce in amplitude, while the initial bump will vanish completely.

Figure 2.8: Point excitation giving rise to spherical waves (left), how light beams in optics do not look like (middles) and a Gaussian beam (right).

If we want to derive the motion of light rays from the wave picture, some problems occur: A light ray should correspond to waves with the wave-crests orthogonal to the direction of the ray. How do they look in the transverse direction (the directional parallel to the crests)? If we want narrows beams, the naïve idea might be to just cut them off as shown in the middle of Figure 2.8. The wave shown there however does not look like a real water wave at all! Remembering the dynamics of the bump we can also understand why: in a situation as displayed in Figure 2.8, a new wave would form and move orthogonal to the steep sides of the waves just as it does on the steep slopes of a bump. If we want to have something resembling narrow beams of rays from geometric optics, we therefore must slowly reduce their amplitude in the transversal direction as depicted in the right of Figure 2.8. We have also learned something from the initial naïve approach above: whenever we start manipulating a beam in the transverse direction – e.g., by forcing it through narrow slits – we will get additional waves propagating in directions different from the initial propagation direction of the wave.

This can be made more formal by the Huygens-Fresnel principle, which gives a rather surprising way to predict the propagation of a wave starting from a surface S where the wave is passing through: look at the amplitude of the wave at each point on this surface and assume a spherical wave (the bump from the beginning of this section) emanating from this point. Intuitively we have decomposed the wave on this surface into a lot of bumps and each one is producing a spherical wave. To get the amplitude of the wave at a point P away from the surface S, we just must sum up all the amplitudes of these spherical waves at P. If the wave-profile is chosen in just the right way – e.g., for the situation in the right of Figure 2.8 – they will all “conspire” to give new waves with almost the same shape, i.e., the situation can be interpreted as a beam with almost constant width.

When we start to drastically alter the surface *S*, the resulting wave-pattern behind *S* will also change substantially. The simplest such situation is the case, where *S* is a surface that blocks all radiation except for two narrow slits at distance d apart. Looking at this situation in the plane perpendicular to the long side of the slits, we get the situation depicted in the left of Figure 2.9. From the left, waves with constant amplitude perpendicular to their direction of propagation arrive. At the two slits, they produce two spherical waves which start with the same phase. At all other points on *S* the wave is completely blocked, so to the right of *S* only the spherical waves from the two slits contribute.

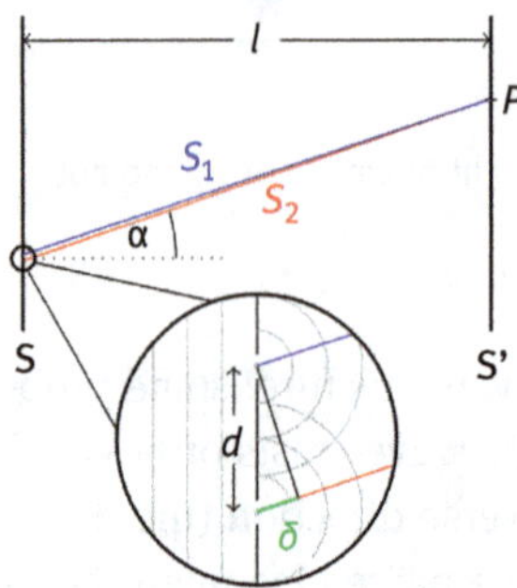

Figure 2.9: Wave pattern behind a double slit. *S* is the surface comprising the two narrow slits. Their distance and width are orders of magnitude smaller than the distance to *P*. The circle below is a magnification of the double slit region.

If we take a point *P* at a macroscopic distance l to the right of *S* – say a few centimeters – then even for IR radiation this is many thousand wavelengths away. As we will see in the following, the slits should have a separation *d* of a few wavelengths, so l it will be much larger than *d*. When drawing s_1 and s_2 from the slits to *P* we thus get the two sides of a very, very narrow triangle. Looking at the base of the triangle, they are basically indistinguishable from parallels, which run at an angle α to the normal of *S*. But this means, that s_2 is $d \sin \alpha$ longer than s_1.

Starting at $\alpha = 0$ the two spherical waves therefore must travel the same distance and arrive with the same phase at *P*, interfering constructively. If we increase α then the wave from the second slit will lag more and more behind the one from the first. When the difference in distance travelled reaches half the wavelength, the two waves arrive with opposite phase at *P* and cancel each other out. Increasing α further, the difference in distance traveled will at some point reach one wavelength and the waves will again arrive in phase (although one is an entire period "behind" the other) and interfere constructively. We thus end up with a pattern of bright and dark stripes on the surface *S*'. By the argument just given, the stripes left and right of the central one can be found at an angle α determined from $\sin \alpha = \lambda/d$.

But this means, that our double-slit can be used to separate different wavelengths! Excluding the forward direction, the angles at which radiation is detectable depend on the wavelength. If we want to have a reasonable change in α with the wavelength this means, however, that d should be of the order of λ, i.e., the two slits should be roughly one wavelength apart.

While the double-slit is conceptually simple, it suffers from several drawbacks:

1. The slits were assumed to be very narrow – so very little light will pass, leading to an instrument with very poor light throughput
2. The stripes vary like a sine from intensity zero to maximum intensity. Even when passing light from a monochromatic source through the slit, we will not get sharp lines but blurred regions, leading to very poor wavelength resolution

To fix 2., one replaces the double slit by a much larger number of slits. Ignoring some subtleties, the basic argument concerning constructive interference remains the same. So, the first maxima can still be found at the same angle. Destructive interference however takes place much faster when moving away from the position of the maxima. So, the broad stripes get replaced with narrow lines, which in addition have a higher intensity. However, even in this case some intensity between these lines is unavoidable. This effect is one of the sources of straylight and will be discussed later on.

To fix 1, one replaces the narrow slits by slits of width approaching d. This does modify the resulting intensity on S', but only in a moderate way and can be controlled by other techniques. The most important one is to use a ruled mirror instead of a (transmission) grating consisting of slits. By "tilting" the surfaces between the lines on the mirror – a procedure known as blazing – one can move the effect to angles which are not used in wavelength separation applications.

2.1.4.2 Monochromator

The magnitude (one could call it "quality" as well) of the separation of the radiation depends on the dispersive element [35] . In most AAS and UV monochromators, flat gratings are used. The light of a single wavelength in a standard grating at normal incidence is diffracted to the central zero order and successive higher orders at specific angles, defined by the grating density/wavelength ratio, and the selected order. The angular spacing between higher orders monotonically decreases and higher orders can get very close to each other, while lower ones are well separated. The intensity of the diffraction pattern can be altered by tilting the grating. With reflective gratings (where the holes are replaced by a highly reflective surface), the reflective portion can be tilted (blazed) to scatter a majority of the light into the preferred direction of interest (and into a specific diffraction order). For multiple wavelengths the same is true; however, in that case it is possible for longer wavelengths of a higher order to overlap with the next order(s) of a shorter wavelength, which is usually an unwanted side-effect. The grating is used under a relatively small angle to the incoming light. The entire spectrum is generated in only one order. The grating is turned to image the selected wavelength onto the exit slit and to the detector.

The line density is high and the distance between the lines is in the range of the wavelength of the incoming light. A higher line density results in better angular dispersion. The second important parameter for the spatial resolution of different wavelengths is the distance from the slits to the collimating mirror expressed as focal length: the longer the focal length, the better the linear dispersion (nm wavelength difference per mm) of the monochromator. High-performance monochromators have 1,800 or more lines/mm and a focal length of about 300 mm. One more parameter of importance for the efficiency of the grating, and therefore the light throughput of the system, is the so-called blaze angle [36]. To improve the light throughput, the grooves are shaped such that dispersion takes place predominantly in only one direction of the grating. As the angle of the grating relative to the incoming light changes with the selected wavelength, there is only one optimal blaze angle possible. Each grating has therefore a wavelength-dependent efficiency. The gratings are usually optimized toward a maximal efficiency at about 250 nm. A typical monochromator for AAS or UV is displayed in Figure 2.10.

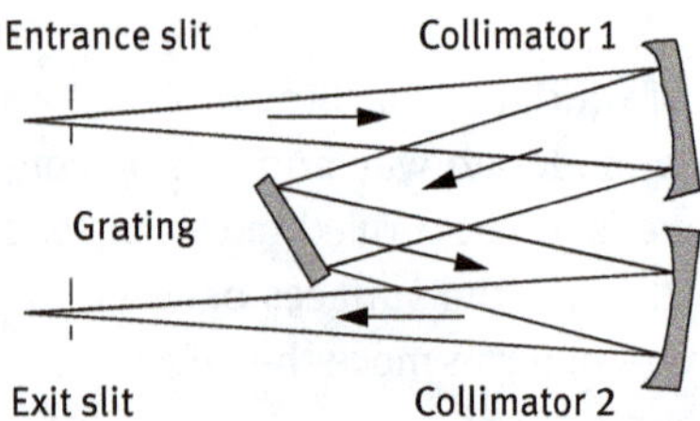

Figure 2.10: Czerny-Turner monochromator. The light passes through the entrance slit, falls on a parabolic mirror and is collimated onto the grating. After dispersion, the light is collected on a second collimating mirror and focused onto the exit slit. The focal length is the distance from the slits to the collimating mirror.

In a second frequently used arrangement, the Littrow monochromator, the same mirror is used to collimate the light from the entrance slit onto the grating and to focus the dispersed light from the grating on the exit slit. This arrangement is simple, rugged, very efficient, and economic. It generates a little bit higher straylight levels than monochromator types with more than one focusing mirror. Other monochromator types (e.g., Ebert) use different positions of the focusing mirror relative to entrance slit, exit slit and grating but this has only a minor influence, if any, on the optical performance of the spectrometer.

These classical monochromators used in AAS and UV have the following characteristic in common:

- Non-collimating planar gratings
- Small angular dispersion
- Blaze angle optimized for wavelengths close to 250 nm
- Only the first order of diffraction is used

Most AAS and UV instruments still make use of such classical monochromators because of their simplicity and ruggedness. High light throughput, adequate optical resolution, and very low straylight levels characterize them. They are usually combined with photomultiplier tubes as detectors, which can be easily positioned behind the exit slit. Regardless, whether only one line or wavelength bandwidth is selected, or whether the instrument is scanning a selected wavelength range, the accuracy and the speed of scanning and acquisition of the data at a specific wavelength is an essential quality criterion for the spectrometer itself. The optical base, the components to fix and adjust grating, slit, and mirror, the quality of the mechanical drive define cost and quality of this part of the instrument. As important as accuracy and stability of wavelength is the stray light generated in the monochromator itself. Straylight influences the signal to noise ratio of the spectrometer. Even more important, it is the most important source of non-linearity in absorption techniques (see sections about AAS and UV).

2.1.4.3 Polychromators

The grating used for dispersion can be spherically shaped. This will act as a focusing optical component as well and will generate the spectrum onto a circle. If the illuminating source, the grating, and the detector or exit slit are all positioned on this circle we talk about a Rawland circle. This configuration is one of the oldest and most powerful arrangements used for spectrographs. The spectrograph is called Paschen-Runge system. The Rawland circle configuration is sketched in Figure 2.11.

Paschen-Runge spectrographs provide high light throughput and good but not exceptional optical resolution. The system was perfectly matched to a detector which could be easily spherically shaped as well (chemical film). The mechanical setup is relatively straightforward as it does not require moving gratings and offers highest mechanical stability. The positioning of electronic detectors is more complicated and requires higher technical effort though.

A fundamentally different way of dispersing light is to use a relatively coarsely ruled grating with about 100 lines per mm, a so-called Echelle Grating. The grooves are shaped for a high angle of incidence of the incoming light in the range of 60–75°. This grating generates high dispersion of the light but also generates a partial overlay of many orders of diffraction. The intensity maxima for a wavelength of interest appear in different orders. For order separation, a second dispersive element, a classical grating or a prism is used which generates dispersion rectangular to the Echelle grating. The resulting spectrum is two-dimensional. A specific wavelength has its exact mathematical predictable position on a plane. It is usually read out by an active pixel sensor, an assembly of diodes, a Charge Couple Device (CCD) or a Charge Injection Device (CID). The layout of an instrument using an echelle polychromator for graphite furnace AAS is sketched in Figure 2.12.

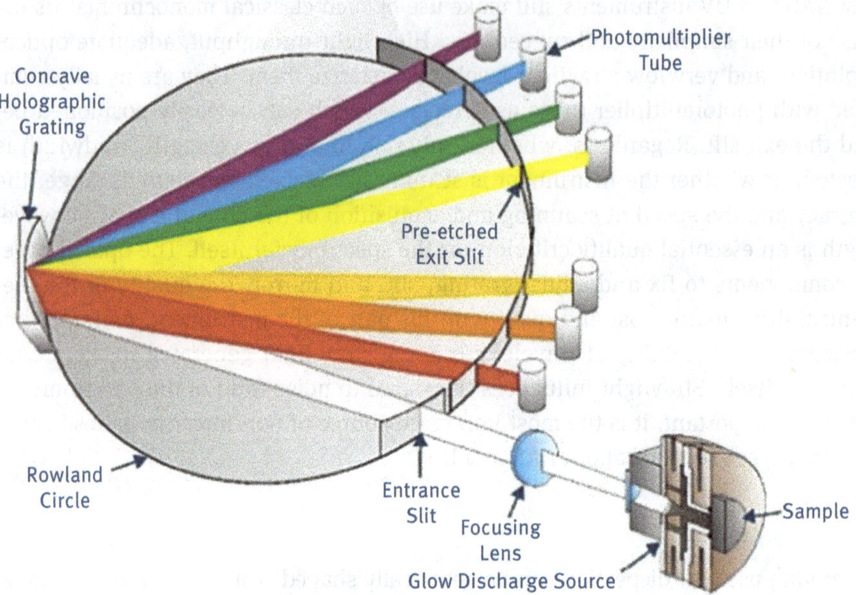

Figure 2.11: Paschen Runge configuration for imaging of a spectral range. The source in this example is a glow discharge emission source. Courtesy of LECO corporation. Michigan, USA, Dr. Zdenek Weiss.

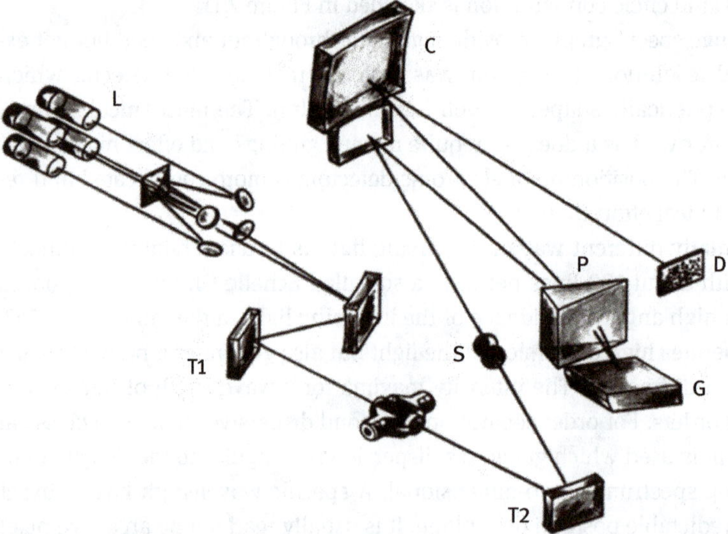

Figure 2.12: Optical system for simultaneous multi-element atomic absorption spectrometry (Perkin Elmer, SIMAA 6000) L: lamp holder with up to 4 lamp positions; T1 and T2: focusing mirrors; S: slit assembly; P: quartz prism; G: grating; C: camera mirror D: detector.

The light from up to four lamps is combined and guided through the atomizer. The sources are then imaged by a second toroidal mirror onto a slit assembly with two entrance slits, which stigmatically limits the light bundle for the two dispersing components, the Echelle-grating for the main dispersion and a quartz prism for the cross-dispersion. The Echellogram is generated by passing collimated light through the prism to the grating and back through the prism onto a mirror which focuses the spectrum on a monolithic solid-state detector. Rectangular photosensitive areas corresponding to the image of the entrance slit are positioned for – in this case – 60 wavelengths.

Echelle systems can be used as polychromators, just as described above, as monochromators. In the latter the Echellogram can be shifted in two dimensions relative to the detector by moving one or both dispersing elements so that a small portion of an order is selected and imaged onto a slit or an active pixel sensor. An example of a high-resolution Echelle-monochromator is imaged in Figure 2.13. This

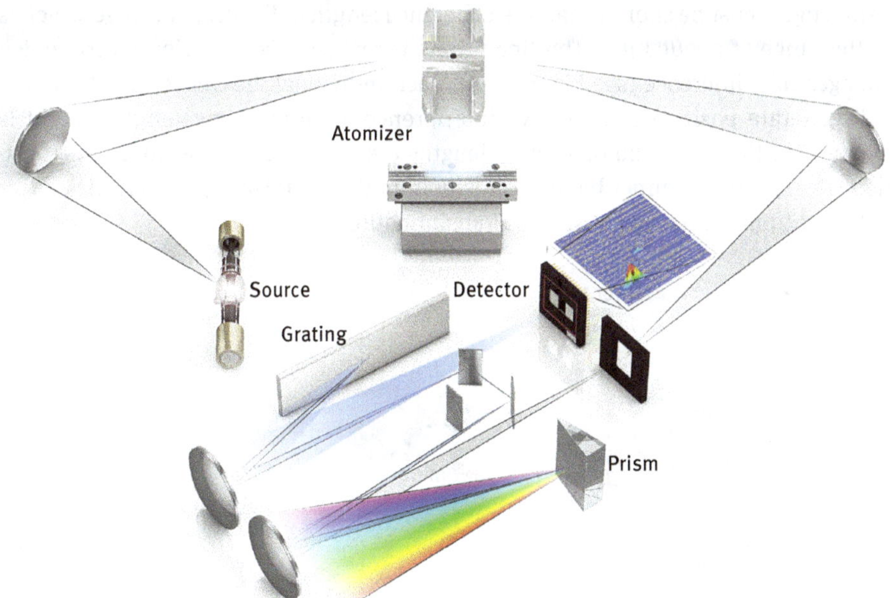

Figure 2.13: DEMON (ISas-Berlin, Germany) Double-Echelle monochromator used in continuum source AAS. The radiation from the xenon-arc is focused on the atomizer (e.g., flame), on the entrance slit of the pre-monochromator. This is a prism in Littrow mount which is dispersing the radiation twice. The selected wavelength range is passed through an intermediate slit and imaged on the Echelle part of the system, also in Littrow mounting. After dispersion by the Echelle grating the radiation is imaged on a linear CCD detector. Both dispersing elements, the prism as well as the Echelle grating, can be moved in very small angles to provide access to the entire spectrum from 180 to 900 nm.

monochromator is used for continuum source AAS. The element specific line sources are replaced by a single high photon output white light source. The monochromator must provide resolution in the range of a line source profile, i.e., about 2 pm to meet the criteria of profile overlap in AAS.

Echelle monochromators or polychromators offer very good resolution within a compact design. They offer high photon throughput but require sophisticated algorithms and instrument control to select the optimal order and position of reflection.

Yet another completely different principle of generating a spectrum is the interferometer. An interferometer is quasi-tunable filter. Interferometers make use of interference of waves to detect the magnitude to be measured. Effects which change the intensity of the active wavelength of one of the beams within the system can be recorded with high sensitivity and precision. Light from one source is separated into at least two beams. The two beams have a slightly different optical path length. They are superimposed again with the help of special mirrors. The difference in path length must be shorter than the coherence length of the original wave which is in the range of about 1 µm. The fine-tuning is realized via a moving mirror which changes this difference (see Figure 2.14). Each individual difference in path length will generate positive constructive interference at certain wavelengths. Light is maximal at this spot. The other wavelengths are weakened by negative superposition. The spectrum cannot be read directly from the recorded light intensities but is obtained from them using a Fourier Transformation.

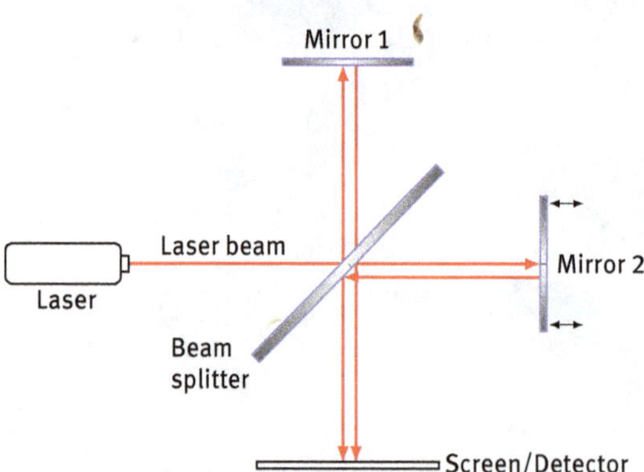

Figure 2.14: Michelson interferometer.

Light from a coherent source (e.g., Laser) is focused on a semi-permeable mirror (Beam Splitter); half of the light is transmitted to the moveable mirror 2, half is reflected to the fixed mirror 1, reflected to beam splitter and brought to interference with the transmitted light of mirror 2. Part of this mixed waves is decoupled via the beam splitter and sent to the screen (usually the detector) Substance for measurement is resident in a cuvette between beam splitter and screen.

An interferometer does not need small slits like classical monochromators. Light throughput is therefore significantly higher. The entire spectrum can be recorded rapidly which provides better signal to noise compared with a scan. If the spectrometer is simultaneously calibrated with a very narrow and exact wavelength of, e.g., a laser source, accuracy, and precision of the recorded wavelengths are extremely high. Interferometers are mainly used in IR and Raman spectrometries.

2.1.4.4 Filter photometer

Often it is necessary only to select a certain wavelength range for illumination of the sample. In this case, optical filters are used to cut off the wavelength toward the shorter and the longer wavelength range. The rather complex optics of a spectrometer becomes a simple setup of a photometer. This principle is often used in atomic or molecular fluorescence. In most cases the filter element is placed in front of the sample to be illuminated. Filters are often used in reflection mode. The surface of the optical component (mirror) with filter function is illuminated. Only a part of the wavelength spectrum is reflected in the direction of the sample to be illuminated. Transmission filters have the same operating principle. In this case the light is passed through the component. The unwanted part of the wavelength spectrum is absorbed. The quality criterion is the wavelength characteristic of the filter. This is predominantly the ratio between intensity of the light of the wanted wavelength to the intensity of the blocked wavelength. Light of unwanted wavelengths acts as stray light for the measurement. If the filter passes a narrow part of the spectrum only, it is called dichroic filter. If it blocks the spectrum in one direction toward the shorter wavelength range or the longer wavelength range, providing a wide range of the spectrum in between, it is named edge filter. If it blocks out a certain short wavelength range and is permeable to the left and to the right side of the spectrum, it is called notch filter.

Whereas the transmission of the useful wavelength is often above 90% of the incident light, the blocking characteristic is often the important criterion for analytical quality. The performance of a filter or of a combination of filters can be determined with the help of a transmission spectrum (see Figure 2.15).

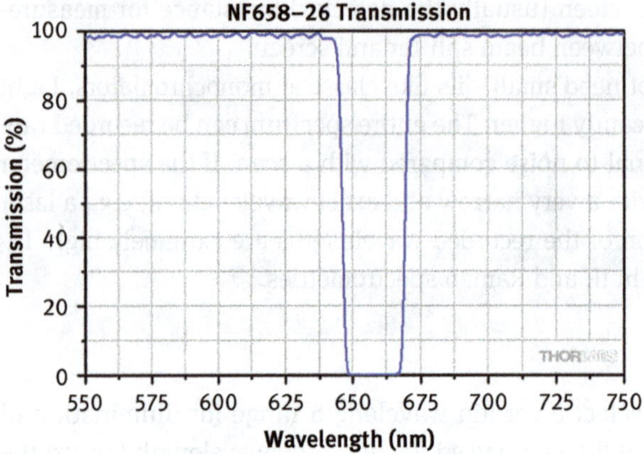

Figure 2.15: Transmission spectrum of a notch filter used in spectroscopy. Courtesy of Thorlabs Inc. Newton, New Jersey, USA.

2.1.4.5 Optics for X-rays

For wavelength below 100 nm, beyond the wavelength range of far UV radiation, matter becomes more and more impermeable. Lenses for radiation shaping must be very thin to keep the energy loss small. The refractive index of matter is very close to and slightly smaller than unity. In optical systems described above the refractive index of quartz is about 1.5. It is obvious that shaping of X-ray beams follow different principles. Still diffraction, refraction and reflection can be used for beam focusing although the shape and illumination of the optical components differ strongly from classical optics.

In XRF analytical instruments filters are used to separate the characteristic radiation of the anode material (see above) from bremsstrahlung. Filtering this way is, to some extent, limiting the usability of the technique for elements which are used as anode material in the X-ray tube. Shaping of the X-rays to a parallel beam or spatial limitation of the X-ray beam is usually realized by metallic tubes or fin-type stops. Just like in optical spectroscopy wavelength (line) separation is essential for analysis. The X-rays must be diffracted. Single crystals providing a lattice spacing in the range of a nm, which matches the wavelength of X-rays, are suited for this purpose. Single crystals are therefore used for wavelength separation in X-ray spectroscopy. LiF-single crystals are an often used as main optical component. Today mirrors coated with multiple layers are in use as well.

2.1.5 Electrical current or voltage: indicator for analytical information

The task of a detector is to transform radiation into an electric current. The sensitive element may be a photosensitive area behind the exit slit of a monochromator providing intensity information without optical resolution. Sensitive elements may be positioned on a line providing one-dimensional wavelength resolution, or on a plane providing two-dimensional resolution. Read out of the generated current may be continuous or discontinuous. Collecting and read out functions must be fast enough to follow the dynamic processes taking place in the measurement cell or volume which usually requires read out times of <10 ms. The number of electrons generated per number of photons shall be as close to one as possible and shall be similar over the entire wavelength.

In early years of spectroscopy, the chemical film was the detector of choice. Just like in photography, the photons generate a blackening of the photographic active substance which is proportional to the number of photons falling on the respective spot. Photographic films are flexible, can be bent and thus, e.g., positioned around a circle of a Paschen-Runge spectrometer. Films carry multidimensional information, but the information is not directly available. The illumination is separated from read out. Illumination requires time and the information on the film does not contain any information on the temporal development of the registered process. In most analytical instruments photographic films are no longer in use.

A widely used tool and one of the oldest detector types for analytical instrumentation is a photomultiplier tube (PMT). In this device, photons striking a photosensitive cathode, release photoelectrons which are accelerated in an electric field to a dynode, where several secondary electrons are released and accelerated to a second dynode, etc. The electron flux is amplified further in a chain of about 10 dynodes (see Figure 2.16).

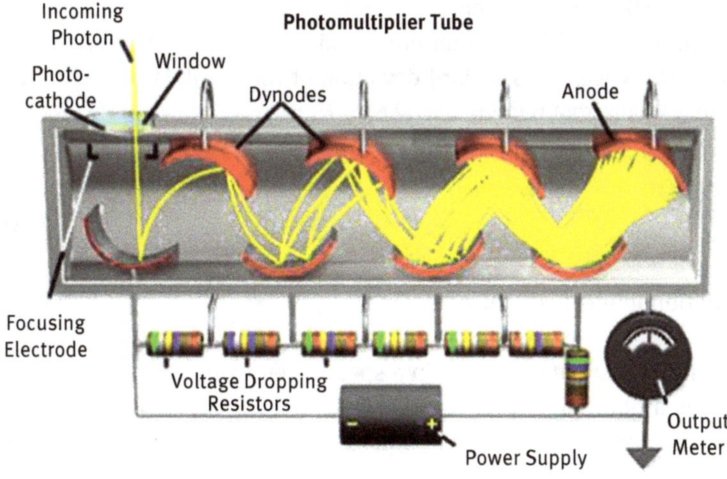

Figure 2.16: Photomultiplier tube for transfer of photons into electrons. Courtesy of Hamamatsu learning center, Japan.

The amplification of a photomultiplier depends strongly on the kinetic energy of the photoelectrons and therefore on the voltage between the photocathode and the last dynode. This is adjustable in the range of up to somewhat above 1,000 V. At this voltage, amplification is typically a factor of between 10^6 and 10^7. The efficiency of detection depends on the type of the cathode material and is expressed as quantum efficiency (that is the number of electrons released by 100 photons). The quantum efficiency is wavelength-dependent and usually between 10% and 30% in the wavelength range typical for elemental absorption or emission, atomic fluorescence, or UV spectroscopy. The output current of a photomultiplier is converted to a voltage and amplified in conventional solid-state circuitry yielding an analytical signal. The combination of photomultiplier and amplifier should provide linear response of signal to light intensity over at least three decades. The gain of the photomultiplier is adjusted by an instrument-inherent algorithm so that a certain predefined electric current is flowing to the amplifier. If large numbers of photons reach the photomultiplier cathode per unit time (e.g., in the case of a bright radiation source or very small absorbance), the statistical variation in the signal generating process (the so-called shot noise) is small. The standard deviation of a series of measurements is often below 0.1% transmission. When the absorbance increases, fewer and fewer photons reach the detector. The amplitude of statistical noise increases relative to the signal and the baseline noise therefore increases. The lowest standard deviation for an absorbance reading is therefore obtained at close to zero absorbance. This is exactly when it is needed: at the lowest concentrations, close to the detection limit! If the statistical noise in absorption techniques is defined by the detector noise at zero absorbance, the system is said to be shot noise limited. The statistical noise is dependent only on the number of photons per time unit hitting the photo multiplier. Shot noise limitation must be the goal in instrument design. In this case, the standard deviation of the blank is related to the light throughput in a square root relation (eq. (2.6)).

Equation 2.6:

$$\text{s.d.} \sim \frac{1}{\sqrt{i}} \tag{2.6}$$

where i is the radiation intensity on the detector.

The relative intensity of a radiation source in a spectrometer with a given amplifier can be expressed as a photomultiplier voltage. This value is roughly comparable between instruments of the same design. It is, however, no quality criterion for instrument components, such as lamp or optical system, as the layout of the amplifier board may be different from instrument to instrument. The layout of the amplifier electronic should be designed such that the PMT is working in an optimal range of about 500–1,000 V. The relation between voltage and number of incident photons is very non-linear. As a rule of thumb, 40 V decrease in the photo multiplier voltage corresponds to a doubling of the incident light energy. This type of detector seems to be the ideal photon capturing system for optical spectroscopy. It is fast enough to follow fast changes in photon intensity. However, the device is relatively bulky and provides no wavelength selective reading. The wavelength must be limited by the monochromator exit slit. Thus, light with wavelength within only one spectral window will be measured. In AAS, e.g., this usually is a window between 0.1 and 1.5 nm.

The trend in chemical analysis, however, is to obtain as much information per measurement as possible. The information content of a measurement can be increased if a solid-state detector, providing many photosensitive spots or areas on a semiconductor chip, is used. The technology for solid state detection has progressed dramatically in the last few years.

The techniques used are single photodiodes, multiple photo diodes, charge couple device detectors (CCD) used in linear and two-dimensional read-out arrangement. A modification of the CCD is the CID (charge injection device) and the active pixel sensor (CMOS). CMOS detectors in specific are used in most commercial cameras today. It is just appearing in the first commercial analytical instruments but will certainly become very popular in the years to come.

Fundamental aspects to consider for each detector type are usable wavelength range, quantum efficiency as a function of wavelength, speed of read out, read out noise and dark noise. In general, detectors in modern instruments represent an important part of the instrument cost.

A CCD solid-state detector basically consists of one, several, or an array of photodiodes located on a semiconducting material. The electrical charge is generated by photons which excite electrons from immobile bound states (from the valence band) to a conductive band. The electric charges, electrons, and holes, are transported rapidly in an electric field, collected below electrodes, transported to the edge of the detector and measured there as a current. After reading out, the detector is empty, again. The schematic of such a detector chip is sketched in Figure 2.17.

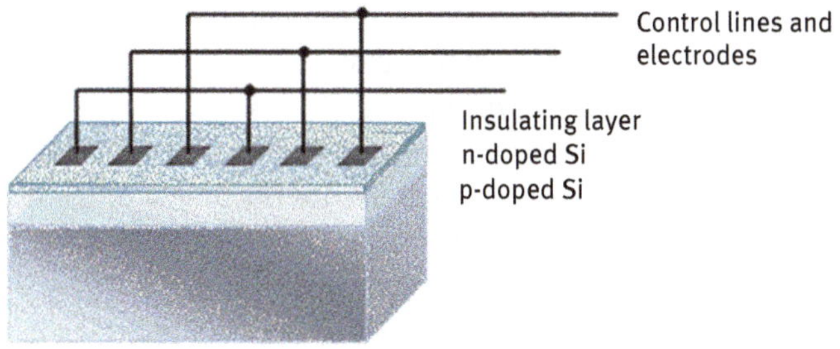

Figure 2.17: Charge-couple-device.

In atomic spectroscopy, the charge coupled device is usually imaged from the back side (the side without electrodes) to obtain highest quantum efficiency down to the UV range of the spectrum between 160 and 200 nm.

The CID used in atomic spectroscopy consists of an X-Y addressable array of photosensitive capacitor elements. Every pixel in a CID array can be individually addressed via electrical indexing of row and column electrodes. The number of charges after illumination is read out in form of a displacement current of individually selectable pixels. Readout is non-destructive, because the charge remains intact in the pixel after the signal level has been determined. Flexible readout and processing options are made possible by addressing each pixel individually. On the other hand, CID detectors are usually illuminated from the front side. The spectral range in the low UV region is limited and CID detectors usually use a fluorescence surface coating to improve quantum efficiency below 220 nm. These coatings may wear and limit the detector lifetime.

The latest detector technology is the active pixel sensor (APS), often called CMOS technology (complementary metal oxide semiconductor) because of the architecture used. Like the CCD, an APS detector is based on the photodiode principle and converts incident light (photons) into electronic charge (electrons).

Other than in CCD detector chips, circuitry at each pixel determines the charge. It is the active circuitry that gives the active-pixel device its name. The performance of this technology is comparable to many charge-coupled devices (CCDs) and allows also for a larger image array and higher resolution.

The operation of an instruments equipped with a solid-state detector is like that of those using a photomultiplier. The output current, however, is no longer defined by an adjustable voltage. The output is linearly proportional to the photon flux for about 3–4 orders of magnitude.

X-ray radiation cannot be transferred into an electrical current by photomultipliers or active pixel sensors. The short wavelength radiation is transferred into fluorescence and is afterward detected with a photomultiplier. This conversion is usually realized with a suitable doped crystal – often a NaI crystal doped with thallium. The assembly is called scintillation counter. It is working for element detection of heavier elements with an atomic number higher than 26.

A complementary detector is the proportional counter. This is basically a tube filled with an inert gas. The X-ray radiation interacts with the inert gas by ionization of the gas atoms. The generated electron is accelerated toward the anode in the assembly. On its way it generates secondary ion-electron pairs. The positive ions move to the cathode. This way short-term voltage impulses are generated which are proportional to the X-ray energy reaching the detector. The radiation counter detects radiation of a longer wavelength which is emitted by elements with shorter atomic number up to 25.

An important magnitude of all types of detectors is dark noise. This is the signal generated by the device when no radiation is received. The standard deviation of the dark noise is often the limiting factor for the detection limit possible with the device. The dark noise of detectors, in particular of semiconductor-based detectors, is temperature dependent. The higher the temperature, the higher the dark noise. To lower the dark noise or to keep the signal at least stable, these detectors are often temperature controlled or even cooled below zero centigrade. Temperature dependence becomes more pronounced if longer wavelengths are registered.

Toward the other side of the spectrum, at wavelengths longer than 1,500 nm (1.5 µm), the photon energy striking the detector becomes small. On the other hand, long wavelength radiation generates heat which strongly influences the dark noise of semiconductor-based detectors. Still these types of detectors are frequently used in IR spectroscopy. Doping with mercury cadmium telluride (MCT) allows tuning of the wavelength from 1 µm up to about 30 µm. To keep the dark noise low the detectors are cooled. If best signal to noise ratio is required, liquid nitrogen is used as coolant.

Heat can be used as means to detect IR photons. The heat generated by the infrared radiation falling on these detector types is changing an electromagnetic property of the material which can be translated into an electrical signal. The sensitivity, i.e., the change of the property should be high, and the detector should cover a wide wavelength range with a similar sensitivity. The change shall be fast, and the response should be proportional to the change in radiation. As an example, a widely used detector, the DTGS (deuterated triglyceride sulfate) detector, shall be described. The crystalline substance is ferro-electric showing a permanent polarization. Temperature changes due to infrared radiation changes the distances in the lattice and with-it polarization. The change in polarization is measured as a voltage surge which is proportional to the change in temperature.

2.1.6 Generation of atoms for elemental analysis

Elemental analyzers which are not based on electroanalytical methods or X-rays make use of optical interaction between atoms or ions and an electromagnetic field. These must be freed from their chemical compounds. Generation of these atoms/ions is a main technical component of the respective instruments.

2.1.6.1 Nebulizers

Most analyses are performed out of liquid samples. Liquids must usually be converted into an aerosol of fine particles mixed with the medium which carries the energy required to generate free atoms or free ions (see schematics in Figure 2.18). The atoms are resident in a clearly defined zone for measurement or extraction.

The primary device for aerosol generation is pneumatic. This is due to simplicity and cost. Small volumes of sample (1 µL to 100 µL/s) are mixed with a suitable

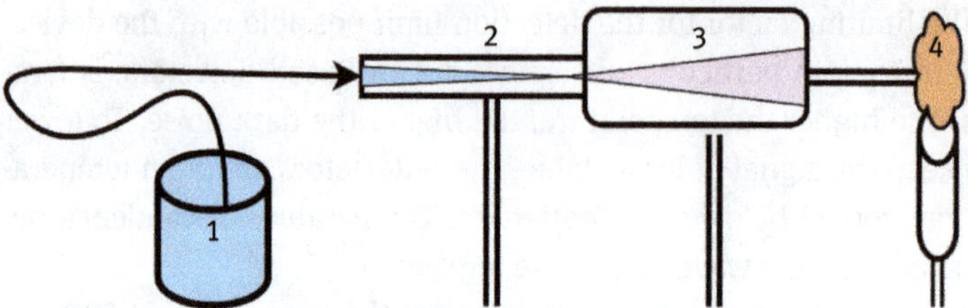

Figure 2.18: Generation of free atoms or ions out of liquids.
1, sample for analysis dissolved in a suitable solvent; 2, nebulizer for the generation of aerosols; 3, harmonization of particle size; mixing and disposal of waste; 4, generation of atoms or ions.

gas in a sprayer. The droplet size should be in the range of few μm to allow fast volatilization and temperature increase. The efficiency of use of sample depends on the flow of the nebulizer gas and sample flow. Various technical principles provide optimal suitability for the various types of samples and specific applications.

Nebulizers for sample introduction into flames do not differ much. They use a gas flow of about 100 mL/s through a concentric nozzle. The nebulizer gas flow consists of the oxidant used in the flame. The sample flow is in the range of 100 μL/s. The primary aerosol is usually further split by an impactor. These nebulizers work with a gross efficiency of up to 25% of the initial sample. The high flow of nebulizer gas makes pumping of the sample to the nebulizer unnecessary. In flame AAS/OES nebulizers are usually self-aspirating. However, pumps might provide exact control of the nebulization parameters by generating low pressure or overpressure at the nebulizer entrance.

Nebulizers for ICP work with a significantly lower nebulizer gas flow of 20 to 40 mL/s. The gas is usually identical to the plasma generating gas (Ar) which flows at a rate of at least 200 mL/s concentrically around the sample channel. Nebulizers can be self-aspirating as well, but the sample is usually pumped with a pre-defined flow. The flow must be steady with a minimum of pulses. Several different principles of nebulization have been applied to make different specifications possible. These are ruggedness against salts and solid particles, low sample volumes, organic solvents, etc. In general, higher flow rates of sample and gas result in better stability against complex matrix at the cost of a low usage of sample. Lower flows allow more efficient use of sample but require samples with low matrix concentration and extremely low sizes of particles, if any. An overview of nebulizers and their typical characteristic is listed in Table 2.1.

Most of the nebulizers are used together with a buffer volume, the mixing chamber. Mixing chambers are basic requirement for flames where the burner gas (ethine) is mixed with the nebulizer gas, additional oxidant, and the sample to generate the required combustible aerosol. The mixing chamber often uses impactors to remove larger droplets and guide the aerosol in a laminar flow to the burner. Mixing chambers in ICP are used almost exclusively to remove larger droplets and generate a gentle steady aerosol flow into the plasma. As plasma torches are much less rugged against high load of solvent and matrix than flames, aerosol formation becomes a crucial element

Table 2.1: Nebulizers for AAS and ICP systems with their typical characteristic.

Nebulizer	Characteristic	Use
Pneumatic, self-aspirating	Rugged, high matrix tolerance, limited use of sample	Standard for AAS
Pneumatic with impact bead	High efficiency, lower matrix tolerance	Most AAS application, limited use for N_2O-C_2H_2
High-pressure pumped nebulizer	High efficiency, low sample volume	Special applications in AAS, coupling with columns
Meinhard concentric	Good efficiency, self-aspirating	Clean samples in ICP-OES and ICP-MS
Micro-concentric (MCN), pumped	Small sample volumes, low flow	Low matrix, mainly used in ICP-MS
High-efficiency nebulizer (HEN), pumped	Small sample volumes, low flow	Low matrix, mainly used in ICP-MS
Cross-flow nebulizer, pumped	Limited efficiency, high ruggedness	Standard for ICP-OES, good matrix tolerance
V-groove, pumped	Limited efficiency, high ruggedness	mainly for ICP-OES, high matrix load
Cone spray, pumped	Limited efficiency, high ruggedness	mainly for ICP-OES, high matrix load
Direct injection nebulizer (DIN)	Lowest sample volume, no mixing chamber	mainly for ICP-MS
Ultrasonic nebulizer	Highest efficiency, desolvation required	Application for lowest detection in ICP-OES and ICP-MS

for good analytical repeatability. Just as in the case of the nebulizer, various technical principles are used to satisfy the various analytical needs such as clean samples at low flow, samples with high concentrations of dissolved salts fed with high flow, etc. In ICP direct injection behind the nebulizer is made use of for micro injection cases. Both nebulizer and mixing chamber are made of chemically inert material. Depending on the type of main application (matrix, solvent, etc.) different materials with different chemical stability character are offered.

A completely different and very efficient way of nebulization is ultrasonics. This type of nebulizer device is used mainly in ICP-OES. The sample is pumped with a flow rate of about 20 µL/s onto a transducer plate vibrating at about 1 MHz. The carrier gas is fed orthogonally on the plate. Droplets of about 2 µm diameter are formed and transferred into an aerosol with an efficiency of about 30%. This aerosol would be too rich to be introduced into an ICP plasma torch directly. The solvent would blow out the plasma or cool it to non-tolerable conditions. The solvent must be removed to a large extent before the aerosol is entering the plasma. This is achieved by consecutively heating and cooling the aerosol in a heater and a condenser leg. The dry aerosol is around 10 to 50 times richer in analyte and this factor can be directly translated into limits of quantitation. Elements with borderline detection limits according to, e.g., the drinking water regulation can be detected comfortably with OES this way. Limitations of this accessory are the matrix load, which should be low, memory effects, and carry over in the device and fairly high cost.

One more important device of liquid transfer into vapor and atoms are electrothermal atomizers. As this device offers combined volatilization and atomization it will be discussed in the section on atomizers/ionizers below. The same is true for devices which generate analyte gas with the help of chemical reactions at low temperatures. More detailed information on aerosol formation for elemental spectroscopy can be found in the literature [37].

Vaporization by electromagnetic energy

Solids (liquids in special cases) are directly volatilized by energy rich radiation. Laser sources, arcs, sparks or radiation at extremely short wavelength became popular to directly volatilize small sample volumes, scan across surfaces, run depths profiles in solids, etc. The amount of sample required can be extremely small or destructive processes may be avoided completely.

2.1.6.2 Atomizers/ionizers

Generation of atoms and ions out of a cloud of gaseous, liquid, or solid aerosols requires high energy. Atoms are usually generated between 1,000 and 3,000 K. A high degree of ionization requires significantly higher temperatures of up to 10,000 K. Thermal excitation depends strongly on the analytical principle applied and on the elements of interest. Widely used atomizers are energy rich flames, high-frequency plasma, electric arc and spark, X-ray, laser radiation.

Flames are the method of choice for AAS/AFS.

A burner, usually machined from titanium, 5 to 10 cm in length with a very thin slot having a width of 0.5 to 1.5 mm is used to provide a stable, laminar flame. Using an ethine gas flow of 2–6 L/min and an air or nitrous oxide flow of up to 10 L/min, the maximum temperature of 2,300 °C (for air flames) and 2,800 °C (for nitrous oxide flames) permits the atomization of most elements while most of the electrons are still in the ground state. Only alkaline-earth, alkaline-, and some of the lanthanide and rare earth elements are significantly excited or ionized.

Pressurized oxidant is guided through the nebulizer (A) and the end cap (B) to the mixing chamber (D); in front of the nebulizer may be a device to separate liquid droplets too big for the atomizer (flow spoiler C). In the mixing chamber the oxidant arriving from the nebulizer is mixed with the burning gas (ethine) and additional

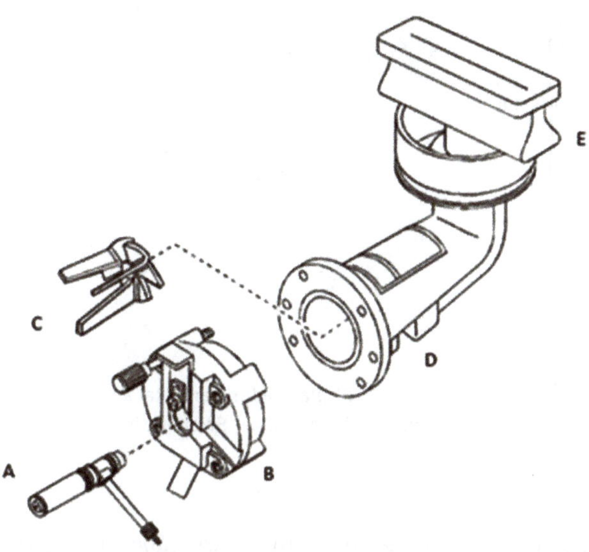

Figure 2.19: Nebulizer, mixing chamber, and burner head (courtesy of Perkin Elmer inc. Waltham, Ma, USA).

oxidant is added. The dispersed droplets are guided to the burner head (E) where the mix is ignited. Bigger droplets are discarded from the mixing chamber (figure 2.19.).

For elements with low ionization potential and high volatility at moderate temperature flames with lower temperature may be preferred to avoid strong ionization of these elements. Slightly modified burner heads are operated with air/propane mixtures at temperatures of roughly 1,300°K.

Low temperature flames are also used to atomize analyte elements which have been transferred into analyte gases by chemical reaction (hydride or cold vapor technique). These flames are predominantly used in atomic fluorescence spectrometry. The burners are small and are operated by hydrogen.

The argon plasma

The most widely used ionizer in the last 40 years is the plasma torch. ICP-based instruments have gained the biggest share in elemental analysis. Volatilization, atomization, and excitation of the sample aerosol takes place in a high-power radio frequency or microwave-induced argon plasma. Argon is passing through an assembly of electrically conductive coils or plates which feed high-frequency radiation or microwave radiation with an energy in the range of 1.5 kW. The plasma is generated by electrons and ions.

At typical RF plasma is sketched in Figure 2.20.

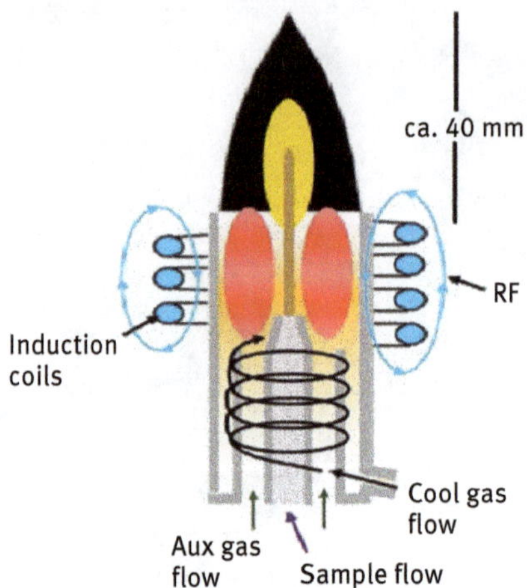

Figure 2.20: RF plasma for elemental analysis Courtesy of Thermo Fisher Scientific inc., Waltham, Ma, USA.

The support for the plasma consists of three concentric quartz tubes. Inside the inner tube the sample gas is flowing with roughly 15 mL/s carrying the aerosol. The tube is shaped ending in the "injector". In the circular orifice between the inner tube and the outer tube gas for maintaining the plasma, the plasma gas, is running with about 20 mL/s. The outer orifice provides the cooling gas at a rate of at least 200 mL/s which protects the quartz support from melting down due to the high plasma temperatures. This outer tube of the torch is longer, reaching into the coil providing the HF frequency. The exact position of the torch in the coil is important to obtain a symmetrical plasma. A bonnet keeps the torch centralized in the coil. For igniting the process, a primary spark is necessary which generate electrons. These oscillate in the HF field and generate argon cations which oscillate as well and form the plasma. Once the plasma is ignited it is kept stable by minor self-adjustment of energy and frequency fed by the generator. The plasma consists predominantly of argon ions in exited states as well as in the ground state and fewer argon atoms in various states. Typical processes are (eq. 2.7).

Equation 2.7: Electronic reactions in an argon plasma:

$$\begin{gathered} e^- + Ar = Ar^+ + 2e^- \\ e^- + Ar^+ = Ar^{+*} + e^- \\ e^- + Ar = Ar^* \end{gathered} \tag{2.7}$$

Excited states of argon, matrix introduced with the sample nitrogen and oxygen by the surrounding air generate a strong emission in the plasma. This is characterized by a rich spectrum of discrete emission lines as well as a broadband continuum generated by bremsstrahlung, due to the change of speed or direction of electromagnetic particles in the electromagnetic field. Thus, a plasma source is a very intense radiation source.

The mechanisms for excitation and atomization of elements are similar. Both electrons and ions contribute to the excitation process of the elements of interest:

$$M + e^- = M^* + e^-$$

$$M + e^- = M^+ + 2e^-$$

$$M + Ar^+ = M^+ + Ar$$

$$M + Ar^* = M^+ + Ar + e^-$$

$$M + Ar^* = M^{+*} + Ar + e^-$$

$$M + Ar^+ = M^{+*} + Ar$$

$$M^+ + e^- = M^*$$

The total energy is expressed as "temperature." Various particles inside the plasma have physically very different temperatures [38]. Still a medium temperature is assumed in a typical ICP for analytical purposes. The highest plasma temperatures of about 8,000 K are observed in the plasma from the coil center to about 10 mm above the coils. Once the sample channel mixes with the plasma gas (mainly by gas expansion) the overall plasma temperature drops by several thousand degrees. The energy to volatilize solvent and matrix is extracted from the plasma. In this zone, about 15 mm above the load coil, the strongest emission is observed. At higher zones where the plasma cools down, recombination of ions and electrons and of atoms takes place.

The aerosol is injected into the base of the central plasma. Introduction into the plasma, however, is difficult. It can be achieved by a viscosity difference between sample channel and the hottest plasma zone (see above). The axis temperature of the plasma is lower due to the gas/sample mix which is cold and energy-consuming. The formation of the central channel is of utmost importance for the correct function of the plasma. It can be checked with an element showing a strong colorful emission when heated.

The size of a typical plasma plume is less than 10 mm in diameter and the hot zone is around 40 mm high. The strongest emission to the sample channel is around 3 mm wide surrounded mainly by plasma background radiation. The observation height is often selected at about 15 mm above the coil. In this region the ions are extracted into the mass spectrometer. One can observe the plasma from the front as well (radial observation). The observation zone is about 10 times longer this way (more photons compared to radial observation). However, there are many more temperature zones and potential matrix compounds which also increase the unwanted background. Both types of plasma observation are commercially offered. Optimal burning conditions, stability, and freedom from contamination by matrix are obtained when the plasma is standing vertically.

The high effective temperature for the atomization/ionization of virtually all elements is certainly one of the main reasons for the straight road of success for this atomizer. Compared with hot flames, the ICP operates with non-hazardous gas and can be easily run unattendedly. The high consumption of Ar is a considerable cost factor. The plasma is less stable against high matrix compared to the flame and the exact settings of operational conditions such as flow of sample and gas, type of nebulizer and mixing chamber, generator power, observation height, and direction of observation require significantly more sophistication than the rugged flame.

The shortcomings of ICP, in specific argon gas consumption, led to the development of mini torches or torches which operate with air as cooling gas or gas mixes. Despite reports on substantial progress in cost saving [39] these systems did not find wide use in laboratories and were not commercialized.

Mixed gas plasmas have been used successfully to change the plasma conditions and increase stability and robustness of the plasma [40]. The idea is predominantly to increase the thermal conductivity of the plasma (hydrogen, oxygen, nitrogen) or to increase the electron density (helium) and thus generate higher plasma stability. The mixed gas is usually entered to the central channel, i.e., the sample gas. Oxygen can remove high load of carbon and reduce carbonaceous residues on the torch or the cones in case of ICP-MS.

A different way of generating a stable plasma has been a topic of research for many years [41] but has been commercialized only recently: the microwave-induced plasma (MIP). Small microwave-induced plasmas have been used for small samples, mainly after separation in chromatographic systems. Microwave-induced plasmas for general purposes have been proposed by researchers at the end of the first decade of the present century. [42]. The commercially available instrument uses a magnetron at a frequency of 2,45 MHz and Nitrogen as a plasma gas. The plasma operates at a power of about 1 kW. Under these conditions seeded electrons can be accelerated such that they can ionize enough neutral atoms to maintain a stable plasma. Although the excitation temperatures are reported to be significantly lower than in an inductively coupled plasma (2,000 to 6,500 K), most elements accessible to ICP-OES and flame AAS can be ionized or atomized efficiently and are thus accessible to elemental optical emission spectrometry [43]. This method is marketed as a highly productive alternative to flame AAS. Only one gas is needed (nitrogen) which can be generated on site. The technique is therefore independent of bottled gas, does not require burnable and potentially dangerous ethine and expensive N_2O, and can be operated unattendedly.

The electrothermal atomizer

The energy required for element atomization can be generated in a small graphite cavity which is heated by ohmic resistance up to 3,000 K. The graphite must be protected from air by an inert gas (argon). The small size of the atomizer (about 6 mm diameter and 30 mm length) provides a cell of less than 1 cm^3 volume which defines an evenly hot cell for generation and moderate time of residence for atoms. It is made of ultrapure high-density graphite with a cold resistance of around 15 mOhm. The electric power applied is of low voltage – less than 10 V – and high currents – up to 600 A.

This device revolutionized AAS about 50 years ago and made it the most powerful elemental analytical technique concerning limits of quantitation for about 20 years starting from 1975 (see figure 2.21). The second most important feature is the unparalleled low sample consumption of this atomizer type. At the turn of the millennium ICP-MS took over a lot of the applications in big laboratories. Nonetheless, graphite furnace AAS is one of the simplest and reliable analytical techniques if quantitative analysis of metals is performed. The graphite atomizer is not only an atom source, but it serves as well for drying liquid samples and separating analyte from matrix. Many solids can be volatilized directly, and mixtures of solids and liquids can be run. The system can be operated automatically and unattendedly. The way of operation includes the following steps:

- Introduce a small, exactly defined, and reproducible volume of sample into a chemically inert and stable reservoir.
- The solvent and the bulk of the matrix is removed at various temperatures programmed and controlled by a computer.
- A few nano gram of stable compounds including the analyte elements are left after the pretreatment steps
- The device is heated to the atomization temperature in a very short time so that all the atoms are experiencing the same temperature and are introduced into the light beam enclosed by the inert walls of the device.
- No gas is flowing through the system and the analyte atoms simply diffuse out of the tube at their own pace.

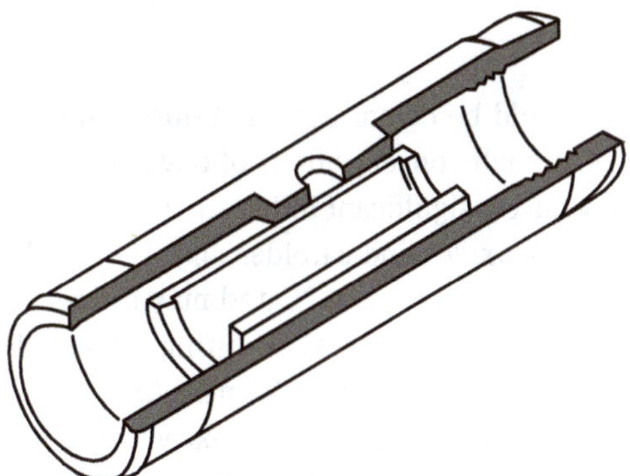

Figure 2.21: Graphite tubes with integrated L'vov platform; cylindrical design with contacts at the ends of the tube. Electric current flows along the tube axis.

Drying and pyrolysis steps usually require between 1 and 2 min. The measurement step requires only 2–5 s at temperatures between 1,600 and 2,800 K. The time to bring the tube from pyrolysis to atomization is only in the range of a second. The measurement cycle is completed by a cleaning step for another 6–8 s at 2,800 K with a full gas flow to purge the remaining debris out of the cavity. The tube is then cooled by a water-cooling system to temperatures below 100 °C within about 20 s. For protection of the graphite tube from the ambient air, the tube is carefully shielded by graphite contact cylinders. An argon gas flow inside the contact cylinders around the graphite tube is constantly flowing and protecting the tube from the outside. An independently controlled inner gas flow is used to direct the removal of matrix toward the tube center and the dosing hole. The mechanical design of a graphite furnace based on the Massmann principle [44] is shown in Figure 2.22. Argon absorbs almost no radiation in the wavelength range used in AAS.

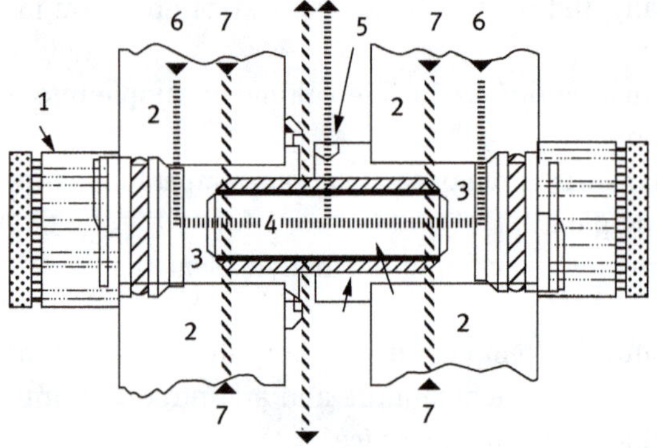

Figure 2.22: Sketch of a Massmann-type graphite furnace.
1, sealed quartz windows; 2, metallic cooling chamber; 3, graphite contact cylinders; 4, graphite tube; 5, hole for sample introduction; 6, flow of internal purge gas; 7, flow of external inert gas which protects the tube from ambient air.

The atmosphere inside a graphite furnace should be chemically inert during the atomization. This is de facto not the case. Very small concentrations of oxygen and/or gaseous carbon compounds and matrix may have a significant influence on the atomization efficiency of elements which form stable oxides and carbides [45]. The graphite surface itself becomes reactive at temperatures above 700 °C and may influence chemical reactions during the thermal pretreatment (pyrolysis) or during the atomization steps. The buffering effect of the combustion gases in flame AAS is often substituted by a chemical additive (modifier) which, if present in excess over the other compounds, will effectively define the chemical atmosphere in the graphite furnace [46].

During the atomization step the internal gas flow is switched off so that the gas inside the tube is hardly moving at the beginning of the atomization step. The first second of the atomization step is characterized by an increase of the gas phase temperature by more than 1,000 °C and an expansion of the gas volume by a factor of 4. This results in a forced convective gas flow from the tube center toward the tube ends and out of the dosing hole. The atoms should be released into the gas phase only after it has almost reached its final temperature and gas volume. Therefore, platform atomization has become popular [47]. A thin graphite body inside the tube has minimal physical contact to the tube. It is mainly heated by radiation from the tube wall. The sample (about 20 µL) is pipetted onto this platform. The heating of the platform is delayed relative to the wall. Under these conditions the atoms are removed from the absorption volume mainly by diffusion into a gas atmosphere which is hotter than that of the platform. It usually takes 1–4 s from the appearance of the first to removal of the last atoms from the measurement beam. The mean residence time of the atoms is about 3 orders of magnitude longer than in flames and this explains the excellent absolute sensitivity of the graphite furnace.

Unfortunately, "there ain't no free lunch" and a very high matrix density in the light beam often accompanies the high analyte atom density, explaining the need for an excellent background correction system (see the section on AAS). After the introduction of the first commercial graphite furnace in 1970 [48] the first ten years were characterized by a rapid development of the technique and the methodology to reduce the effects of the numerous interferences that occurred in the non-optimized systems of the early years. The physicochemical and technical requirements of GFAAS are now well described and documented.

GFAAS today is characterized by a good freedom from interferences and relative detection limits in the range of 0.1 µg/L (2 pg for 20 µL sample volume) for most elements. Depending on matrix and analyte elements, a single graphite tube can be used for from 300 to 1,500 atomization cycles which, under favorable conditions, represent 100 to 500 duplicate sample measurements including calibration and quality control samples. The furnace operates automatically during analysis with limited requirement for operator interaction.

The graphite furnace is used as well as efficient vaporizer for the ICP atomizer/ionizer source. In this case the µL small amounts of sample are pipetted, dried, pyrolyzed, and finally brought to a temperature where the analyte elements are efficiently volatilized. This dried and clean aerosol is transported with the sample gas to the plasma and the elements are determined with ICP-OES or ICP-MS. These systems are commercially available (see Figure 2.23).

Low temperature atomizers: chemical vapor generation

Chemical reactions are applied, which transfer elements, dissolved in the solution for measurement, into gaseous molecular compounds or into gaseous free atoms. These compounds or free atoms are separated from the matrix and transported by an inert gas to the atomizer. Gaseous molecules shall be decomposed such that free atoms can be determined by AAS/AFS/OES. The analyte-containing gas is transported to a heated cell, a hydrogen flame, an ICP or to a graphite furnace. The paramount function of the process is enhancement of specificity and analyte preconcentration.

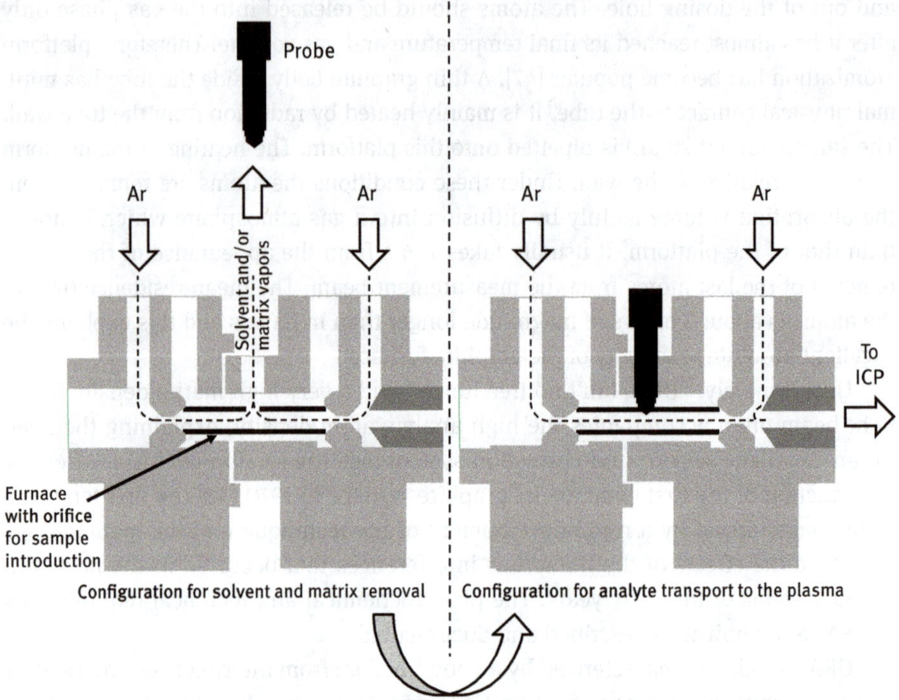

Figure 2.23: Electrothermal vaporizer for ICP-OES. from literature [122].

Central unit for the transfer of dissolved elements into gaseous reactions is a batch-type reaction flask or a flow-through unit. The reagent is added in liquid form to the solution for measurement. The reaction kinetics should be fast, and the process should lend itself to automation. Most units offered on the market today are on-line flow systems. The acidified solution for measurement is pumped with a flow rate of a few mL/min to a merging point with the reagent (usually a reducing agent). Often the carrier gas, which supports separation of the gaseous analyte from the liquid matrix, is added at this merging unit as well. The reaction takes place downstream in a tube within seconds. The mix of gas and liquid is guided to a separator where the liquid is discarded, and the gas is transported to the measurement cell. The classical atomizer is a heated quartz cell at close to 1,000 °C. The diameter of the cell is between 5 and 8 mm. The speed of the gas is strongly reduced in the atomizer to about 1 mm/s. The cell is mounted into the light beam and forms the absorption volume in AAS. The hydrides are thermally cracked but – depending on the element – atoms are not stable at these temperatures. At this point in time, the radicals play an important role to contribute to atom formation for most of the hydride forming elements [49].

A much better atomizer would be a graphite furnace. However, the analyte element from the reaction gas has first to be trapped to the moderately heated graphite surface.

This works remarkably quantitative if the graphite surface is treated with an involatile noble metal, e.g., Ir or Pd [50]. The trapped hydrides are subsequently atomized at their optimal temperatures, i.e., around 2,000 °C. This technology combines the advantages of ideal sample pre-treatment and atomization conditions. It is therefore the analytical method with the best detection limits and highest selectivity. The analyte bearing compound is atomized by temperature or by chemical reactions and quantitated by AAS or AFS. The sample may be added as a defined volume of a few microliters up to about 1 mL (volume-based flow injection or time-based flow injection), or it may be added continuously until a steady state signal is obtained. As an example, the setup of a Chemical Vapor generator with volume-based flow injection is sketched in Figure 2.24.

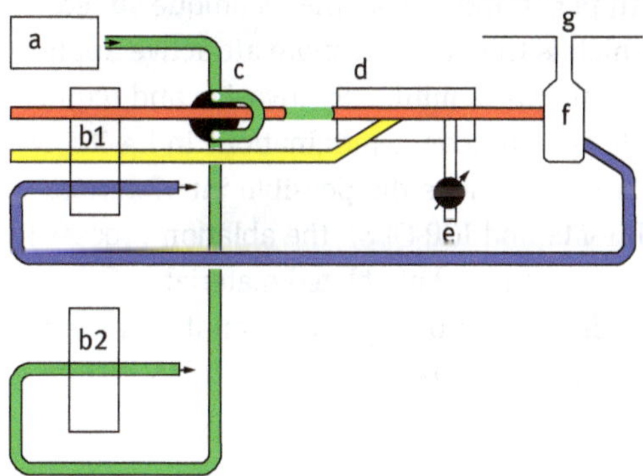

Figure 2.24: Chemical Vapor Generation unit; volume-based Flow Injection.
a, sample for measurement in auto-sampler; b1 and b2, peristaltic pumps; propulsion of sample (green line), carrier solution (red line), reagent (yellow line), waste (blue line); c, 4-port-valve switching between sample channel (green) and carrier channel (red); d, mixing block, mixing of carrier/sample with reagent and carrier gas (e); e, carrier gas with control unit; f, gas-liquid-separator; g, device for spectroscopical measurement or sample enrichment.

If the gaseous sample is transferred to ICP it may be introduced into the sample gas. In this case only the gaseous analyte elements are determined. It can be mixed with aerosol from a conventional mixing chamber (see above) and the gaseous elements can thus be determined with the other elements in the aerosol. In any case the limits of quantitation for gaseous elements can be enhanced in ICP-OES or ICP-MS substantially.

Atomization with high-energy radiation

High-energy laser beams with short pulse duration can be used to volatilize small amounts of material from mainly solid surfaces. This technique is called laser ablation. It is mainly used in combination with ICP-MS but could be used with ICP-OES as well. The aerosol which is produced in this process is transferred into the ICP, as final atomization/ionization source, by means of a carrier gas, usually Ar. The amount of the material ablated is dependent on the wavelength of the laser, the energy of the laser beam, and the size of the focal point (usually in the range of less than 0.001 mm^2. The penetration depth of one shot is usually in the range of 0.1 µm. The amount of sample ablated per shot is therefore in the sub-ng range. If repetition rates of about 50 Hz are applied, the material sent to the plasma is roughly in the range of 10 ng/s. which is similar to the mass introduced by an aerosol generated from a solution with 0.1% dissolved solids, however, without solvent.

The exact location for ablation of the sample can be selected via precision steering of the ablation chamber and is video monitored. The high lateral resolution of between 1 and 100 µm and the low depth penetration make the technique suited for lateral and depth profiling of surfaces which is the probably more attractive application as compared to bulk analysis. The technique is minimally invasive and requires extremely small sample quantities [51, 52]. Quantitative determinations in LA-ICP-MS require sophisticated strategies for calibration Besides the possible interferences of the analytical techniques (see sections ICP-MS and ICP-OES), the ablation process itself may be strongly influenced by the matrix, the kind of ablated material.

Calibration with solids standards as similar as possible to the analyzed sample seems to be a prerequisite. A schematic setup of a laser ablation ICP-MS system is sketched in Figure 2.25.

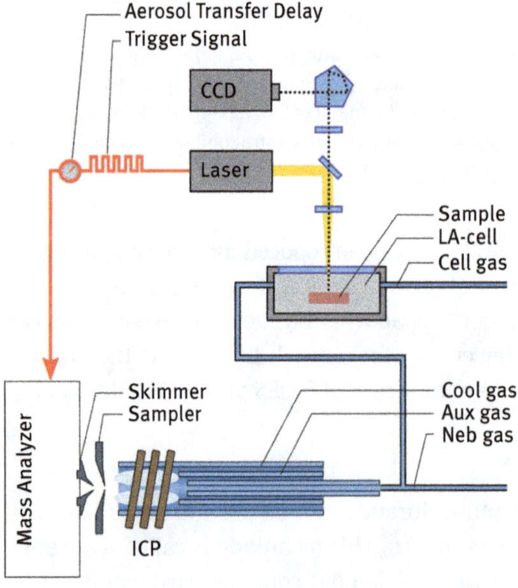

Figure 2.25: Representation of an LA–ICP–MS setup (courtesy of TOFWERK AG, Thun, Switzerland).

Direct measurement of the excited sample

When a high-energy laser pulse is directed on a surface (typically a solid), a plasma is formed which, after the pulse has decayed, is quickly fading as well. The emission of this pulse contains a lot of white light (bremsstrahlung by recombination of electrons and ions) but it contains wavelength specific emission of the elements ionized and atomized by the high-energy photons of the focused laser as well. The emitted light can be focused into a spectrometer and can be evaluated analytically. As the white light would negatively influence the detection of the specific emission lines, the measurement is triggered several hundred nanoseconds after the laser pulse and is usually only very few microseconds long. To improve signal to noise, usually many individual pulses and measurement cycles are averaged. The technique is called laser-induced plasma spectroscopy (LIPS) or, as the plasma breaks down prior to measurement, laser-induced breakdown spectroscopy. Compared to ionization/atomization in a steady burning plasma, LIPS requires a very subtle timing of the laser and the measurement system. On the other hand, the technique can be used with portable instrumentation in the field as it does not require supply of bottled gas. It is mainly used for qualitative and semi-quantitative analysis of solids, but the atomizer is able to stimulate emissions in liquids, gases, aerosols and gels. A LIPS system was working at the Mars Science Laboratory Rover Curiosity.

Sources to generate analytical response without volatilization of matter

If one looks at ionization, atomization, and volatilization by photons (see laser ablation described above) it becomes obvious that the wavelength of radiation plays an important role in the kind of activation of the sample. High-energy radiation at longer wavelength, e.g., 1,064 nm of an Nd YAG laser, results in strong heat dissipation in the ablated matter, melting and volatilization. If the frequency of the Nd YAG laser is quintupled to 213 nm, the penetration of the beam into matter is much more focused with minimal warming but efficient ionization.

If much shorter X-ray radiation is applied to matter it will penetrate the sample without volatilization. If the energy is sufficient, electrons of the inner atomic orbitals can be removed.

The inner orbits are filled with outer electrons, and radiation, characteristic for the material, is released. This radiation is detected and used for qualitative and quantitative analysis. It will be discussed in the section of X-ray fluorescence spectroscopy.

Ionization of molecules

Optical molecular spectroscopy is based on direct interaction of light with liquids or gases. Mass spectrometry, however, requires ions for the measurement process. Whereas in elemental mass spectrometry, described above, the target of elemental release is the free ion, the ideal in molecular mass spectrometry is ionization of an intact molecule. This is often impossible or, at least, difficult. Several methods have therefore been applied to generate ions for mass spectrometric separation.

The oldest and most widely used method is electron impact ionization (EI). Electrons are fired on the strongly diluted vapor of the analyte inside a small chamber. In practice the analyte is volatilized in high vacuum of about 10^{-4} Pa. Electrons are generated at temperatures above 2,000 °C, usually from a heated tungsten wire and are exposed to about 70 V in an electric field. The molecule is usually losing an electron according to

$$e^- + M = M^+ + 2\,e^-$$

Molecules have invariably paired electrons. The result of electron loss is a radical molecule cation. Toward the mass spectrometer the ionization chamber is equipped with a negatively charged hollow electrode which accelerates the molecule cations toward the mass spectrometer.

Electron bombardment leads to extensive fragmentation of molecules. These will be discussed in the section on molecular mass spectrometry. Methods of milder ionization have therefore been of utmost interest for the spectroscopist. The idea of “buffering” has been an early attempt to keep the molecules intact. Buffer gas (CH_4) is introduced into the ionization chamber which leads to a pressure increase to about 100 Pa.

A small portion of CH_4 is ionized by electrons; the rest is present as neutral molecules.

The methane cation radicals hit methane and generate new chemical compounds with very short life time. Among the compounds is CH_5^+ which acts as an extremely strong acid. It will protonate molecules with free electron pairs easily and produce cations en route the chemical reaction. If argon is used as buffer gas, the heavy argon atoms are partially ionized. They are accelerated in an electrical field toward a chamber also filled with argon at reduced pressure. Elastic collision between Ar^+ and neutral argon leads to fast and heavy atoms which, upon impact with the analyte generate both cations and ions. Both ion types can be run through a mass spectrometer for analysis. Often, the analyte is embedded in matrix as well. The matrix will obviously appear in the mass spectrum.

Novel methods of volatilization and ionization of molecules gained utmost importance when mass spectrometry became the gold standard in macromolecular analysis. Proteomics is the most dynamic field of research in analytical chemistry of the twenty-first century. Methods for volatilization of macromolecules became accessible 20 years earlier and were decorated with the Nobel prize of chemistry in 2002 [53]. When a small volume of dissolved analyte is pushed through a charged capillary, aerosol-like charged droplets are sprayed toward a counter electrode. The spray is usually supported by a chemically inert gas. The solvent is gradually evaporating, and the charge remains on the surface of the droplet. Finally, free charged protein molecules remain. The polarity of capillary and counter electrode determines the charge (positive or negative) of the molecule. The ions may carry one or several charges when reaching the mass spectrometer. The method is called electrospray ionization (ESI) and is used extensively in modern mass spectrometry.

Lasers can be used to effectively ionize matter. This has been discussed in the section on LIPS above. However, matter is decomposed into single atoms and ions. Soft laser desorption was proposed at the end of the nineteen eighties by Koichi Tanaka. The sample must be embedded in a solid matrix. The entire matrix absorbs the energy of a high-energy UV laser in the form of a blast. Small, charged bits and single charged analyte molecules are formed. These are accelerated and analyzed in a mass spectrometer, often a time of flight design. A similar process developed by Karas, Bachmann and Hillenkamp [54] was called matrix-assisted laser desorption (MALDI). Again, the analyte is embedded with suitable matrix to form the ionized analyte molecule of interest upon laser impact. MALDI MS became the gold standard for macromolecule determinations today.

Specific methods of ionization are used for surface analysis. These methods excite e.g., metal surfaces with heavy ions (e.g., Ar) which are generated in a separate chamber. The ion beam is then accelerated and focused onto the surface which itself is in a vacuum chamber. Secondary ions from the samples (anions and cations) are released from the surface and analyzed in a mass spectrometer.

2.1.7 Quantitative detection of atoms: atomic absorption spectrometry (AAS)

2.1.7.1 Basic principles of atomic absorption spectrometry (AAS)

In analytical atomic absorption spectrometry (AAS) we consider the process which describes resonant absorption. A photon, interacting with a bound electron is absorbed completely by the electron. Incident radiation with wavelength λ (frequency ν) transfers energy to an electron which is thus excited to a higher energy level.

$$E = h \cdot \nu = h \cdot c / \lambda$$

This interaction is unique and specific for each electron transition and thus as well for each element. If the lower level is the ground state, the absorption is said to take place at a primary resonance line. In AAS we consider the wavelength range between 190 nm and 800 nm which corresponds to frequencies between $15 \times 10^{14} \cdot s^{-1}$ and $3.7 \times 10^{14} \cdot s^{-1}$ and an energy between 11×10^{-19} J and 3×10^{-19} J. In this energy range, only electrons from the outer shell of atoms can be excited. Light which can be seen by the human eye is within the range used for AAS. The lifetime of the electron in the upper energy level is short, lasting only a few nanoseconds. The electron then returns spontaneously to its original level or to another energy level, emitting the excitation wavelength or a different distinct longer wavelength.

The spectral line width of the absorption profile is initially defined by the uncertainty relation (Heisenberg, 1927). The line has the form of a Lorentz curve, and the broadening is therefore called Lorentz broadening." The energy uncertainty of the excited state ΔE is related to the residence time in the exited state Δt.

Equation 2.8:

$$\Delta E \cdot \Delta t = h/2\Pi \tag{2.8}$$

The energy uncertainty of a transition with a lifetime of 10^{-9} s is in the range of 10^{-25} J, which translates to a frequency uncertainty Δv of 1.5×10^8/s. If, for example we consider a resonant wavelength at $\lambda_0 = 2 \times 10^{-11}$ m, we will calculate a line width of about 4.5×10^{-14} m or 0.045 pm.

Both the emission profile of the source and the absorption profile of free atoms are usually significantly broader than the Lorentz profile. Several effects are inducing additional broadening:

Atoms are in rapid thermal motion, with their velocity depending on the temperature. Their speed is in the range of 10^3 m/s. Atoms which are moving toward the source will absorb photons of slightly longer wavelength, atoms moving away from the source will absorb slightly shorter wavelength. Atom movement results in a line broadening in the form of a Gaussian peak. The effect is called Doppler broadening. The Doppler broadening is described by eq. (2.9).

Equation 2.9:

$$\Delta\lambda = \lambda_0/c \cdot \sqrt{8kT \ln 2/m} \tag{2.9}$$

As the absorbing atoms are experiencing a temperature of around 2,000 K, the Doppler broadening is in the range of 1–3 pm, significantly wider than the Lorentz broadening.

The atoms of interest are not isolated in the absorption volume but are surrounded by combustion products of the burning gas in flame atomizers, by argon and matrix products in electrothermal atomizers and by argon, hydrogen, and other reaction products in the chemical vapor generation technique. The entire atmosphere is mostly at high temperatures in the range of 2,000 K or above. The atoms of interest are frequently colliding with other particles. The duration of this process is very short. The collision changes the excitation/emission process. The profile is defined by a Lorentzian profile and is called collisional broadening or impact pressure broadening. The magnitude of this broadening effect depends on the frequency of collisions v_{col}. The frequency depends on the density of collisional particles, their speed, and their cross section for collisions (their probability to react during a collision). At the conditions in AAS, it is in the range of about 1 pm.

Besides these dominant effects of line broadening, there are other effects which lead to small line broadening or line shift effects. The line broadening effects result in a convolution of the Gaussian curve with a Lorentz curve. The resulting profile can be expressed in the form of a Voigt function.

For the absorption process will be important that the profile emitted by the radiation source is less blurred than the absorption profile of the analyte atom. The processes described above are effective for the line width of the emission as well. Parameters which can be technically influenced are temperature and pressure inside the source. Line sources are designed such that the temperature of the emission is low, and the number of collisions is reduced (see Section 2.1.3). The emission line width of the sources used in AAS is in the range of 1–3 pm, slightly narrower than the absorption profile. If a continuum source is used and radiation is selected by a high-resolution spectrometer, the spectral bandpass should be narrow enough to resolve the absorption profile completely; i.e., it ought to be in the range of about 1–3 pm.

The absorption process

A characteristic wavelength is isolated from the spectrum of a radiation source which contains the element to be analyzed. The mean line width of the emitted radiation shall be narrower than the absorption profile of the analyte element. All photons emitted by the source should be absorbable by the absorption profile.

Depending on the element of interest, a lot of transitions are possible which can be visualized in a Grotrian-diagram or energy level schematic. The more complex the electronic structure of the element, the bigger is the number of transitions. Out of the large number of possible transitions only a part is statistically permitted. The lower level of the transition needs not inevitably be the ground state of the atom. Elevated levels may be thermally excited, but these cases are rather the exception than the rule in AAS. Usually, transitions are selected which are starting from the ground state. An example is sketched in Figure 2.26 for Li, an atom with a simple electronic structure [55]. In the diagram a few transitions only are depicted. The ground state for Li is the 2s shell. From there transitions to 2p (670.8 nm) and 3p (323.2 nm) are observed. The line at 670.8 nm is the primary resonance line, the most sensitive transition in AAS.

The atoms, specific for a certain transition, reduce the original radiation intensity I_0 by a certain amount which depends on the absorption coefficient. This is specific to the electron transition. Important for the atomic absorption process is the awareness that the absorption coefficient is different from element to element. A certain number of free atoms "*a*" in the absorption volume will therefore result in a different reduced intensity I after the absorption process compared with atoms "*b*." If we determine the intensity decay I/I_0, we can calculate the number of atoms in the absorption volume from theoretical physical data. Vice versa, if we know the number of free atoms in the absorption volume, we can predict I/I_0.

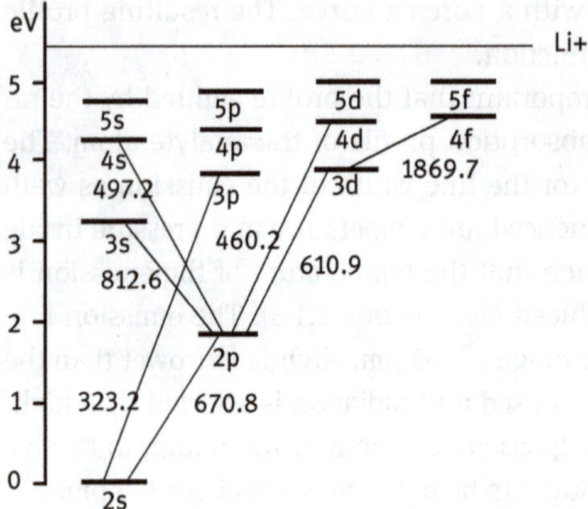

Figure 2.26: Grotrian diagram for Li. The ordinate shows the energy of the respective term in eV. The ionization energy is 5.392 eV. Selected transitions are depicted as lines.

Mathematically, the original radiation intensity is reduced by a logarithmic function. For practical reasons, the logarithm of the absorption, the absorbance *A* is used for calculation. *A* is directly proportional to the number of atoms (*n*).

Technically, free atoms must be generated and locked into a defined "room" for a certain amount of time. The room will allow a certain length of interaction between free gaseous atoms and the light beam (I_0). The generation of atoms in the absorption volume, and the decay of atoms out of the absorption volume, will result in a static or dynamic atom density or concentration (*c*). The fundamental analytical law of photon/atom interaction can be simplified in the following eq. (2.10).

Equation 2.10:

$$A = \log I_0/I = k \cdot c \cdot l \tag{2.10}$$

This so-called Bouguer-Lambert-Beer law relates the instrumentally determined photon intensities *I* and I_0, to the mean concentration of analyte atoms *c*, the absorption coefficient *k*, and the length of the absorption path l.

The prerequisites for an accurate AAS determination of the analyte concentration in the sample can be derived with the help of this simple relationship:

- The concentration of atoms in the measurement beam must be proportional to the concentration of atoms in the original sample; under optimum conditions, all analyte atoms in the sample are atomized and determined.
- The radiation source should emit no wavelengths outside of the absorption profile of the analyte atoms. Any other wavelength cannot be absorbed and reach the detector as, so-called, stray light.
- At high absorbance values the fraction of the radiation which has not been absorbed by the absorption profile becomes small. Eventually it becomes insignificant

compared to the stray light intensity. The function $A = f(c)$ will consequently bend toward the abscissa and will finally reach a maximal absorbance value A_{max}. At this point, a further increase in analyte concentration will no longer result in higher absorbance.

Optical emission spectroscopy with AAS instrumentation

Excitation of electrons to elevated states or to ionization may also be stimulated by thermal energy. For most elements, the probability of excitation at the typical flame temperatures is not high. The emission generated from thermal excitation is therefore generally weak in flames. Alkaline, earth-alkaline and a few other elements such as lanthanum, as an example, are thermally excited to an extent that optical emission spectroscopy in flame AAS becomes attractive though. Determinations using the flame emission technique in these cases may even provide better figures of merit compared to flame atomic absorption. Most AA spectrometers are therefore technical capable of operation in the emission mode as well and methods are described in the users' manuals and are resident in the instruments' software. Even modern publications describe the merits of this technique for selected applications [56]. Optical emission spectroscopy will be described in Section 2.1.8 of this textbook.

2.1.7.2 Technical means to facilitate AAS

Qualification for a product, meeting the expectations of the end-user, are specifications which transfer the fundamental idea into a description of the requirement for the technical component. The specification often describes an ideal conception of the device. The final product will always be a compromise between the idea, the technical feasible, and the limitations set by expenditure.

General layout

In atomic absorption spectrometry (AAS), radiation of a suitable source (section 2.1.3) is focused on a volume where the absorption process ought to take place. This volume is called "absorption volume." The light collected after passing the absorption volume is then focused on the entrance slit of a discriminating optical element which is selective for separation of the absorption line width. Radiation of the selected spectral bandwidth is focused on a detector which can distinguish between the radiation intensity before and after interaction with the element of interest. The light flux is measured and mathematically evaluated.

The schematic setup of an AA spectrometer is sketched in Figure 2.27.

It should be emphasized that the emission source is usually a line source (hollow cathode lamp, electrodeless discharge lamp, boosted hollow cathode lamp), but it may be a continuum source as well. The wavelength selector in AAS is usually a monochromator with medium resolution. It may be a high-resolution monochromator

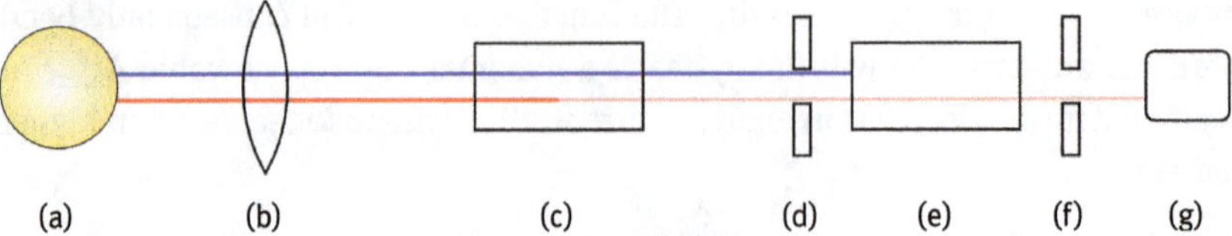

Figure 2.27: Schematic setup of an AA spectrometer: Radiation of a source (a) is focused by an optical element (b) through an absorption volume (c). Only the suitable wavelength range is partially absorbed (red line). Radiation is guided through an entrance slit (d) into the wavelength selector (e) and through an exit slit (f) to the detector (g). Stray light (blue line) is separated from the analytical spectral bandwidth in the wavelength selector assembly.

or a polychromator as well. The combination of technical components must be carefully matched to get a balanced combination of performance and system cost.

The light source in conventional AAS should emit an intense line spectrum of the element(s) to be determined. Under favorable conditions, the half width of the emitted line is in the range of 1–3 pm which is about 5 times narrower than the half width of the absorption profile and about a factor of 100 narrower than the typical resolution of the monochromator or polychromator in an AA spectrometer. The light source most commonly used for AAS is the hollow cathode lamp (Section 2.1.3). A spectral profile of a lamp showing pronounced self-absorption is sketched in Figure 2.28.

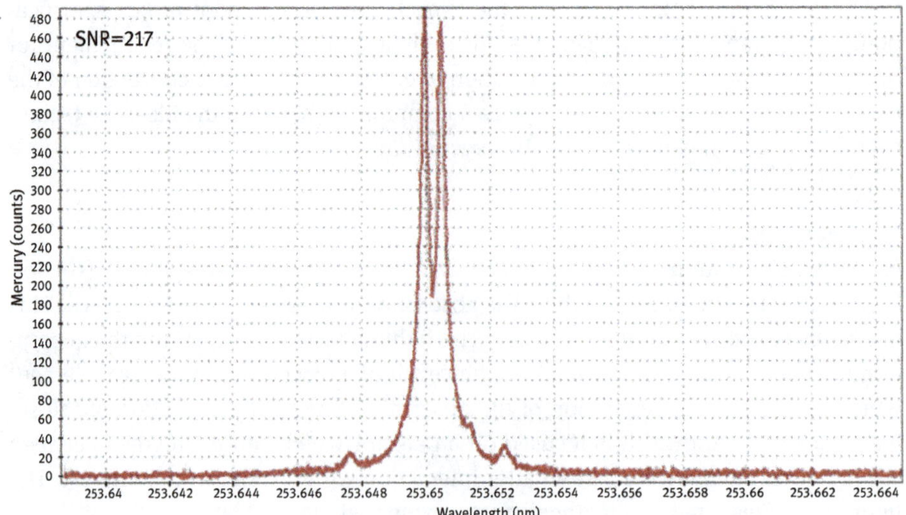

Figure 2.28: Spectral emission profile of a mono-isotopic mercury lamp, recorded with a high-resolution Echelle spectrometer. The resonance line at 253.7 nm is split and shows pronounced self-reversal (courtesy of ISAS Berlin, Germany).

The analytical lifetime of a hollow cathode lamp is defined in mA hours. It is usually in the range of 5,000–10,000 mAh. Thus, a lamp operated at 5 mA should theoretically live twice times longer than a lamp operated at 10 mA. The operating currents, recommended by the manufacturers, usually provide optimum signal to noise ratio but often not the best lifetime.

For elements with a high excitation potential electrodeless discharge lamps or boosted hollow cathode lamps are applied to increase emitted intensity and decrease line broadening. In general, the sensitivity of the determination and the detection limit are extremely dependent on the age of the lamp and its actual operating condition. Spectroscopically important is the active size of the emission spot. The photon discharge should be emitted by an emission spot as small as possible. Lamps are often more important for the analytical performance than sophisticated details of photometer or spectrometer design!

An alternative, to line source AAS is the continuum source AAS (CS-AAS) [33]. This principle has been discussed since the early days of AAS, but its commercial realization took until 2005. CS-AAS uses a strong white light source to illuminate a high resolution spectrometer. As discussed above, the resolution must be narrower than the line width of an absorption profile, i.e., approximately 2 pm.

The source used to illuminate the spectrometer must have a high intensity over the entire wavelength from 190 to 900 nm. In particular in the short wavelength range the white light emission must be extremely hot to provide intense radiation in the UV-C range. The source used in the commercialized instrument is a Xenon short arc lamp operated in hot-spot mode (GLE, Berlin, Germany, refer to Figure 2.3). The lamp is operated at 300 W. The photon output per pm is comparable to a HCL at short wavelengths up to about 220 nm and becomes significantly stronger toward higher wavelengths. The emission spot of the Xenon short-arc hot spot source is significantly smaller than that of a hollow cathode lamp. This is generally favorable for the layout of photometer and/or spectrometer.

Photometer and spectrometer

As pointed out earlier, the selectivity in line-source AAS – the spectral resolution – is due to the emission profile of the source. If this light is focused through the atomizer by means of mirrors or lenses, and imaged a second time on a photosensitive cell, atom specific absorption could be detected. In some cases, if the source emits a simple spectrum, and if only vapor of the analyte element without concomitant particles, molecules, and atoms is generated in the atomizer, the simplest atomic absorption photometer possible can be designed. This principle is used, for example, for the determination of mercury with the cold vapor technique [57]. Atomic fluorescence spectrometers are following this simple principle as well. Figure 2.29 sketches a cold vapor technique spectrometer which can be used in AAS mode and in AFS mode.

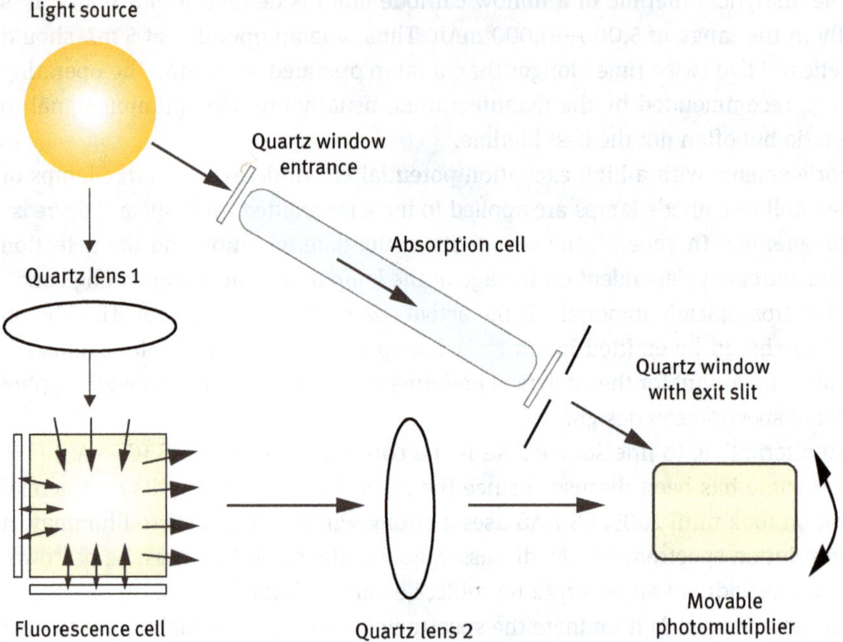

Figure 2.29: AAS/AFS photometer "mercur" Courtesy of Analytik Jena AG, Jena, Germany.

The photometer part of the instrument, in general, is imaging the emission source on the atomizer and onto the detector or, as described below, onto the spectrometer.

AAS, if not limited to a single element, is more complicated: the lamps emit a spectrum of resonant and non-resonant lines, mixed with lines from other elements and from the fill gas of the lamp. In the atomizer, several atoms, molecules, and particles are generated, the analyte atoms usually being only a small fraction – less than a thousandth – of the total absorbing species. Furthermore, a lot of light is emitted in the atomizer which reaches the detector. The line used for the measurement must therefore be isolated by a monochromator or polychromator. As the emission profile of the selected resonance line is sharp, the resolution of the monochromator or polychromator for AAS can be limited to about 0.2–2 nm, 100 times wider than the absorption profile. It is advantageous to be able to select the actual bandpass depending on the spectrum in the neighborhood of the measurement line. This is sufficient to reduce the stray light level for all elements to less than 1% and still provide enough photons at the detector for good counting statistics.

Typical monochromators for AAS are Littrow or Czerny-Turner monochromators (see Section 2.1.4.2). The gratings are usually optimized toward a maximal efficiency at about 250 nm.

Most AAS instruments make use of such classical monochromators because of their simplicity and ruggedness. High light throughput, adequate optical resolution, and low straylight levels characterize these monochromators. They are usually combined with photomultiplier tubes as detectors which could be easily positioned behind the exit slit. Changing from one analytical line to the next requires a change of lamp and the setting of the new wavelength with the help of a stepper motor driven grating and, in some cases, the change of the width of the entrance slit. This procedure takes between 10 and 30 s.

A fundamentally different way of dispersing light for AAS is the Echelle monochromator or polychromator (Section 2.1.4.2). They have been and are used with specific line sources as well as with the continuum source technology. In the latter case, the Echellogram can be shifted in two dimensions relative to the detector by moving one or both dispersing elements so that a small portion of an order is selected and imaged onto a slit or an active pixel sensor (Figure 2.13). This monochromator is used for continuum source AAS. The monochromator must provide resolution in the range of a line source profile, i.e., about 2 pm in order to meet the criteria of profile overlap in AAS.

Echelle monochromators or polychromators offer high resolution within a compact design, or, alternatively an extremely high photon throughput, if moderate resolution is the target. They require sophisticated algorithms and instrument control to select the optimal order and position of reflection.

Detectors used in AAS

A widely used tool for AAS is the photomultiplier tube (PMT) (see Section 2.1.5). Amplification is typically in the range of 10^6 and 10^7. The quantum efficiency is wavelength-dependent and usually between 10% and 30% in the wavelength range typical for AAS. The combination of photomultiplier and amplifier should provide linear response of signal to light intensity over at least three decades. The highest light flux in AAS is measured in the case of zero absorption. Under these conditions, the gain of the photomultiplier is adjusted by an instrument-inherent algorithm so that a certain predefined electric current is flowing to the amplifier. If many photons reach the photomultiplier cathode per unit time (e.g., in the case of a bright radiation source or very small absorbance) the statistical variation in the signal generating process (the so-called shot noise) is small. The standard deviation of a series of measurements is often below 0.1% transmission (0.00044 A). When the absorbance increases, fewer and fewer photons reach the detector. The amplitude of statistical noise increases relative to the signal and the baseline noise therefore increases. The lowest standard deviation for an absorbance reading is therefore obtained at close to 0 absorbance. This is exactly when it is needed: at the lowest concentrations, close to the detection limit. If the statistical noise in AAS is defined by the detector noise at zero absorbance, the system is said to be *shot noise limited*. The statistical noise is dependent only on the number of photons per time unit hitting the photo multiplier. Shot noise limitation must be the ultimate goal in instrument design. In this case, the standard deviation of the blank is related to the light throughput in a square root relation (eq. (2.11)).

Equation 2.11:

$$\text{s.d.} \sim 1/\sqrt{i} \tag{2.11}$$

i is the radiation intensity on the detector.

The relation between gain and light intensity of a PMT is non-linear. As a rule of thumb, 40 V decrease in the photomultiplier voltage corresponds to the doubling of the light energy. This type of detector seems to be the ideal photon capturing system for optical spectroscopy. However, the device is relatively bulky and provides no wavelength selective reading. The wavelength must be limited by the monochromator exit slit. Thus, light with wavelength within only one spectral window will be measured. In AAS, this usually is a window of 0.1 to 1.5 nm.

The trend in chemical analysis is to obtain as much information per single measurement as possible. The information content of a measurement can be increased if a solid-state detector, providing many photosensitive spots or areas on a semiconductor chip, is used. The technology for solid state detection has progressed dramatically in the last few years.

The techniques, used in AAS are single photodiodes, multiple photo diodes, charge couple device detectors (CCD) used in linear and two-dimensional readout arrangement. Detectors such as the active pixel sensor type (CMOS), used in most commercial cameras, or charge injection device detectors (CID), are not yet used in AAS. However, CMOS detectors will certainly become popular in the next decade.

Modern AAS instrumentation is gradually changing to more modern optical systems and detector technologies. These may improve analytical performance. The criteria for analytical quality, however, are not the technical design layout but the figures of merit (FOM).

Zero absorption in AAS

The original idea of AAS, described by Walsh and Alkemade, is simple: light of an exactly defined wavelength is passed through an atom reservoir and is exclusively absorbed by the analyte atoms. If the light flux would be constant in intensity, no other photons would reach the detector, and the atoms of interest would be the only species to absorb photons, AAS would be the ultimate, interference free method for elemental analysis. As usual, this ideal can only be approximated. Several technical means must be implemented to correct for unwanted effects:

1. as smallest changes in absorbance must be detected, it is important to keep the light flux of the source (representing the “0” absorbance baseline) stable to within less than 0.1% during measurement.

2. radiation emitted by thermally excited atoms or by the atomizer would shift the light flux to values higher than baseline (transmission higher than 100% or negative absorbance). This radiation must be detected and separated from radiation emitted by the light source.
3. attenuation of radiation by processes other than analyte absorption, such as scattering by particles, absorption by molecules or absorption by non-analyte atoms, must be distinguished from the analyte absorbance.

The effect of these three phenomena for the analytical result is similar. A shift of the baseline will eventually change the measured absorbance and may lead to an erroneous analytical result. The theoretical and the technical means to correct for the effects, however, are of different nature.

1. Shifts of the photon flux of a radiation source is a relatively slow process. Lamps just ignited, will usually change their intensity and – to a minor extent – the width of the emission profile. The intensity drift is most pronounced during the first few minutes of operation until the temperature of the lamp has become constant. Even then, the change from minute to minute can be significant. When will the drift negatively influence the analytical performance of the system? A measurement cycle in AAS usually requires about 3–10 s in flame AAS or graphite furnace AAS and up to 30 s in chemical vapor generation AAS. If the change in lamp intensity within this time frame is similar or even larger than the shot noise, it will undoubtedly influence the quality of the analytical result.

Lamp intensity drifts can be compensated by splitting off a part of the light, guiding it around the atomizer and using it as a reference for the radiation passing through the atomizer. This process is called double-beam technique. Variations in the primary source intensity can be easily distinguished from absorbance within a negligible time difference between the two readings. The baseline can be easily stabilized within the limits set by the shot noise. On the other hand, a portion of the photon integration time and a part of the total light intensity must be sacrificed for the second beam. As compared to a single-beam system using the same optical and electronic components, the baseline noise (shot noise) of an optical double-beam system is expected to be higher. In Figure 2.30, the optical schematics of a single-beam (a) and of a double-beam instrument (b) are shown.

The splitter and combiner elements may be semi-permeable mirrors or rotating chopper wheels. Usually, one of the elements is a rotating wheel making temporal and spatial separation of the two beams possible.

In single-beam optical systems, the reference measurement can be performed shortly before or after the actual absorbance reading of the sample. This process is often called baseline offset compensation. It will prolong the measurement cycle per sample by a few seconds. On the other hand, no photons are sacrificed for referencing. Accurate double beaming requires an ever-identical light transmission

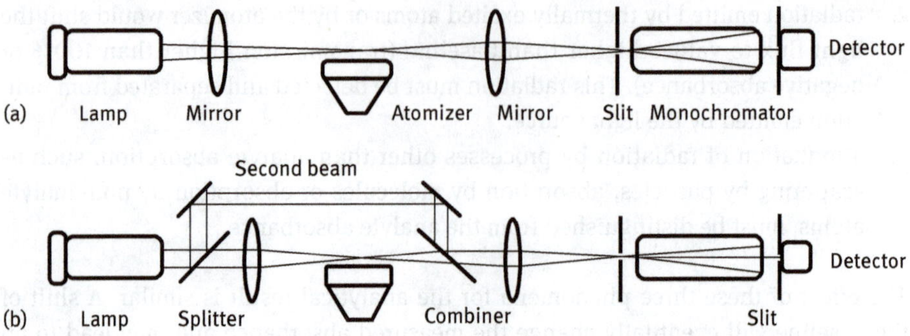

Figure 2.30: Schematic diagram of an optical single-beam (a) and an optical double-beam (b) AA spectrometer. In the optical double-beam system, the light is passed alternately through the atomizer (sample cell) and around the atomizer.

during referencing. In case of a graphite furnace, the absorption volume is only filled by Ar when correcting for the photon flux. An accurate referencing is easily possible. A similar situation is obtained in chemical vapor generation AAS. Time based double beaming is therefore the method of choice for these AAS techniques. The situation is more complex in flame AAS. Absorption by a flame may change when a liquid rather than air is aspirated through the nebulizer and the source intensity reference reading through the flame may therefore be biased. Instruments designed for flame AAS are therefore still often optical double-beam systems. In some of the instruments, means are provided to guide the light beam around the flame with the help of flipping mirrors to obtain a reference reading prior to the actual measurement.

Background correction techniques (see below) may simultaneously correct for light flux variations introduced by the lamp as well as absorption not caused by the analyte element. The discussion about optical single-beam instruments and optical double-beam instruments has lost much of its significance. Various techniques are used nowadays to minimize lamp drift errors.

2. Light emitted by the atomizer or by thermally excited atoms usually changes rapidly with time. Correction for this light is as essential for the analytical accuracy of the reading as is the correction for non-analyte-specific absorbance. The emission, not originating from the light source, is measured, while the lamp is switched off (usually about 50 to 100 times per second for a period of 1 to 3 ms). Radiation measured during this dark period is subtracted from the light measured with the ignited lamp. Usually, the measurement rate is fast enough to correct even for emission effects which change rapidly with time. If the speed is not fast enough, a baseline drift toward negative absorbance can be observed as an indication for incomplete correction for emission by the atomizer. Depending on the temperature of the atomizer, the wavelength selected, the baffles or filters used in the optical system, this type of emission can be more intense

than the intensity of the specific line source. In some rare cases emission radiation may even saturate the detector. The effect may be exemplified with a photographic light measurement of an object close to the sun. The sunlight is so bright that the true exposure for the other object cannot be resolved by the light meter of the camera.

3. Non-specific absorption.
Absorption which is not caused by analyte atoms must be identified and corrected. Technical means to correct background absorption should not deteriorate repeatability, shall have a minor influence on the sensitivity of the determination, and shall be as close as possible to the measurement of analyte absorbance with respect to wavelength and time.

Background absorption occurs primarily when a large number of microscopic particles is present in the absorption volume. These scatter light from the radiation source and generate non-specific absorption or background absorption. The magnitude of scattering is proportional to the number of particles N and to the square of the particle volume V. At longer wavelengths, scattering is much less pronounced than at short wavelengths. The amount of light lost by scattering decreases with the fourth power of the wavelength when going from the short UV toward the visible wavelength range (equation 2.12). Light losses due to scattering for the As wavelength at 193.7 nm should be 12 times more pronounced than for the Cr wavelength at 357.9 nm.

Equation 2.12:

$$I_d/I_0 \sim N \cdot V^2/\lambda^4 \tag{2.12}$$

Background absorption occurs also when molecular vibrations and rotations are stimulated by the source radiation. This effect is used to quantitate molecular compounds in UV/VIS or infrared spectrometry (see Section 2.1.10).

Atoms other than the analyte should absorb only rarely in the narrow wavelength window defined by the light source. When it does occur, however, this type of background absorption is very difficult to correct for.

Background can be made visible using a high-resolution continuum-source spectrometer. As an example, the wavelength resolved spectrum of a dilute nitric acid solution aspirated into an ethine/air flame is depicted in Figure 2.31; the selected wavelength range is close to the Cd resonance line at 228.8 nm. The structured type of background is strongly wavelength-dependent and may reach 0.015 A. At the analyte line (red line in the plot) the baseline is slightly negative.

The technical means used to correct for unwanted absorption make use of physical differences in the absorption process of background absorption and specific absorption. Light scattering, for example, although wavelength dependent, is a broad band effect. Molecular absorbance in many cases is also resonant with radiation over a broader wavelength range than atomic absorption. Non-specific atomic absorption is narrow band but occurs always at a wavelength different from that of

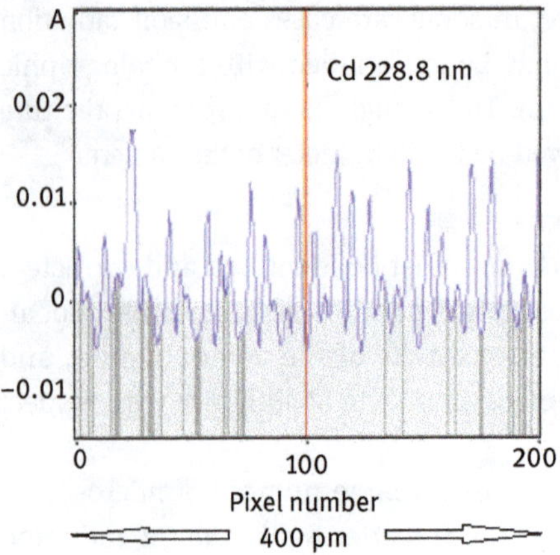

Figure 2.31: Spectrum of an ethine/air flame close to the Cd-resonance line at 228.8 nm.

the analyte resonance line. Usually, it is not known which type of background must be corrected for, but the analyte properties are known. The background correctors so far in use in line source AAS therefore apply a method by which the analyte absorption is changed (minimized) periodically to distinguish it from the background absorption. In continuum source AAS, the spectral range near the analyte line under consideration is inspected with nearly the resolution of the analyte line width and conclusions on the magnitude exactly at the line width are drawn mathematically (see Figure 2.31). As the density of analyte atoms and background species may change rapidly in the absorption volume, measurement of absorption and background should be performed within as short a time interval as possible.

In line-source AAS, basically three methods for quantifying the non-specific absorbance have been used routinely and commercially:

1. Background correction using a continuum source
2. Line-reversal background correction
3. Zeeman-effect background correction

All three systems have found extensive application over more than three decades and the advantages and limitations are well known and described in the literature. Some of the AA-spectrometers offer more than one background correction system.

1. Background correction using a continuum lamp as a second source

Continuum radiation from a deuterium lamp is passed through the absorption volume in rapid alternation with the radiation from the narrow line source [58]. The wavelength range of this radiation is defined by the spectral resolution of the monochromator (0.2 to 2.0 nm) and is about 2 orders of magnitude wider than a typical analyte atomic absorption profile. This light is attenuated by broad band molecular absorption or by light scattering at particles, but it is absorbed only to a negligible extent by the narrow profile of the analyte atoms. The line source is affected by both types of absorption in the same way. Thus, background absorption (BG) – determined with the deuterium arc lamp – can be subtracted from total absorption (AA + BG) – determined with the line source. A schematic view of the two measurement cycles is displayed in Figure 2.32.

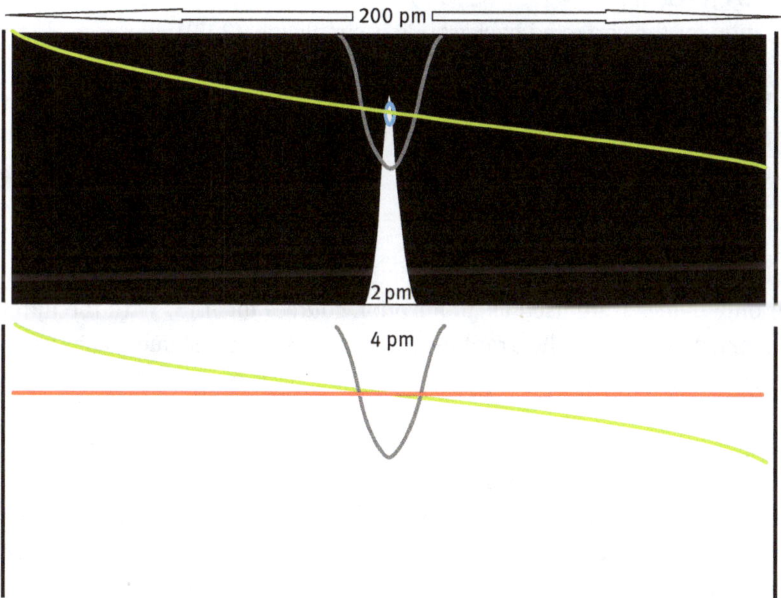

Figure 2.32: Deuterium background correction: The monochromator entrance slit is illuminated alternately with and a line specific lamp, e.g., a hollow cathode lamp (upper plot) and a continuum source, e.g., a deuterium lamp (lower plot). The photon flux of the two sources is matched optically and electronically. The bulk of the photons is distributed over a wavelength $\Delta\lambda$ of about 5 pm in the upper case and over a wavelength $\Delta\lambda$ of about 500 pm in the lower case. Moderate analyte specific absorption (gray profile) is detected by the line specific source but is negligible for the deuterium source. Broadband background absorption is detected by both sources (green line).

The result of this calculation is the correct specific or analyte absorption (AA), if the following requirements hold true:

- The optical beams for the two sources have the same spatial distribution and intensity throughout the absorption volume. This is not necessarily the case.
- The background absorption is homogeneous over the optical bandwidth of the monochromator. That is usually the case for background originating from scattering but not always for background generated by molecular absorption and never for atomic absorption. This can be easily seen in Figure 2.32. The background absorption seen by the hollow cathode lamp (mind the blue circle) is

slightly different from the averaged background by the continuum source (indicated by the red line).
- The intensities of the two light sources are similar enough that they can be handled by the instrument electronics. If the difference in intensities is outside the limit set in the spectrometer software, the lamp current of the line source has to be reduced or increased until the intensities of line source and continuum source are equal again. The D2 compensator usually is limited to wavelengths below roughly 350 nm.
- The equation for the corrected background is simple (eq. (2.13)).

Equation 2.13: Analyte specific absorption (A_a) corrected by background absorption (A_b). The term in brackets is determined with the line source, A_b is determined with the continuum source:

$$A_a = (A_a + A_b) - A_b \qquad (2.13)$$

Background correction with two different radiation sources was found to work perfectly when only flames were used as atomization sources for AAS. With the introduction and extensive use of the graphite furnace in AAS it became clear that it would be much better to use only one radiation source to determine total absorbance and background.

2. Reverse-line background correction

It has been mentioned that a hollow cathode lamp, operated at currents much higher than the standard current, results in lesser and lesser absorbance at a given analyte concentration. The line profile first broadens and then a dip evolves in the center of the profile due to self-absorption by atoms emitted from the cathode. The more volatile the element and the higher the lamp current, the deeper will be the dip in the emission profile (see Figure 2.28).

The analyte absorbance will decrease with increasing depth of the dip due to less and less complete overlap with the absorption profile. The stray light level increases at the same time. The background absorption, however, should not be influenced as it usually is broad band compared to the width of the self-absorbed profile.

The application of a normal current and a boost current at the lamp results in a standard reading of analyte absorption (A_a1) plus background absorption (A_b) and a reading of reduced analyte absorption (A_a2) plus background absorption (A_b). If these two readings are subtracted from each other the result is no longer distorted by background absorption and represents only analyte absorption, with reduced sensitivity (eq. 2.14).

Equation 2.14: Absorbance readings with normal lamp current A_a1, A_b, and boosted lamp current A_a2, A_b:

$$(A_a1 + A_b) - (A_a2 + A_b) = A_a1 - A_a2 \tag{2.14}$$

In comparison with the two-source method, the determination of background absorbance is performed with similar spectral resolution in the two measurement phases. There are requirements for proper operation of this principle.

- The lamps must be manufactured specifically for operation at high currents. In general, the lifetime of any hollow cathode lamp is reduced, when currents much higher than the standard operating current are used.
- Line profiles for moderately refractory and refractory elements show only a minor dip; the remaining absorbance (A_a1-A_a2) is significantly smaller than A_a1. Thus, the detection limits are likely to deteriorate when this background corrector is used.
- The significantly higher stray light level results in curvature of the calibration function at lower absorbance values.
- Finally, the background must not be structured to such an extent that it cannot be considered constant over the spectral width of the broadened emission profile.

Instruments with a background corrector based on the line-reversal or Smith-Hieftje principle [59] are therefore used only for some analyte elements and are always equipped with an additional continuum source background corrector.

3. Zeeman-effect background correction. Magnetic field at the atomizer (inverse Zeeman effect).

Analyte absorbance can also be minimized, if the absorption profile is shifted to higher or lower energy values so that it is largely outside the spectral bandpass of the emission line. This can be achieved by generating a strong magnetic field of about 1 Tesla at the absorption volume. Depending on the electronic structure of the element, the ground state may remain unchanged (normal Zeeman effect) or split (anomalous Zeeman effect), the upper levels are split into three or more components (see Figure 2.33).

During electron excitation, three or more transitions become possible. In the first case the transition energy is the same as without the magnetic field, in the other cases the energies are higher (shorter wavelength) or lower (longer wavelength). The magnitude of the split depends on the magnetic field strength. The wavelength difference between the transitions is small, but in the case of a 1 T field, large enough to shift the so-called σ-profiles far enough that the overlap of these components with the emission profile is minimal. The absorbance of the central profile(s) (the so-called π component) is still half of the absorbance without the magnetic field. The essential feature to make the Zeeman-effect efficient for AAS is polarization. The split profiles will interact with radiation of a certain polarization direction. The type of polarization depends on the direction of the propagation of the light beam relative to the magnetic field. For background correction it is essential that components

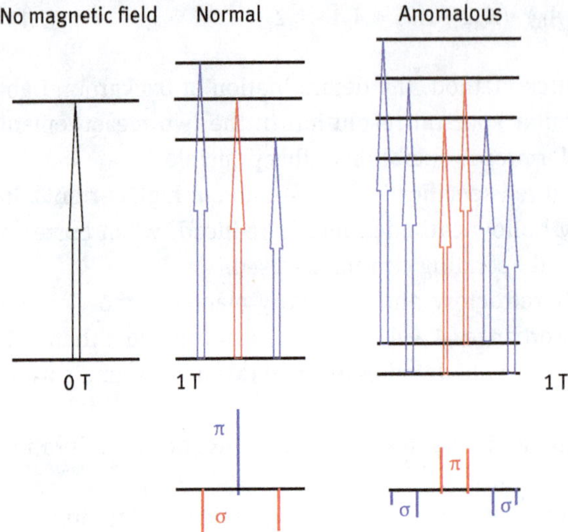

Figure 2.33: Splitting of the transitions without and with magnetic field. Normal Zeeman effect and anomalous Zeeman effect.

close to the original line center are polarized differently from those with a bigger wavelength distance from the original line. Zeeman-effect background correction is technically more complex than the other methods described above. It is almost exclusively applied for graphite furnace AAS. An exception is the DC Zeeman effect at the flame (see number 3, below).

3 technical implementations of the Zeeman effect are used in AAS:

1. Alternating magnetic field at the furnace perpendicular to the light beam
2. Alternating magnetic field in direction of the light beam
3. Constant magnetic field perpendicular to the light beam

The technical principle of background correction will be explained using the alternating magnetic field at the atomizer, perpendicular to the light beam (1).

Under the influence of the magnetic field, the lines are split into transitions which are stimulated by light exclusively polarized parallel to the magnetic field (π-components), and transitions which are only stimulated by light polarized perpendicular to the magnetic field (σ-components). π-Components are at or very close to the position of the emission line of the source, σ-components have a distinct wavelength difference. With the help of a polarizer, the light plane parallel to the magnetic field is rejected. At zero magnetic field, the absorption profile is neither split nor polarized. The usual atomic absorption takes place. The emission line (*a*) is absorbed by the analyte profile (*b*) and by background (*c*). This is classical AAS with polarized radiation (Figure 2.34, left side). The magnetic field is now switched on (Figure 2.34, right side) and the absorption profile is split and

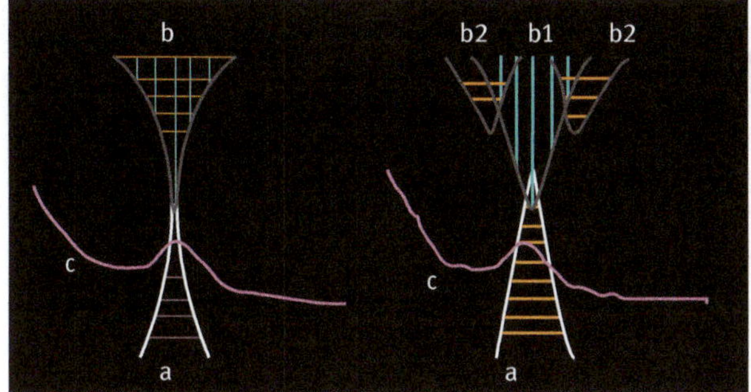

Figure 2.34: Absorption of analyte atoms (b) and background (c) without magnetic field (left) and with magnetic field (right side). The emission line (a) does not contain light polarized parallel to the magnetic field. The interaction of a with b1 is blocked out once the magnetic field is active.

polarized into components *b*1 and *b*2. *b*1 is π-polarized, *b*2 is σ-polarized. The background originating from particles or molecules (*c*) is neither split nor polarized. The transition *b*1 is blocked by the polarizer and analyte absorbance is taking place only by the overlap of *b*2 with the emission profile. This is small or very small, for most elements if the magnetic field is 0.8 T or higher.

The calculation is similar as in the other cases (eq. 2.15):

Equation 2.15:

$$Aa = (b + c)\text{-}(b2 + c) = b\text{-}b2 \text{ with } b \gg b2. \tag{2.15}$$

It is obvious that the background is measured at the wavelength and with the line width of the emission source. Only one source is used which is operated at constant current and intensity conditions. Small changes in the light intensity are compensated as if they were background by the subtraction of 2 readings. The magnetic field is modulated with 50 Hz which, in most cases is fast enough to follow fast absorbance changes in the graphite furnace. However, as for every background correction system discussed thus far, the total absorbance and the background are determined not simultaneously. This is a potential source of error in few special cases.

The Zeeman-effect principle works only if the background absorption is unaffected by the magnetic field. This is true for light scattering in general and, apart from a few well documented examples [60], it is true for molecules as well. It is true for metals only, if the split of the matrix atomic line profiles in the magnetic field does not result in an overlap with the analyte emission profile [61]. In these cases, the optical resolution is limited by the Zeeman principle itself. In general, however, no interference has ever been reported which was not observed with a D2 or a line-reversal background corrector. Many examples have been documented, on the other hand, where interferences observed with a D2 system cannot be found with a

Zeeman-effect background corrector [62]. One characteristic of ZAAS measurements has already been emphasized: the σ absorption profiles are not shifted completely outside of the spectral bandwidth of the emission profile. The more complex the interactions of the magnetic field with the magnetic and spin properties of the resonant electron, the more energy levels will be created. This eventually results in an analyte absorbance by the σ profiles of between less than 10 and up to 50%. In other words: ZAAS is slightly less sensitive than conventional AA. On the other hand, there is a significant gain in signal to noise, as only one source is used. Generally, the detection limits in ZAAS as compared to conventional systems are equal or better in cases of simple or no background absorption and much better in cases of complex or high background absorption. More important is the much lower risk of analytical errors in this system.

4. Specific and background absorption in high-resolution continuum source AAS. High-resolution continuum source AAS, as realized commercially, is measuring simultaneously the light intensity over a spectral range of about 400 pm, using 200 individual detectors (pixels) of a CCD. The optical resolution is about 2 pm. These pixels will experience different photon intensity due to analyte absorption and background. Usually the spectrometer is set, such that the analyte line is in the center of the window for measurement. In Figure 2.31 the analyte absorption on the Cd- line is zero. The background may be quantitated by selected pixels to the left and/or to the right side of the analyte line only, it may be averaged over the entire window, or it may be treated by mathematical models. Instead of the sequential measurement of cycles with different information, the entire spectrum is treated to calculate the analyte absorption. Usually, the background absorption is measured in a distance of 5 pm to the analyte line. Practical experience shows that the analytical quality of background correction is comparable to that of the Zeeman-effect technology.

Sample introduction and principles of atom generation in AAS

Elemental analysis is undertaken with liquids, solids, slurries, and gases. Most samples in AAS are introduced into the instrument as liquid. Solids and slurries have been predominantly analyzed using the graphite furnace technique. The chemical vapor generation (CVG) is applied to generate analyte bearing gases from liquids with suitable chemical reactions. Samples from separation techniques can be coupled to AAS as well. However, as this is a typical domain for ICP-MS, the latter topic will be handled exclusively in the mass spectrometry part below.

AAS is based on the interaction of light with the outer electrons of free atoms in ground state. Atoms dissolved in liquids in ionic or molecular bound form, bound in solids, or present as gas or transported in gases, must be first atomized, and kept stable for the time of the measurement process in a confined volume, completely enclosed by the probing radiation beam. The temperature must be high enough to keep the atoms stable for the time of measurement, but it should be low enough not to thermally excite ground-state electrons.

Greatest significance must therefore be granted to the atomizer of an AAS instrument.

Flame atomizers (see Section 2.1.6)

A test solution is transformed into an aerosol of droplets and burnable gas. The droplets shall be small enough to be dried rapidly at elevated temperatures. The size distribution shall be narrow enough to ensure vaporization within a small time window. The mix of gas and sample must be heated to temperatures high enough to ensure atomization of most elements of the periodic system. The flame atomizer is the simplest and most widely used system for atomization in AAS. Most elements are in the ground state experiencing temperatures between 2,300 °C and 2,800 °C. Only alkaline-earth, alkaline-, and some of the lanthanide and rare earth elements are significantly excited or ionized. The ionization, however, can be easily suppressed by chemical additives (ionization buffers). The chemical reactions of the analyte elements are often defined or strongly influenced by the composition of the flame gases. Important parameters such as atomization efficiency or chemical reactions with the matrix, are strongly dependent on the reducing or oxidizing properties of the flame. Compared to the analyte atoms and to the matrix, the flame gases are always the bulk compound, minimizing reactions between analyte and matrix. The number of analyte atoms present per unit time in the light beam is diluted as well. The few gas phase interferences possible, can be easily controlled by optimizing the gas flows or the observation height in the flame, or by changing from an air flame to a nitrous oxide flame. The velocity of the gases in AAS flames is very high: about 200 cm/s in air/ethine flames and 700 cm/s in nitrous oxide/ethine flames. The time in which the compounds pass from the cold zones inside the burner head through the hot zones of the flame lasts only a few milliseconds. Obviously, the observation height is therefore of major importance for parameters such as atomization efficiency or chemical interferences. As the atoms and the matrix can be observed and quantitated for a short time only, the power of detection of flame AAS is limited on the one hand but very moderate background absorption is experienced on the other. Non-specific absorption can be easily corrected. Flames are transparent at wavelengths longer than 230 nm. In this range, flame flicker contributes little to the measured baseline noise. At shorter wavelengths the flame absorbs more and more radiation. About 50% of radiation is lost at 200 nm. Changes in the flame conditions will therefore significantly contribute to the overall noise. This noise can usually be minimized by the background corrector. In the long wavelength range above 350 nm the flame may contribute to the baseline noise by emission radiation.

The sample is introduced dissolved in acidified or basic aqueous solutions or in the form of organic liquids with a pneumatic nebulizer (see Section 2.1.6) as a fine aerosol, which is mixed with the combustion gases in the mixing chamber. The droplets diameters are distributed about a median of 10 µm. The aerosol is desolvated, vaporized and finally atomized in the flame. Usually, the liquid sample is aspirated at a rate of about 5 mL/min by an oxidant flow of about 5 L/min. Only around 20% [63] of the sample is transformed into a fine aerosol, the rest is discarded. The aerosol is introduced into the flame until a constant analyte flow through the light beam is observed. This will take about 2 s without automatic sampling device and about 5 s with automatic sampler. The steady state signal is integrated two or three times for about 1–3 s, representing several hundred individual photon counting cycles of a few milliseconds each. A typical triplicate flame reading will therefore require about 10 s without autosampler and about 15 s with autosampler. For this type of sample introduction 1 ml of sample is required. The constancy of the sample flow and the stability of the flame will limit the repeatability of the flame determinations at higher absorbance readings. An optimized system can achieve a repeatability as good as 0.1–0.3% relative standard deviation.

The flame conditions usually change between aspiration of air only (no sample introduced) and aspiration of solvent. This may cause small changes in the baseline. Simple drift compensation (double beaming; see above) by measuring the lamp light directly through the flame, shortly before or after the absorbance reading, is therefore possible only if the solvent flow is kept constant. As this is usually not the case in routine analytical work, flame AA spectrometers are often optical double-beam instruments where two light paths – through the flame and around the flame – are compared.

The merits of flame AAS are well documented and described. Besides the unmatched sampling speed and the simple and rugged instrumentation, there is vast knowledge on reliable methods for essentially interference free trace element analysis. Flame AAS is the cheapest atomic spectroscopy technique. The limits of detection range from a few µg/L up to about 100 µg/L based on element and concomitants in the solution analyzed. The maximal concentration of dissolved solids which can be introduced depends strongly on the method of nebulization selected. It is about 1–5% for classical nebulization and close to saturation for high-pressure nebulization or microsample injection. The detection limits referred to the solid sample are therefore in the range of 50–1,000 µg/kg. The sampling efficiency and the detection limits are probably the most serious limitation in flame AAS and are among the reasons for the success of the graphite furnace as a means of sample vaporization and atomization.

Graphite furnace AAS (see Section 2.1.6.2)

A small, exactly defined, and reproducible volume of sample is introduced into a chemically inert and stable reservoir. The solvent and parts of the matrix are removed at various temperatures programmed into the computer of the system. Finally, only very few µg of stable compounds including the analyte elements are left. The atomizer is heated to a very high temperature in a short time so that all the atoms are experiencing roughly the same temperature and are introduced into the monochromatic light beam enclosed by the inert walls of the device. No gas is flowing through the system and the analyte atoms simply diffuse out of the tube at their own pace. All these parameters can be calculated from physical equations. The analytical signal can be approximated using chemometrics.

The combination of an electrothermal vaporizer with AAS is in fact the instrumental technique coming closest to the ideal of absolute (standardless) analysis.

The buffering effect of the combustion gases in flame AAS is often substituted by a chemical additive which, if present in excess over the other compounds, will effectively define the chemical atmosphere in the graphite furnace.

During the atomization step the internal gas flow is switched off so that the gas inside the tube is hardly moving at the beginning of the atomization step. The first second of the atomization step is characterized by an increase of the gas phase temperature by more than 1,000° and an expansion of the gas volume by a factor of 4. This results in a forced convective gas flow from the tube center toward the tube ends and out of the dosing hole. The atoms should be released into the gas phase only after it has almost reached its final temperature and gas volume. Therefore, platform atomization has become popular [57]. A thin graphite body inside the tube has minimal physical contact to the tube. It is mainly heated by radiation from the tube wall. The sample (about 20 µL) is pipetted onto this platform. The heating of the platform is delayed relative to the wall (see Figure 2.35). Under these conditions the atoms are removed from the absorption volume mainly by diffusion into a gas atmosphere which is hotter than that of the platform. It usually takes 1–4 s from the appearance of the first to removal of the last atoms from the measurement beam. The mean residence time of the atoms is about 3 orders of magnitude longer than in flames and this explains the excellent absolute sensitivity of the graphite furnace.

High matrix density in the light beam often accompanies the high analyte atom density, explaining the need for an excellent background correction system. During the atomization step, even refractory matrix such as oxides can be vaporized almost completely. After the introduction of the first commercial graphite furnace in 1970 [48] following suggestions by Massmann [44], the first ten years were characterized by a rapid development of the technique and the methodology to reduce the effects of the numerous interferences that occurred in the non-optimized systems of the early years. The physicochemical and technical requirements of GFAAS are now well described and documented.

GFAAS today is characterized by a good freedom from interferences and relative detection limits in the range of 0.1 µg/L (2 pg for 20 µL sample volume) for most elements. Depending on matrix and analyte elements, a single graphite tube can be used for from 300 to 1,500 atomization cycles which, under favorable conditions, represent 100 to 500 duplicate sample measurements including calibration and

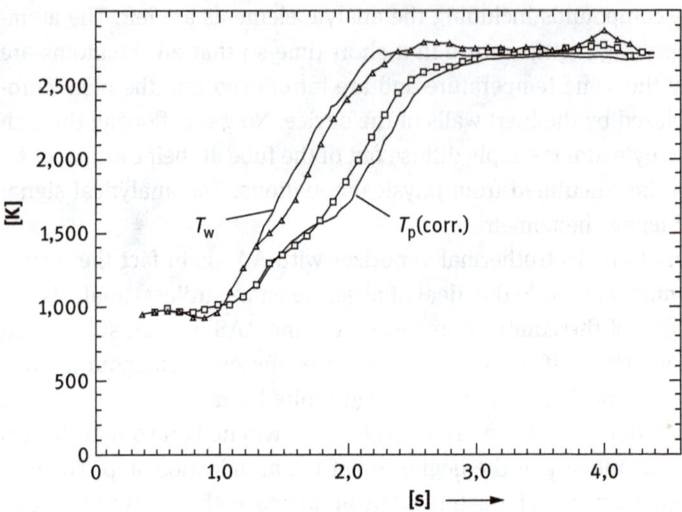

Figure 2.35: Temperature evolution in a Massmann-type graphite tube with L´vov platform. The temperatures of wall and platform are obtained with a fast reading pyrometer and plotted as a function of time. The gas phase temperature near the wall and near the platform is determined with CARS methods [64]. The wall temperature is retarded from the platform temperature by up to 700 °C.

quality control samples. The furnace operates automatically during analysis with very limited requirement for operator interaction.

Chemical vapor generation (see Section 2.1.6.2)

Chemical reactions transfer elements, dissolved in the solution for measurement, into gaseous molecular compounds or into gaseous free atoms. These compounds or free atoms are separated from the matrix and transported by an inert gas to the absorption volume. Gaseous molecules are decomposed such that free atoms can be determined by AAS. The paramount function of the process is enhancement of specificity.

As an example, the determination of Arsenic (As) using the Chemical Vapor Generation AAS is described (Figure 2.24.)

The sample for measurement is completely dissolved. It is acidified and As is present in oxidation state +3. Blank, standards, samples and quality control solutions are resident in the auto-sampler for automatic operation.

The instrument control software will activate the peristaltic pump for propulsion of the carrier solution (usually 3% HCl), the reagent (0.5% $NaBH_4$ in 0.2% NaOH) and removal of the waste. The second pump will be activated to transport a small volume (approximately 0.5 mL) of the sample to the valve.

The valve will be turned, and the measured sample will be injected into the carrier stream. Carrier, sample, and the reagent will be brought to reaction in the mixing block. The reaction will take place downstream the line and will be supported by an Argon gas flow of about 100 mL/min. Reductant and acid will release nascent hydrogen which will reduce As^{3+} to AsH_3. The combined gases, Ar, H_2 and AsH_3 will be separated from the liquid and transported to the reaction cell. The reaction cell is often a heated quartz cell at about 950 °C. AsH_3 is decomposed by heat, H-, OH-, and O- radicals to As atoms which can be determined by AAS.

After separation of liquid and gases, the analyte compound is accompanied by the carrier gas (Ar) and reaction gases (e.g., H_2) only. Specificity and sensitivity of the measurement is therefore exceptionally high. Instead of transfer to a heated atomization cell, the gases can be preconcentrated and atomized in a graphite furnace, guided to an ICP-OES or ICP-MS or can be transported to the atomization cell of an atomic fluorescence spectrometer. Due to the high specificity, the detection limits of elements readily forming analyte vapor, are lowest when processed through a chemical vapor generator.

The atomizer for CVG in AAS is usually a heated quartz cell. The cell is about 5–10 cm long and is embedded into an electrically heated oven, with programmable temperature control. This system is stabilized at a constant temperature between 700 and 1,000 °C before the measurement process is started.

Mercury is the only element which is released as metal vapor which persists in atomic form at room temperature. The reduction of the cation can be accomplished with mild reductants so that separation from matrix is even more complete than for the other elements discussed above. Analyte atoms mixed with Ar are directed to the non-heated atomization cell for AAS or to a special small cuvette for AFS determinations. Mercury is among the elements with the lowest possible detection limit in optical spectroscopy.

In Table 2.2 the elements suitable for CVG are listed. Only few of these are determined with CVG in practice.

Table 2.2: Analyte vapor for determination in AAS or AFS.

Element	Volatile compound	Spectroscopy technique	Standards
Hg	Atoms	AAS, AFS, QC, GA	Yes
As	AsH3	AAS, AFS, QC, GA	Yes
Sb	SbH3	AAS, AFS, QC, GA	Yes
Bi	BiH3	AAS, AFS, QC, GA	Yes
Se	SeH2	AAS, AFS, QC, GA	Yes
Te	TeH2	AAS, AFS, QC, GA	No
Ge	GeH4	AAS, QC, GA	No

Table 2.2 (continued)

Element	Volatile compound	Spectroscopy technique	Standards
Sn	SnH4	AAS, QC, GA	No
Pb	PbH4	AAS, QC, GA	No
Cd	Atoms	AAS, QC, GA	No
Cu	CuH	AAS, GA	No
Ag	AgH	AAS, GA	No
Au	AuH	AAS, GA	No

Column 2 lists the chemical compound released after reduction. Column 3 defines the techniques usually applied for determination; QC=QUartz cell atomizer; GA=preconcentration and atomization in the grapite furnace. In column 4 the availability of international standards is indicated.

Solids and slurries

Elements can be determined directly from solids or mixtures of solids with liquids using AAS. A degradation of the analytical figures of merit are tolerated up to the point where the method can be considered as exploratory analysis. As explained above, introduction of samples into a flame is subtle and the sample residence time in flames is short. Solids, even when mixed with liquid, can hardly be decomposed under these conditions. The atomizer of choice for solids or liquids inclosing particles (so-called slurries), is the graphite furnace. Numerous publications [65] prove the suitability of GF-AAS for the direct analysis of solids or slurries [66]. It has been proven that this option may enable unmatched detection limits or unparalleled speed of analysis. Still in routine laboratory life, solid sampling is rare.

Several prerequisites must be matched to obtain good analytical quality from solid sampling analysis:

- Very small masses of solids (µg-range) must be weighted and introduced reproducibly into the atomizer
- The small mass must be representative for the sample (homogeneous distribution of the analyte in the sample)
- The analyte element must be released from the solid completely or at least reproducibly
- The matrix effects must be controllable
- The solid must be removed from the atomizer after the measurement cycle

These propositions, in fact, are often met and the power of solid or slurry sampling in AAS is generally greatly underestimated.

In Figure 2.36 an automatic solid sampling device is imaged. The sample for measurement is loaded on a graphite platform and automatically balanced on a

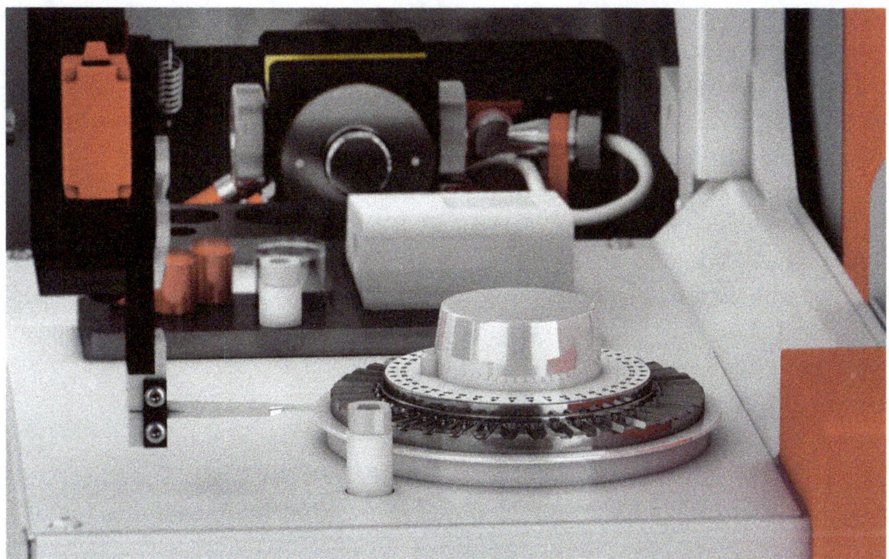

Figure 2.36: Automatic solid sampling device (Analytik Jena, Germany). Courtesy of Analytik Jena, Jena, Germany.

microbalance. The platform is then inserted into the graphite tube where it acts like a platform. The sample is handled like a liquid described above.

Equally important as direct solid sampling is the slurry technique. About a mg of a powdered solid is balanced into the cup of an autosampler. A mL of an acidified liquid is added and the solid is evenly distributed into the liquid. An aliquot of the slurry is then pipetted into the graphite furnace. The advantage over direct solid sampling is that bigger masses of sample can be handled, and automation of sample introduction is much simpler. The disadvantage is that the samples need to be powdered such that they can be evenly homogenized into a liquid. It is known that a part of the analyte is dissolved into the solution used to prepare the slurry. The repeatability of the determination is therefore usually higher than in the direct solid sampling technique.

Automated handling of samples

The auto-sampler is a central accessory of the analytical system, managing the course of action of an analytical method. This holds true for most of the analytical techniques described so far and in the future text. In the following section we will call all types of reference solutions, quality control samples, etc. solution for measurement: SFM. The functions of an auto-sampler are or may be:

- Store SFM and feed them to the analytical instrument in a pre-defined sequence
- Run cleaning sequences between SFMs

- Dilute SFM
- Add reagents to SFM
- Mix or homogenize SFM
- Feed SFM, assess the analytical result, interrupt the automatic process, and initiate an extra action

In AAS the auto-sampler holds between 1 mL (Graphite furnace) up to 50 mL (flame, CVG) vessels. Depending on the sample volume, between 20 and 100 samples can be handled without manual interaction. SFM and reagents are usually pumped by a precise piston pump. In a simple device the sample may be only aspirated by the pneumatic nebulizer. As an example, the combined auto-sampler for Flame and GF-AAS is sketched in Figure 2.37.

Figure 2.37: Auto-sampler for flame and graphite furnace AAS Courtesy of Shanghai Spectrum Instruments Inc., Shanghai, PR China.

Auto sampling is the pre-requisite for an unattended operation of the measurement device and thus for time- and cost-efficient analytical spectroscopy. The sampler also strongly contributes to accuracy and repeatability of the entire system and influences the total time of an individual determination. The parameters that need control are:

- Accuracy of sample feed and sample dilution as a function of the volume
- Repeatability of sample feed and sample dilution as a function of the volume
- Carry over and/or contamination effects by the sampling capillary
- Time required for stable sample feed (in case of flame AAS and CVG-AAS)
- Possible changes or deterioration of the sample by long sampling capillaries

Auto samplers are usually technically mature. Still the influence on the analytical quality must be tested and validated by the analyst.

Just as in the case of chemical vapor generation, flow systems in combination with an autosampler can be used to facilitate sample introduction or even enable automated analyte preconcentration, matrix separation, or species analysis [67]. A widely used system for flame AAS is an injection device for samples. This may be volume or time controlled. The basic idea is to feed nebulizer, mixing chamber and flame constantly with solvent. The aerosol pathway and the burner head remain clean, and the burning conditions of the flame are stable. Corrections, such as lamp intensity (double beaming) can be performed in this cycle. The SFM is then injected for a period of fractions of a second up to 10 s. Absorbance is measured in this cycle followed by solvent again. The injection technique in flames is gaining increased attention nowadays but its analytical potential is still underestimated [68].

2.1.7.3 Chemistry in AAS

Flames

The final aim of sample nebulization, confinement of droplet size, mixing with combustion gas, heating in the flame (see above), is to generate a spatially homogeneous concentration of atoms in the absorption volume. The transfer efficiency of dissolved elemental ions or compounds to atoms shall be maximal. Physical and chemical effects are involved on the way from liquid to free atoms. This has an influence on the analytical quality.

A nebulization rate of 5 mL/min will typically yield a flow of 0.08 mL/s into the mixing chamber. At least 75% of this volume is discarded and the remaining 0.02 mL/s are thoroughly mixed with the burning gases. Most of the sample is water or acid. Assuming that 3,000 mL/min = 50 mL/s ethine (C_2H_2) and 8,000 mL/min = 130 mL/s oxidant (e.g., air) are flowing through the system, it can be easily calculated that about 1 µmol of liquid is homogenized into 2 mmol of burning gases and 6 mmol of oxidant. The chemical environment is thus defined by the burning gases and their ratio.

On the way from the aerosol generation through the radiation beam of the spectrometer, many parameters must be controlled to obtain best analytical performance. The solution for measurement (SFM) is dissolved in water, acid, lye, or organic solvent. The element to be determined is dissolved as cation, present as a complex or as a molecule. The type of acid or the matrix will influence the type of compound from which atomization will take place.

During aerosol formation the chemical form is expected to be stable. The nebulizer and mixing chamber should generate a narrow droplet size range. Bigger droplets should be skimmed. The solvent and higher matrix concentration may influence the droplet formation due to different viscosity and surface tension. Compared to a dilute (<1%) nitric acid solution, for example, a solution of 5% sulfuric acid will result in a slightly lower uptake rate and bigger droplet size. On the contrary, a higher flow rate through the nebulizer and smaller droplets may result, if organic solvents are aspirated. These effects are called physical effects.

About 200 cm^3/s of aerosol move through mixing chamber and the body of the burner through a slit of 0.06 cm width and 10 cm length (in the case of air/ethine flames). The speed of the aerosol thus is around 300 cm/s or 3 mm/ms. The droplets reach the hot zone of the burner head neck and are rapidly evaporated within few milliseconds. The chemical form of the solid, forming from desolvated droplets, will strongly depend on the solvent used and the matrix present. Often it is the chloride or the nitrate which is primarily formed. Directly above the burner head slit the aerosol is reaching its burning temperature of 2,200 °C (air/ethine) and 2,700 °C (N_2O/ethine). The gas volume will expand by a factor of about 8 in the case of the air/ethine flame and about 10 in case of the nitrous oxide/ethine flame.

The transition from the salt particle to the atoms is taking place above the burner head. Chlorides and sulfates are often directly fragmented into atoms, nitrates may be transformed into oxides and afterward reduced to the elements. Molecules are rapidly disintegrated or are first forming oxides before they are atomized.

Quantitation of the atoms takes place between 5 and 15 mm above the burner head. The atom formation takes place within 1 to 4 ms after ignition. From the various reaction pathways described above it is obvious that instrumental parameters such as droplet size and distribution, the exact position of the burner head relative to the measuring light beam as well as the relative mix of burning gas and oxidant will have a big influence on the atom concentration in the absorption volume.

In Figure 2.38, the physicochemical processes taking place from aerosol formation to determination are sketched.

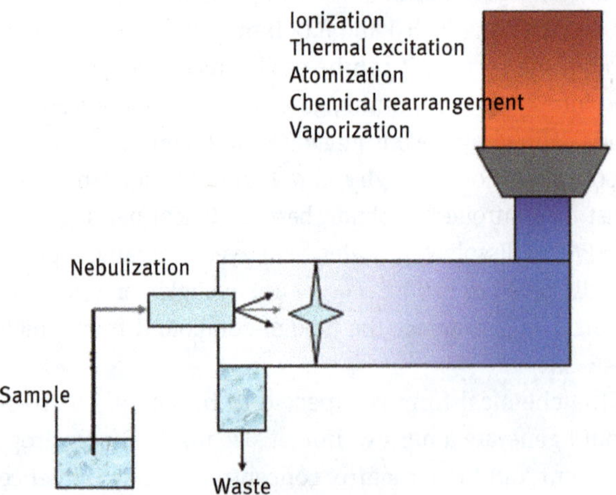

Figure 2.38: Processes taking place in flame AAS from nebulization to atomization/ionization.

The chemistry of the analyte atoms is the first parameter of optimization. Elements which form stable oxides are usually determined in fuel-rich (reducing) flames, elements which are more stable than their oxides are preferentially determined in oxidant-rich flames. The optimization of flame stoichiometry becomes increasingly important with higher atomization temperature. Whereas elements which are usually atomized in ethine/air flames can often be atomized under compromise gas conditions, refractory elements atomized in ethine/N_2O flames require a careful optimization of the fuel/oxidant ratio. As an example, the steady state atomization signals (absorbance versus time) of V, atomized in an ethine/N_2O flame is illustrated in Figure 2.39.

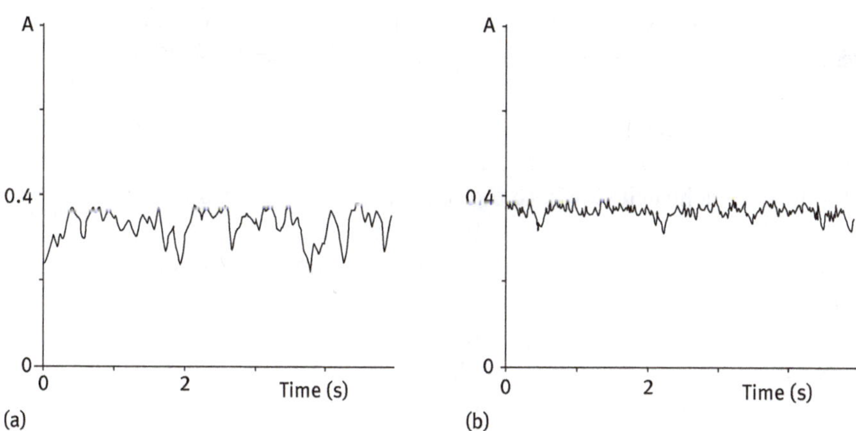

Figure 2.39: V atomized in Ethine/N2O-flames. Left side. Balanced stoichiometry: Poor repeatability, limited sensitivity. Recording on right side. Fuel-rich stoichiometry: good repeatability, optimal sensitivity.

Despite the big excess of burning gas over the gas formed by solvent, matrix and analyte, the chemical effects of the latter cannot be excluded. Volatilization of a medium-volatile element chloride in the flame is faster than that of the refractory element oxide. The concentration of atoms as a function of measurement height above the burner head may therefore be different. Fine-tuning of instrument settings with respect to optimized atom formation is a multi-parameter task. Usually, it is recommended to optimize for ruggedness rather than for highest sensitivity. The optimal conditions for the determination of the individual elements are well described in the literature and are part of the recommended conditions given by the manufacturers. Matrix and solvents generate effects on the analytical quality of the measurement, but these effects are rather moderate in flame AAS.

Graphite furnace

The absorption volume in graphite furnace AAS holds 500–800 mm^3 or 0.5–1 mL, respectively. The inner surface of the tube is 350–550 mm^2. Tube and platform are made of high-density graphite coated with a dense crystalline structure of pyrolytically deposited graphite. This material is chemically widely inert up to temperatures of about 400 °C. About 10 to 30 µL of solvent is usually pipetted into the tube for one determination, which corresponds to roughly 1 mmol of solvent. When the solvent is volatilized, it will generate more than 20 mL of gas. If the solvent contain 1% of dissolved solid, the volatilized solid would still generate a gas volume which is alike the tube capacity. Solvent and matrix therefore play a dominant role in GF-AAS. The sample is introduced into the tube as a liquid, a slurry or a solid. Liquids and slurries often contain dilute or medium concentrated acid, less frequently lye or organic solvent. The analyte element is present as cation, as a complex or as organic molecular compound. The same is true for the matrix which is usually present in a huge excess.

During pipetting the tube is filled with Argon. Once the furnace program is starting, the inside of the tube is purged with a flow of a few mL Ar per second.

The furnace is heated slowly so that the solvent is vaporized within about half a minute. The volume, generated by the solvent, should be less than the Argon purge flow. The dissolved solid is preconcentrated and transformed into a solid compound. Depending on solvent and matrix this may be a salt (nitrate, chloride, sulfate, etc.) or a molecular compound. Reactions may take place which can be actively directed by the addition of modifiers, chemical compounds which help to improve the analytical quality of the determination. During drying, the graphite surface remains chemically inert, however, adsorption phenomena may occur.

Next the tube is heated up to a temperature where the bulk of the matrix (e.g., organic compounds, NaCl) can be volatilized and removed. Many of the compounds are disintegrating, such as nitrates or some chlorides. These are releasing reactive species or radicals which may react with the graphite surface and/or form new bonds with the analyte element. The latter one should be retained quantitatively in the tube. The temperatures during this pyrolysis step are 400–1,100 °C but seldom higher. Many analyte elements cannot be retained in the tube and thus separate from the matrix under these conditions. Chemical modifiers are used to stabilize the analyte element or to make the matrix more volatile. The chemical reactions during pyrolysis are often complex and often not completely understood. An example of the complexity can be found in [69]. The graphite surface becomes active and plays an important role during this process as well. It usually acts as a reductant, but it may as well interact with the modifier and form chemically active spots on the surface which stabilize the analyte elements to higher temperatures. Few modifiers are used nowadays [46] to facilitate analyte/matrix separation, among them the most popular are Pd/Mg modifiers. Most important during pyrolysis is to remove matrix and modifier with controlled speed so that it is removed out of the dosing hole as completely as possible and is not re-condensed at cooler parts of the graphite furnace. In case of a high concentration of organic material in the graphite tube, carbon is often formed during the pyrolysis step. This will settle down inside the tube and gradually fill the tube with amorphous carbon waste which cannot be removed out of the tube again. In this case air is used as an alternative purge gas, acting inside the tube at temperatures between 400 and 600 °C. The organic compound is ashed instead of pyrolyzed this way, whereas the highly structured crystalline tube surface is not significantly oxidized under these conditions. In general, a carefully optimized pyrolysis step is only required if the total content of matrix exceeds 0.1 µg. In all other cases a fast pyrolysis at temperatures around 400 °C, followed by the atomization step will be a safe compromise which will speed up the determination.

During atomization the temperature is rapidly increased to the level where the analyte element is quantitatively atomized. In this step the internal Ar gas flow is stopped. The temperature increases by 1 to 2 °C per millisecond. The increase is not linear but depends on the temperature range to be bridged. If we assume a temperature increase between 1,000 °C and 2,000 °C we will face an expansion of the gas volume in the tube by a factor of 1.8. Between 400 °C and 2,000 °C the factor would be 3.4. The atomization temperature for most elements is between 1,500 °C and 2,600 °C. Temperatures above 2,600 °C are not recommended as the graphite starts to sublime significantly above that level. Even after a thorough pyrolysis step, most often there is still a large excess of matrix which will volatilize, disintegrate, expand and react while the temperature is rapidly increasing. Modifiers help to dominate the chemistry in the atomization volume. At temperatures above 2,000 °C, the partial pressure of carbon increases and the partial pressure of oxygen decreases. The graphite surface now plays an important role in the process of atomization and/or formation of carbides. Good conditions for atomization are obtained when atomization takes place from a platform inside the tube. The expansion of the gas phase is almost complete under these conditions, and the atoms are moving predominantly by their natural diffusive speed.

Cleaning

In the final step, the tube is heated to 2,600 °C and remaining matrix is largely removed out of the absorption volume. This cleaning process is facilitated by the Argon purge gas flow through the furnace. In Table 2.3 chemical processes in the furnace are sketched in a simplified way.

Chemical vapor generation (CVG)

The classical CVG technology will start from completely mineralized samples only. The only exception is mercury where, under certain circumstances, the gaseous analyte can be released from dissolved molecules, particles, and solids as well.

The hydride forming elements are usually reduced from their lowest cationic oxidation state. These cations of As, Sb, Bi, Se, Te can be transformed into hydrides

Table 2.3: Chemical processes during the main steps of a graphite furnace program.

Program Step	Process	Analyte	Matrix	Modifier	Surface
Drying	Desolvation, formation of solids	Present as salt, complex molecule	Partly volatilized, salt or molecule	Salt or partly volatilized, addition as gas possible	Chemically inert, physical adsorption effects
Pyrolysis	Matrix separation	Present as compound or stabilized by modifier	Volatilized, split, chemically activity	Active in forming compounds reduced, active with surface, split	Minor chemical activity, adsorption processes, reducing
Atomization	Atom formation volatilization	Atoms, stable molecules,	Atoms, molecules, radicals	Atoms, radicals	Reducing, carbon donor
Heat out	Cleaning of absorption volume	Volatilization	Volatilization	Volatilization	Reducing

quantitatively and rapidly by strong reducing agents such as $NaBH_4$ in acidic medium. Elements like Ge, Sn, or Pb are reduced in weakly acidified and buffered medium. The transfer of these hydrides is usually not quantitative. If the elements are present in higher oxidation state in the solution for measurement, the reaction with the reductant is slower, not quantitative, or even negligible. The pre-reduction step is therefore a part of the sample pretreatment. The hydrides formed, are transported to the gas-liquid separator by the carrier gas (Ar) and hydrogen (formed by the reaction of acid and boronhydride), water vapor, HCl, and radicals such as H°, OH°.

In a typical flow system, the carrier gas flow is about 150 mL/s. A 1% NaBH4 solution is added to the acidified carrier. The flow rate is about 2 mL/min of the liquid reductant. If we calculate with the molar mass and the volume of an ideal gas, about 500 µL/s of H_2 will be generated and mixed with 2.5 mL/s of Ar carrier (a typical value for a CVG flow system). After gas/liquid separation, the gas mix is flowing to the atomizer. The classical atomizer is a heated quartz cell at close to 1,000 °C. The diameter of the cell is between 5 and 8 mm. The speed of the gas is strongly reduced in the atomizer to about 1 mm/s. The hydrides are thermally cracked but – depending on the element – atoms are not stable at these temperatures. At this point in time the radicals play an important role to contribute to atom formation for most of the hydride forming elements [49]. A much better atomizer would be a graphite furnace. However, the analyte element from the reaction gas must first be trapped to the moderately heated graphite surface. This works remarkably quantitatively if the graphite surface is coated with an involatile noble metal, e.g., Ir or Pd. The trapped hydrides are subsequently atomized at their optimal temperatures, i.e., around 2,000 °C. This technology combines the advantages of ideal sample pretreatment and atomization conditions. It is therefore the element analytical method with the best detection limits and highest selectivity.

As mentioned above, mercury is easily reduced to the element. A much milder reductant can be applied. International standards for mercury determination are based on $SnCl_2$ in a moderately acidic environment. Hydrogen is obviously not generated under these conditions. Hg is transported through the gas-liquid separator by the carrier gas and flushed through the absorption volume, usually a glass cell with small diameter. The measurement cell does not have to be heated. Although the conditions in the absorption volume are ideal for best detection limits, the sensitivity of Hg in AAS is only moderate. This has to do with the transition probability of the electrons. Hg is therefore nowadays mainly determined using the fluorescence method.

2.1.7.4 Instrument suitability

The techniques of AAS allow the determination of elements within a certain working range, a lower limit of quantitation and a certain repeatability. The trueness of the result may depend on the concentration of concomitants in the sample. Limits, induced by the sample preparation, add to the complexity of judgement. Time for analysis and laboratory cost demand using the optimal method for the analytical question.

Basis for the judgement is the cookbook or the recommended condition which is stored in the data base of modern instruments. It lists characteristic concentrations or characteristic masses of each element doable with the respective technique. It suggests standard settings for spectrometer and atomizer, chemical additives such as buffers and modifiers. Experience with many element determinations shows that the standard deviation of an AAS spectrometer under optimal conditions is in the range of about 0.0005 A. This is about an order of magnitude lower than the absorbance at the m_O/c_o level, which is 0.0044 A. Thus, the characteristic data can be estimated as the concentration or mass range where quantitative determinations become possible. Both the characteristic data of a certain element and the standard deviation of the blank must be verified once the technique has been selected.

Example to illustrate the process

The regulations of the European Union require a Cd concentration of less than 1 mg/kg in agricultural soil [70]. The standard treatment of the soil is based on an aqua regia digestion with a total final dilution of 3 g solid in 100 mL of solvent. 1 mg/kg translates to a minimum detectable concentration of 0.030 mg/L in the solution for measurement. The cookbook data for Cd with flame AAS reports a c_O value of about 0.010 mg/L. The estimated limit of quantitation is below the lowest level to be determined. Hence flame AAS seems to be the appropriate technology for this application. The characteristic mass of GF-AAS in this case would be in the range of 1 pg. At 10 µL injected sample volume, this translates to 0.1 µg/L at the limit of quantitation. The sensitivity of this technique would be ways to high for the analytical question. The latter method would certainly be able to run the test but at the cost of an additional sample dilution, with higher risk of contamination, and significantly longer analysis time. Following this type of evaluation, the selected method must be further checked for its suitability.

A software resident automated process is taking place to setup instrumental conditions for the determination of the selected element(s).

It is recommended to run 10 repetitions of the spectrometer baseline measurement under these conditions. The result is the so-called spectrometer blank. The mean value and its lower and upper limit indicate the lowest possible standard deviation of all future analysis under these conditions. The standard deviation should be below 0.001 A for most of the possible analyte elements.

Both the graphite furnace and the chemical vapor generator are predominantly filled with argon while a blank solution is atomized and measured. Argon does not absorb radiation in the AAS wavelength range. Spectroscopically the mean value and the standard deviation of the baseline should not change when these atomizers are operational in the light beam. The furnace, however, is operating at maximum electrical current during the atomization phase and the strong electromagnetic stray field might have a slight influence on the baseline. In addition, tube or contacts may be contaminated or misaligned and the mean value of the blank may be different from zero. It is therefore important, to record blank and standard deviation of 10 replicates under the conditions selected for the intended determination. The mean value and the standard deviation obtained from the atomizer blank will therefore define possible shifts of the origin and the best possible l.o.q. under the recommended conditions.

Flame gases will absorb radiation in the short wavelength range. For ethine-air flames a significant absorbance is measurable only below 220 nm. The effect is significantly stronger in the ethine-nitrous oxide flame. Shifts of the mean absorbance of the flame and an increase in the photometric standard deviation may be observed. Using the recommended conditions for the intended determination, 10 atomizer blanks should be run and mean value, s.d., and possible systematic drifts should be recorded.

The atomizer blanks are important to judge on s.d., l.o.q., and blank of the instrument. The difference between the photometer blank and the instrument blank may help to identify shortcomings of an instrument component. The obvious next step is the blank obtained from a test solution which is close to the acid and reagent concentration of the sample. It should be as low in element concentration as possible. Again the s.d. is determined under the recommended conditions.

The recommended conditions contain valuable information on the sensitivity of the determination in form of the characteristic concentration c_O or characteristic mass m_O. In flames this value is based on a constant feed of the solution for measurement into the nebulizer. The graphite furnace is usually defined with analyte mass introduced into the furnace. The reciprocal normalized value is m_O. m_O is related to

c_0 via the injected volume: $c_0 = m_0/V$, where m_0 is usually defined in pg and V in µL. The resulting pg/µL can easily be converted into µg/L.

We already mentioned that most AAS reference solutions can be run with an s.d. of 0.0005 to 0.001 peak or integrated absorbance units. A concentration or mass of 10 times m_0/c_0 should result in an r.s.d of 1–2% under optimal conditions.

A standard of about 10 times m_0/c_0 is therefore suited to measure the reciprocal normalized sensitivity under the selected conditions and to determine the l.o.d. and l.o.q. under ideal measurement conditions (eq. (2.16)).

Equation 2.16: Limit of detection (3 s.d.) obtained from 10 replicates of the blank, and c_0 determined from a reference solution containing about 10 times the expected c_0 concentration:

$$\text{l.o.d.}_{\text{blank}} = 3 \cdot \text{s.d.}_{\text{blank}} \cdot c_0 \cdot (0.0044)^{-1} \tag{2.16}$$

It is more difficult to define the upper limit of the working range. If we assume a maximal working range of 2 ½ orders of magnitude, we may prepare a reference solution which contains about 2,500 times the l.o.d. concentration or roughly 800 times the calculated value of c_0. Calculation of the real c_0 value at this concentration value will yield a new reciprocal sensitivity which can be compared to c_0 (see Table 2.4). This way evaluation of linearity and r.s.d. will become straightforward.

These basic tests allow to define a few of the important figures of merit for the selected technology and the selected instrument. With these tests the basic function of the instrument is verified. They are significant for the reference solutions but not yet for the application.

Is the instrument suitable for the application?

Unknown samples in instrumental analytics are often very different in composition. In international standard methods (ISO, EN, ASTM, etc.) one will often find limiting concentrations of concomitants under which the method has proven to be functional, or an instrument has proven to be suitable.

Testing the suitability of an instrument requires some knowledge on the worst sample, which should serve as a test solution for the method development. If the worst sample exceeds the limiting concentrations in the external or laboratory internal standard it must be diluted to this limit with direct result on the l.o.q. (see the examples above). The best test solution would be a sample containing all concomitants but no analyte. As this is often not possible, one should select two test solutions: one straight sample and one sample where a known amount of the analyte element is added, approximately in the range of about 10 times m_0/c_0.

In flame AAS the effect of concomitants is usually manageable. The effects induced by matrix can be grouped into sources. In Table 2.4, the effects, the probability of an interference, and its correction are listed.

Table 2.4: Matrix effects in flame AAS.

Source	Influences	Occurrence	Correction
Viscosity of sample	Uptake rate, nebulization, droplet size	Frequent	Standard addition
Vaporization	Observation zone	Sometimes	Burner height, modifier
Matrix/analyte reaction	Free atoms, observation zone	Sometimes	Flame type, stoichiometry burner height
Degree of ionization	Free atoms	Sometimes	Modifier
Light scattering	Spectral background	Sometimes	Background corrector
Molecular absorption	Spectral background	Rare	Background corrector
Absorption by other elements	Spectral background	Rare	High-performance BG corrector
Burner blockage	Flame stability	Frequent	Cleaning

In all these cases, the characteristic concentration in the sample will be different from that of the reference solution. Often the effect can be compensated by instrumental actions like activation of the background corrector, changing of the observation height in the burner, changing the flame stoichiometry to stronger oxidizing or stronger reducing. In many cases, the addition of a chemical modifier or buffer solution will prevent the effect. Calibrating with the method of standard additions will remove the interference in most cases. Care must be taken to assure that the addition curve is strictly linear. In cases of absorbance shifts parallel to the absorbance axis due to non-corrected background absorbance or blanks the method of standard additions will not correct for the error. Concomitants may influence s.d., and r.s.d., and ruggedness as well. This may originate from a less uniform droplet size distribution from the nebulizer, a cooling of the flame due to dissociation energy of salts in the plasma, reduced absorbance due to different flame stoichiometry or burner height.

The solutions for measurement, the worst sample, and the spiked worst sample, is aspirated into the flame for method development. Changes in the color of the flame, in the stability of the flame or in deposits on the burner head can be identified easily. The standard deviations and relative standard deviations should serve as indicators for possible additional noise sources by the matrix. The characteristic concentration, calculated from the spike will show possible effects of the matrix on the efficiency of nebulization, vaporization, or atomization. A long-term aspiration of the spiked sample will provide information on the stability of the characteristic concentration. Non-specific absorption due to matrix can be identified by activation of the background corrector. Non-specific absorption is indicated by the instrument. The corrected absorbance of the sample or the spiked sample will be different without and with background corrector. Optimization of the measurement conditions may involve:

- Measurement with activated background corrector
- Variation of the burner height: a higher observation zone will often reduce matrix effects at the cost of moderately reduced sensitivity.
- Variation of burning gas (ethine) and oxidant. A more robust flame (higher gas flows) will generate flames with higher energy which are less influenced by matrix. The sensitivity of sample and standard may slightly drop, but the interference may be prevented.

As described already earlier, matrix effects can be usually minimized by dilution. A similar effect may be obtained if only a limited volume of sample is introduced to the nebulizer instead of continuous aspiration.

Whereas in flame AAS analyte and matrix are strongly diluted by the flame gases, the tube of a graphite furnace may be quickly filled with concomitants during atomization. These may influence background, characteristic mass, s.d. and r.s.d. and long-term stability. A quick example may illustrate this: 20 µL of sea water pipetted into GF-AAS represent 600 µg of NaCl or about 10 µmol. These would generate a gas volume of about 220 µL at room temperature and 1,700 µL at 2,000 °C, which is a typical atomization temperature in GF-AAS. A graphite tube has a capacity of 500–800 µL. Reduction of matrix prior to atomization is therefore one of the major goals in GF-AAS.

The recommended conditions, developed in many years at institutes and in companies, are based on analyte-matrix separation. The aim is to provide the most promising initial point for method optimization. They are compromise conditions which are usually rugged but may not represent the optimal conditions for a simple matrix with negligible concentrations of concomitants.

Measure for the extent of possible matrix effects in GF-AAS is background absorption. If the expected worst sample generates only negligible background at a low pyrolysis temperature – say, background absorbance below 0.1 – the method development needs not to be focused on analyte matrix separation. If, however, the background of the sample is high (0.5A or above), a careful optimization of analyte absorbance versus background absorbance is required (see Figure 2.40).

Just like in the case of flame AAS, the selected test sample and the selected spiked test sample are injected into the graphite furnace. The sample volume will define the analyte mass in the furnace. It shall be selected taking the target l.o.q. into consideration. In most cases, the high sensitivity of GF-AAS makes it possible to use small sample volumes between 5 and 10 µL which will speed up the time per determination and generally reduce matrix effects. Background absorbance, possible shifts of the baseline and appearance of the absorbance pulse relative to that of the reference solution will provide valuable information on the influence of concomitants. Possible effects and ways to overcome them are listed in Table 2.5.

Table 2.5: Matrix effects in GF-AAS.

Source	Influences	Occurrence	Correction
High mass of concomitants	Background, number of free atoms, residence time	Frequent	Pyrolysis, modifier, platform tube, suitable background corrector
Gas phase interference	Number of free atoms residence time	Sometimes	Atomization temperature, platform tube, modifier
Vaporization interferences	Number of free atoms	Frequent	Modifier, alternate gas, pyrolysis temperature
Carry over	Blank, absorbance drift	Sometimes	Heat out steps, alternate gas
Light scattering	Spectral background	Frequent	Background corrector
Molecular absorption	Spectral background	Sometimes	Background corrector
Absorption by other elements	Spectral background	Rare	Background corrector

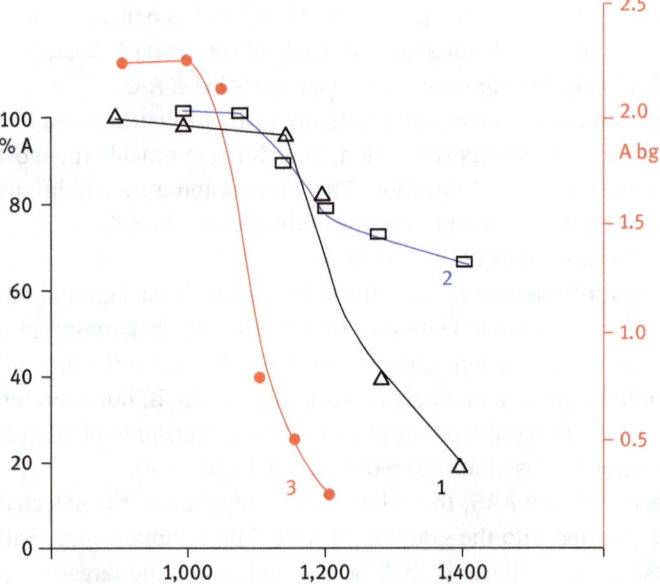

Figure 2.40: Absorbance of Mn in the reference solution and in seawater, and non-specific absorbance as function of the pyrolysis temperature. Left side: relative analyte absorbance (integrated absorbance in %) in the reference solution (1) and in seawater (2). Right side ordinate: non-specific absorbance of 20 μL of sea water (A in peak height). The red line 3 shows the absorbance as a function of the pyrolysis temperature (abscissa, °C).

Vaporization interferences in furnace AAS are frequent. While the potential of in situ analyte matrix separation is unique for GF-AAS, it bears the risk of interferences as well. Matrix components such as reactive chloride atoms, for example, may react with analyte atoms and remove them partly out of the furnace during the pyrolysis step. The analyte loss may even happen during drying, if organically bound analyte is determined. Modifiers which bind to the analyte element very strongly are best suited to avoid these pre-atomization losses. In some cases, a modifier which binds the reactive matrix may be the tool of choice. When, for example, Tl is determined in chloride containing matrix, a mixture of 95% Ar and 5% H_2 gas in the furnace neutralizes reactive Cl radicals as stable HCl molecule.

Gas phase interferences due to a direct interaction of analyte and matrix are rather infrequent in GF-AAS. The higher the actual gas phase temperature, the less probable is the interference. An efficient platform and analyte stabilizing modifier are the best tools to minimize or avoid direct gas phase interferences.

The optimization of the atomization temperature is performed in a similar way as described above. The optimal pyrolysis temperature is kept constant, and the atomization temperature is increased stepwise.

Carry-over is a problem mainly for refractory analyte elements. It may be enforced by concomitants which form refractory molecules such as stable oxides or carbides. Carry-over will only occasionally result in interferences. Usually reference solution and unknown sample will show similar effects: Upward drift of repetitive results from low to high concentration and drift downward when changing from a higher to a lower concentrated standard or a blank. A carefully optimized heat out cycle will often help to minimize carry over. In some special cases a mixture of, e.g., 5% Cl_2 in 95% Ar as alternate gas will help to remove refractory elements or compounds from the furnace.

As pointed out above, high non-specific absorption is frequently observed in GF-AAS. Small droplets and solid particles, molecules and direct elemental overlap are observed and reported in numerous publications. The type of spectral interference, spatially inhomogeneous and occasionally spectrally structured, has fostered the popularity of powerful background correction system such as the Zeeman technology, the high-resolution continuum source spectrometer or the Smith-Hieftje principle. Zeeman-effect GF-AAS is by far the most often used background correction system in high-performance GF-AAS.

The main task for method development in GF-AAS is obviously the optimization of processes in the furnace. The distinguished control page is the temperature/time program in the instrument's software.

CVG-AAS

CVG is essentially a technique which separates analyte from matrix. In a quartz or glass atomization cell the analyte element is often only surrounded by an excess of carrier gas (usually Ar) and reaction or burning gas (usually H_2, H_2O vapor and radicals). In case of Hg, the reaction gas can be avoided as well by making use of a mild reductant, usually $SnCl_2$. Sometimes concomitant elements which are forming volatile compounds as well, will be transported also to the atomizer. This could be, e.g., As in the case of Se or Sb determinations. Still, gas phase interferences in the atomizer are seldom in AAS and quasi non-existent when Hg is determined. The same is true for non-specific background, which is low, if present at all. However, gas phase effects are likely to be observed. They are mainly provoked by incomplete atomization at the relatively low temperatures inside the quartz cell. It has been shown [71] that some of the volatile element molecules, are not stable, when decomposed at around 1,000 °C. Radicals such as OH and H are required to generate atoms at least for the short time of their measurement. If the gas mixture is preconcentrated on a graphite tube or on an amalgamation device, the analyte-matrix separation is complete and interferences in the atomizer are practically not observed.

Chemical effects are expected in the reaction cell, however. The reaction of the cationic analyte to a volatile hydride, metal, or other compound requires a defined chemical environment of acidity, chemical activity of the reductant, volume, and speed of the carrier gas. Analyte elements present in multiple oxidation states in the solution must be reduced prior to transformation into the volatile compound. In the case of mercury, the element determined most often using the CVG technique, reduction is easy. However, the element must be kept in oxidizing environment prior to the reduction to avoid reductive adsorption to the walls of containers and tubing. Concomitants which easily form amalgams may cause interferences for Hg determinations. All these parameters are usually defined in the recommended conditions which may differ from instrument to instrument. Table 2.6 lists the effects which may influence CVG-AAS and AFS.

Method development in CVG-AAS follows essentially the procedure described above for flame and graphite furnace AAS. Optimization parameters are the pump rate of sample or carrier solution relative to the pump rate of reductant and the carrier gas flow. The temperature of the atomizer can be optimized within a small range only if an electrically heated quartz tube is used. The graphite furnace is uncritical concerning its atomization conditions. Coupling with chromatographic methods is often necessary to determine the species of an analyte rather than the total concentration of the element.

Analytical quality versus sample and element throughput

The analytical task of a routine laboratory and a research institute is often different. The research laboratory aims to develop new methods with hitherto unmatched detection limits or specificity. The routine laboratory is often following standard methods and is targeting for ruggedness and low cost per analysis. Generally, determinations can be run with various techniques described in this textbook and often the results will be adequate with respect to the required figures of merit. In this case, the simplest

Table 2.6: Matrix effects in CVG-AAS and -AFS.

Source	Influences	Occurrence	Correction
High mass of concomitants	Formation of vapor, atomization or lifetime of atoms	Sometimes	Buffering of sample, concentration of reductant, cell temperature, GFAAS coupling
Gas phase interference	Number of free atoms, residence time	Sometimes	Cell temperature, limit upper working range, GFAAS coupling
Quenching	AFS signal	Frequent, rare for Hg	Dilute reaction gas in higher carrier gas flow
Carry over	Blank, absorbance drift	Rare, mainly for Hg	Oxidizing environment of sample
Redox effects	Kinetics, generation of analyte gas	Frequent	Pre-reduction of sample, buffered sample
Non specific absorption	Absorbance	Rare	Background corrector

approach will often be the most convenient for the laboratory as well. The decision on the best analytical technique and method, however, will not be based on the instrumental approach only, but on the total flow of analysis, from sample collection via sample preparation to determination and reporting. The technique which will meet the requirements of the total process best will be the selected one.

2.1.7.5 Application examples

AAS is a universally used method in elemental analysis. There are applications in numerous fields in industrial, environmental, food and feed, medical, and other applications. A selection of examples which would cover only an approximately acceptable overview of the applications would by far go beyond the chapter of this book. Two examples are selected which will explain typical but totally different challenges connected with element determination.

1. Contaminated soils: an easy standard flame AAS application.

The EU limits of heavy metals in agriculturally used soil (mg/kg in dry matter) are:

Cd 1 mg/kg; Cu 50 mg/kg; Ni 30 mg/kg; Pb 50 mg/kg; Zn 150 mg/kg; Hg 1 mg/kg.

1/10 of the threshold value shall be the instrumental detection limit. Mercury will not be determined with flame AAS but with the cold vapor technique. The sample preparation is following the aqua regia method with a dilution factor of 3/100. The expected maximal matrix concentration in the solution for measurement will therefore be less than 3%. The d.l. values in the solution for measurement are listed in Table 2.7 together

Table 2.7: Parameters for method development; flame AAS, elements in soil.

Element	Lower limit in test sample	Target l.o.d.	Char.conc.	Comment
	mg/L	mg/L	mg/L	
Cd	0.033	0.003	0.008	Borderline sensitivity
Cu	1.5	0.15	0.020	
Ni	0.9	0.09	0.050	
Pb	1.5	0.15	0.14/0.06	Pb 217 nm preferential
Zn	4.5	0.45	0.007	Dilution and buffer may be necessary

with the characteristic concentration and an expected course of action for optimal measurement conditions.

The only critical element where the target l.o.d. is expected to be close to the limit of the instrument is Cd. The characteristic concentration for Pb on the main resonance line 283.3 nm (0.14 mg/L) and on the more sensitive secondary line at 217 nm (0.06 mg/L) suggests to making use of the secondary line rather than the primary resonance line.

All elements can be determined with the air ethine flame. Gas flows and burner height are similar and should be set according to the recommended conditions. Standards are prepared in 1/3 concentrated aqua regia to match the chemical conditions of the sample after digestion.

The only element where ionization might occur is Zn. As the salt concentration in soil is high enough to suppress ionization for the elements under consideration, the samples can be used without further handling. The standards may be prepared individually for each element or mixed from a multi-element standard. 0.1% KCl should be added to blank and standards to avoid possible ionization interferences. The samples used for the determination of Zn should be diluted 1:10. The concentration of the reference solutions for Zn is based on diluted samples.

Optimal concentrations for the standards are:

Cd: 0.020 mg/L; 0.050 mg/L; 0.2 mg/L
Cu, Ni, Pb: 0.3 mg/L; 1.0 mg/L; 3 mg/L;
Zn: 0.050 mg/L; 0.2 mg/L; 0.8 mg/L

One of the soil samples should be spiked with standard 1 to check the recovery in matrix.

Background correction should be activated for all elements to compensate for possible background absorption due to the matrix. The instrument blank under the individual measurement conditions should be determined from 10 measurements to estimate s.d. and l.o.d. This is particularly important for Cd where the method is approaching the limits of the instrument performance. The Cd lamp should have had a time to warm up for about 10 min. The atomizer should be ignited and run aspirating deionized water for 10 min to warm up of all instrument components.

After having tested the basic instrument performance, calibration and samples can be run using an automatic sequential multi-element run with, e.g., 5 replicates per blank, standard, sample, and spike. If, say, 20 samples and 2 spiked samples and one reference sample with known concentration are run together with a blank and 3 standards, the time for the determination of one element can be roughly estimated:

5 s delay time to stabilize the sample flow and 20 s per sample will amount to about 30 s per sample. The determination of 5 elements in 20 samples will require about 1 h and 15 minutes. This time includes the additional time required for the operation of the autosampler.

2. Determinations in ultrapure materials: an unusual challenge

Contamination control in ultrapure chemicals and ultrapure water has become more and more important. One of the main driving forces is the semiconductor industry where all solvents, leaching agents, photoresist solutions, etc. must be certified to extremely low element concentrations. The same is true for purified water used in power plants. Usually, concentration limits are defined for the various reagents which are acceptable for the industrial process. The most critical elements for a semiconductor process, e.g., Ca, Fe, Na, Zn, are also among the most ubiquitous in the environment, found in the labware, in water and in reagents used for the analysis. One of the most important parts of the analysis is therefore to obtain a blank level which is as low as possible, and – even more important – is stable! As the detection limit of the method is defined by the standard deviation of the blank level, it is evident that contamination and carry over may easily become the limiting factor for the determination. Manipulation of the sample as well as number of solvents, vessels, and reagents involved should therefore be reduced to a minimum.

On the other hand, as the reagents to be analyzed are ultra-pure, the matrix is well known and can often be removed easily prior to atomization. This is true for pure water, pure acids including the less volatile acids such as H_2SO_4, alkaline solutions, organic solvents, and more complex organic compounds such as photoresist dissolved in organic solvents or water. This, of course, does not hold true for inorganic salts or brines which are composed of practically 100% pure medium volatile matrix, which often cannot be removed completely or even partially prior to atomization. For water and pure solvents, the relative detection limits can be reduced by preconcentration in the graphite tube, making use of repetitive pipetting of up to 40 µL with intermediate drying steps. The furnace program in this case is stopped after the drying steps, sample is pipetted again, and drying is commencing a second time. The detection limits can in this way be lowered by up to a factor of 2–5 compared with the published values, provided that the standard deviation of the blank remains unchanged. In the case of matrices with high organic content (e.g., photoresist) the total mass of carbon which must be removed from the furnace during the pyrolysis step is the limiting factor for the relative detection limits achievable. In the case of salts and brines, the detection limit will be degraded by a necessary dilution factor of more than one order of magnitude. An attractive alternative in this case, is the separation of the matrix from the analyte using an analyte selective sorbent, which may be coupled on-line to graphite furnace AAS [72].

The elements, commonly determined in ultrapure reagents by graphite furnace AAS, are the alkaline elements Na and K, Mg and Ca from main group 2 of the periodic table, Al, Pb and the transition elements Cr, Mn, Fe, Ni, Cu and Zn. Ultrapure water and acids are usually analyzed after a simple pyrolysis at a temperature which is just high enough to completely remove the solvent. The atomization temperature is selected from the conditions recommended by the manufacturer. If pyrolysis temperatures of less than 500 °C are sufficient, no modifier is required for the elements listed above. The addition of modifier should in such cases be avoided to minimize the risk of contamination. In the case of photoresist, organic matrix is carbonized during the pyrolysis step and needs to be removed in an additional step, usually supported by an internal air flow at about 550-600 °C. Modifier should be added in this case, if the volatile elements K, Na, Pb, and Zn are to be analyzed.

The elements Fe, Ni and Cu were determined in concentrated hydrofluoric acid (40%), concentrated H_2O_2 (30%), and 10% H_2SO_4 [73]. Multiple injection was used for one of the acids (HF) to improve the relative detection limits.

The spectrometer is placed underneath a small laminar flow bench which protects particularly the autosampler and the furnace from ambient air.

The spectrometer is used exclusively for the determination of ultrapure solutions. Autosampler vessels made from fluorinated hydrocarbon (PTFE/PFA) should be used, after overnight treatment in 5% ultraclean nitric acid and after rinsing with ultrapure water. The autosampler washing bottle should be made of cleaned PTFE/PFA or polycarbonate and should be filled with ultraclean water or a highly diluted (e.g., 0.1%) ultrapure nitric acid. If the analysis is being performed for the first time, the entire autosampler tubing system should be cleaned by flushing 5% ultraclean nitric acid through the diluent pump several times, followed by a rinse with ultrapure water. Graphite tubes are sometimes contaminated with elements such as Ca or Fe, even though the manufacturing process includes a high temperature cleaning step in a halogenated hydrocarbon atmosphere. The contamination is usually on the surface of the tube. It can be removed by repetitive heating of the tube to 2,500 °C, or by execution of the program optimized for the determination of the elements. The absorbance for the furnace blank and the blank due to ultrapure water or to a reference blank should be as small as possible, statistically distributed around 0 or around a small mean value of a few milli-absorbance. Even more important than the absolute blank level is the stability of the blank value. If no downward drift can be detected for repeated blanks, the standard deviation of the blank should be defined by the photometric noise and the detection limits should therefore be optimal.

All matrices can be removed at temperatures lower than 300 °C. The individual solutions, HF, H_2O_2, and H_2SO_4 have different boiling points, however, and the sample volumes pipetted are different. The graphite furnace program must therefore be adjusted to each sample individually or be optimized for the matrix, which is expected to be most difficult, in this example, sulfuric acid. An example program for 30 µL of H_2SO_4 (10% w/v) is listed in Table 2.8. In steps 1 and 2, primarily water is removed from the liquid and the sulfuric acid is preconcentrated. Steps 2 and 3 are optimized to remove sulfuric acid in a controlled way. These steps would not be required for H_2O_2 or HF but would not be detrimental to this determination either. Step 5 is a pyrolysis intended to remove adsorbed SO_2 or SO_3 from the graphite surface before the atomization and heat out takes place in steps 6 and 7. In programs for the more volatile acids or for organic solvents, steps 1 and 2 would be similar or identical to those shown, while steps 3 and 4 would not be required. The pyrolysis, atomization and clean out steps would also be identical to those listed.

Table 2.8: Graphite furnace temperature/time program for the determination of Cu, Fe, and Ni in 10% H_2SO_4.

Step	Temperature	Ramp	Hold	Internal gas
#	°C	s	s	
1	120	1	30	Max
2	140	20	300	Max
3	200	20	60	Max
4	250	10	15	Max
5	400	30	30	Max
4	2,300*	0	5	Stop
5	2,500	1	5	Max

*The optimal atomization temperature for Zn would be approximately 2,100 °C for Cu, 2,300 °C for Fe, and 1,900 °C for Zn.

In the case of HF, the relative detection limit may be improved further, by pipetting the sample twice. First, 30 µL of the sample are pipetted, and steps one and two (the drying steps) are run. The furnace heating is then stopped, another 30 µL of sample pipetted and the whole furnace program carried out.

Standards added to the samples can be recovered with practically identical characteristic masses from sulfuric acid and hydrogen peroxide. In hydrofluoric acid, the characteristic masses for the three elements are higher by about 20% for Fe, 25% for

Cu, and 30% for Ni. All three elements form stable volatile fluorides which cause either a volatilization or a gas phase interference. The interference may be minimized by the addition of 5% H_2/95% Ar during the pyrolysis step. In this example the concentrations were quantified by the method of additions calibration.

The relative detection limits obtained in concentrated HF, concentrated H_2O_2 and 10% w/v H_2SO_4 range between 0.03 and 0.1 µg/L and are better than those previously published for the same elements in similar samples [33]. The improvement can be explained by higher sample volumes and by a reduction of the variation in blank levels.

2.1.8 Quantitative detection of atoms: atomic fluorescence spectrometry (AFS)

2.1.8.1 The fluorescence process

The atoms excited in the absorption process (see above) undergo a de-excitation within a very short time, usually within nanoseconds. This process is accompanied with radiation emitted at the same wavelength as absorption, at longer wavelength, or even at shorter wavelength. A simplified schematic of excitation and de-excitation is sketched in Figure 2.41.

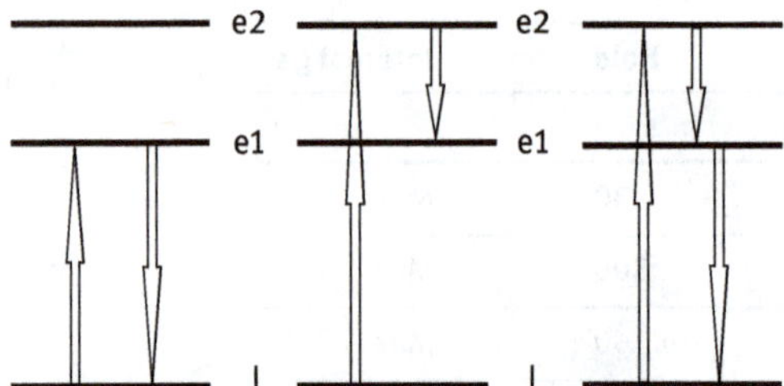

Figure 2.41: Transitions in fluorescence; l = lower level (ground state), e1 and e2 are excited states.

Electrons resident in a lower level (l), in AAS and elemental AFS often the ground level, may be excited to various upper levels (e1, e2) depending on the wavelength (energy) of the exciting radiation. From the excited level, they may directly return to the lower level (left side in Figure 2.41) or they return to lower energy levels in a stepwise process. If the energy of absorption and emission is identical, the process is called resonant. In elemental AFS only resonance fluorescence is of practical importance. It should be mentioned, but not further explicated, that fluorescence can be stimulated between excited states (excited fluorescence) or from thermally excited energy states as well (thermally assisted fluorescence).

The basic expression for light emitted by fluorescence (fluorescence radiance) is expressed in eq. (2.17) (according to Winefordner [74]).

Equation 2.17: Basic equation for radiance originating from a fluorescence process:

$$B_F = \left(\frac{I}{4\pi}\right) Y_{21} \cdot E_{\nu_{12}} \int_0^{\infty} K_\nu \, d\nu \tag{2.17}$$

Fluorescence is isotropic. For the energy collected, the path length in direction of the optical axis is the important magnitude which is expressed by the term in brackets. The second term, Y_{21}, factors in the quantum efficiency of the process fluorescing power/absorbed power. $E_{\nu_{12}}$ is the intensity of the radiation absorbed at the line of interest (spectral irradiance of the source). The expression under the integral is the absorption coefficient analogous to that in AAS.

Equation (2.18) shows the absorption coefficient for the fluorescence process.
$h\nu_{12}$ is the energy of the exciting photon. c is the speed of light.
B_{12} is Einstein´s coefficient of induced absorption.
g_1, g_2 are the statistical weights of the two energy states under consideration and n_1 and n_2 are the concentrations of atoms in these states.

Equation 2.18:

$$\int_0^{\infty} K_\nu \, d\nu = n_1 \left(\frac{h\nu_{12}}{c}\right) B_{12} \left[1 - \frac{g_1 n_2}{g_2 n_1}\right] \tag{2.18}$$

The integrated absorption coefficient is proportional to the concentration of atoms in the absorption volume and to the energy of the exciting photons. Einstein’s coefficient of induced absorption B_{12} is the number of transitions normalized wrt time, spectral energy density and absorbing species. The coefficient in brackets finally includes the population of the energy levels and the statistical weights of the energy niveaux. In the steady state, which must be assumed for AFS, the excitation rate equals the de-excitation rate.

For the practical use in AFS the following statements can be made:

1. The fluorescence radiance depends on the geometric structural conditions of the absorption volume (which becomes the fluorescing volume) in the direction of the optical path collecting the fluorescing radiation.
2. The radiance is proportional to the atom density in the cell, provided the optical density is low.
3. The emitted radiation is proportional to the irradiated intensity if the upper energy level is not saturated. This holds true for irradiation with the line sources used in AAS/AFS. It would be different if lasers would be applied for irradiation.

2.1.8.2 Instrument setup

In Atomic fluorescence spectrometry (AFS), the incident radiation passing through the absorption volume is not evaluated. Instead, the light emitted after the absorption process is quantitated. In resonance fluorescence the emitted light has the same wavelength characteristic as the absorbed photons. As pointed out earlier, the selectivity in line-source AAS – the spectral resolution – is due to the emission profile of the source. If this narrow line is focused through the atomizer by means of mirrors or lenses, and imaged a second time on a photosensitive cell, atom specific absorption could be detected. Atomic fluorescence spectrometers are following this simple principle as well (see figure 2.42).

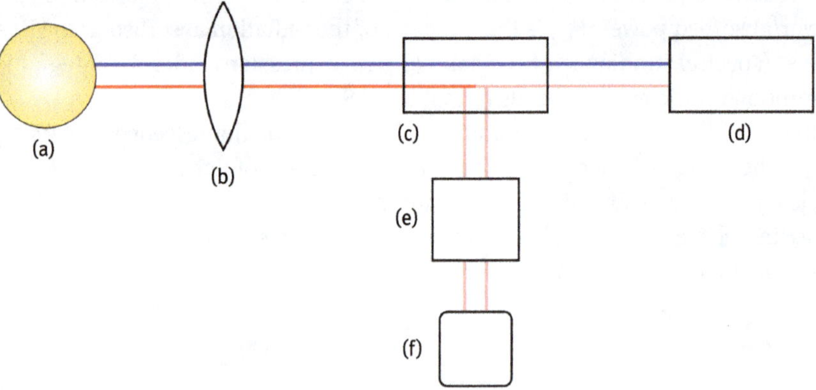

Figure 2.42: Schematic setup of an AF spectrometer.

Radiation of the source (a) is focused by an optical element (b) on the fluorescence volume (c). The suitable wavelength range is partially absorbed. The degraded incident radiation is trapped by a suitable component (d). The excited electrons will emit radiation when falling back to their original niveaux. This light is guided through a wavelength separator assembly (e) and quantitated by a suitable detector (f).

The photometer part of the instrument, in general, is imaging the emission source on the atomizer and onto the detector or, as described below, onto the spectrometer. The emission source has a certain spherical shape with an area (F) emitting the photons under an angle (Ω). The image of the entire emission spot must be guided through all limiting geometrical restrictions, such as windows, lenses, mirrors, atomizers, and optical slits to obtain the maximum possible Etendue (see eq. (2.5)).

Quenching in atomic fluorescence

The chemical reactions described in AAS concern the number of atoms generated for measurement and the stability of the atoms in the absorption volume. These reactions do not influence the processes of electron excitation. This is different in atomic fluorescence spectroscopy. Reactions between the excited atom (say Hg) and gases may result in an energy transfer between the analyte atom and the matrix, usually in the gaseous state. The dominant effects reported in the literature [75] are:

1. Transfer of energy from excited analyte electrons to matrix electrons.

2. Transfer of energy from excited electrons to vibrational modes of the matrix molecule
3. Transfer of energy from excited electrons to translational energy of matrix atoms
4. Chemical reactions which deactivate excited states.

The number of possible interferences from these processes in flame and graphite furnace atomizers turned out to be serious. Despite the wider dynamic range and potentially better detection limits, AFS never became commercially successful for most AAS applications. An exception are all applications which are based on analyte-matrix separation. Additionally, the number of species in the absorption/fluorescence volume must be strictly limited. This is possible only using the chemical vapor generation and, even more, the cold vapor technique. In these fields AFS has found important applications.

While mechanism 1, listed above, is rather rare in CVG, mechanism 2 may reduce the fluorescence intensity in the presence of gases such as H_2O, NO, N_2, CO_2, H_2.

An inert gas like the carrier gas Ar may cause quenching as well by mechanism 3. Finally, chemical reactions with hydrogen may likewise result in quenching. One example is the reported [75] reaction:

$$Hg^{*}+H_2 \rightarrow HgH + H \rightarrow Hg + 2H$$

Gaseous organic compounds may contribute to quenching as well. In all cases the result is a decrease in fluorescence intensity or, with other words, the observed analytical response decreases. If this effect can be calibrated it will negatively influence the signal to noise ratio but will not introduce an analytical error. If the effect is stronger or weaker in the sample than in the reference solution, it will generate an analytical error.

Quenching is the more probable, the more different gaseous compounds are present in the measurement cell. Water vapor must be reduced carefully from the reaction gas. A mild reductant ($SnCl_2$), as used in the case of Hg, will avoid H_2 generation. Hydride formation will be accompanied by the generation of Hydrogen. This may be one of the reasons why AFS became most popular only for the determination of mercury. CVG systems based on atomic fluorescence for the determination of hydride forming elements are commercially available, but their distribution is limited, and the number of international standards based on CVG-AFS is limited as well.

Method development in AFS will start from the recommended conditions and is following the procedure described above for CVG-AAS. The H_2 flame atomizer in AFS systems is uncritical with respect to atomization conditions. AFS used for the determination of Hg with the cold vapor technique must mainly be optimized for perfect chemical conditions regarding the mercury stability. The cation is gently reduced to the metal without the necessity to use nascent Hydrogen. The Hg vapor is transported by the carrier gas into the non-heated or only moderately heated measurement cell. Concomitants volatilized with the mercury, or reactions which force quenching are not expected under these mild conditions. The technique allows to determine Hg with an approximately 10 times superior detection limit compared to AAS. The disadvantage compared to AAS is that the zero point of the baseline is not defined as it is in AAS. As an emission method, the intensity is displayed in arbitrary units. The spectral linearity is up to 5 orders of magnitude, provided the detector is set to the right amplification conditions. Method development should start with a measurement of the carrier gas alone. The intensity recorded at high detector amplification is a measure for stray light limitation close to the detection limit. The next step is related to the setting of an optimal detector amplification. A reference solution shall be selected which defines the upper range of expected maximal sample concentration plus spike. The reading can be used to set the detector amplification for best dynamic conditions. With these settings the repetitions of blank solution are recorded as well as a standard above but close to the estimated l.o.q. Afterward the worst sample alone and with a suitable spike is run as described above. The difference to AAS is mainly that sensitivity must be defined in arbitrary intensity units. Characteristic masses or concentrations cannot be used.

2.1.8.3 Application example: the determination of mercury with the cold vapor technique and atomic fluorescence spectrometry (AFS)

It was explained earlier that AFS is a specifically stimulated emission technique. Other than in AAS the measurement units are arbitrary intensity units, and the linear dynamic range is much wider and is usually not limited by the fluorescence process but by chemical and physical limitations in the reactor or atomizer. Close to the detection limit, the number of photons to be counted is close to zero and not 100% as in AAS. The technique is therefore able to provide a significantly better signal to noise ratio compared to AAS. Hg is reduced from the cationic state in the solutions for measurement with a mild reductant, usually $SnCl_2$. After analyte-matrix separation, it is almost exclusively gaseous elemental mercury which is transported to the measurement cell by the carrier gas argon. Quenching becomes very scarce. For further focusing of the element, the element can be quantitatively trapped on a noble metal containing adsorber and heated out as purified metal vapor (amalgamation technique [76]). It is not surprising, that AFS became the most important technique for the determination of extreme ultra-traces of Hg.

The generation and the separation of Hg is very similar to the CVG-AAS procedure. The determination of ultra-traces of mercury with AFS is described painstakingly in the EPA method 1,631 which is based on amalgamation, or in the European equivalent DIN EN ISO 17,852 [77] which is based on direct determination after cold vapor separation. The instrumental detection limit of AFS is in the range of 1 ng/L or even below. This is in many cases below the blank level of reagents, container blanks, or laboratory environment. A significant part of the standard method is therefore concerned with blank control. On the other side, Hg at low concentrations is easily adsorbing to container walls and instrument tubing. If not in cationic state, it is lost through the walls or caps of polymer containers. The second most important task for the analyst is therefore the stabilization of the analyte in the cationic state. A mixture of $KBr/KBrO_4$ in hydrochloric acid has proven to be clean and strongly oxidizing. The salts can be cleaned of Hg by heat before preparation of the oxidant. The reductant, a solution of 10% $SnCl_2$ in 3% HCl, can be cleaned by blowing high purity Ar though the reductant solution for a few minutes. All reagents and containers should be cleaned with the blank solution containing the bromine oxidant.

The chemical vapor generation part is similar if not identical to the CVG technique described already. About 1–2 mL sample is required to obtain lowest d.l. without amalgamation. The flows of carrier, reductant and sample may be adjustable by changes in the pump speed or in the type of tube used for these solutions. The recommended values of the manufacturer are usually optimal but can be fine-tuned by the user. The flow rates are in the range of 5 mL/min (carrier) and about 2.5 mL/min (reductant). The waste can be adjusted as well by the same parameters. The rate of removal should be slightly higher than the sum of carrier/sample and reductant but not too high to keep the loss or reaction gas, including the analyte, minimal. The flow rate of the carrier gas is somewhat higher than in CVG with hydrogen generation and will be optimal in the range of 100–200 mL/min. In particular for the determination of Hg with AFS it is essential that the analyte/argon mix is dry. Often a special drying unit based on membrane technology is used as an additional drying line between gas liquid separator and measurement cell.

The lamps in mercury-specific instruments are usually simple low-pressure U-type mercury lamps. They are operated at their optimal power condition and cannot be adjusted by the user. Their photon output can be measured in a specific mode in most instruments. If the test fails, there will be a warning message displayed by the instrument. The test may run automatically in the background when the instrument is switched on. Other than in AAS, the intensity of the lamp has a direct influence on the fluorescence intensity and hence on the sensitivity of the measurement. The fluorescence process should provide zero light on the detector when no fluorescence is taking place. Because of straylight, this is not the case. The background emission of lamp and cell can be measured at high detector amplification. This value can be calculated and used as a reference for instrument and cell integrity. Some of the instruments in the market provide this information as a check parameter. Out of range straylight will then be automatically indicated as a warning message.

The working range of AFS can be influenced and shifted by the amplifier settings. Higher gain will usually result in a better signal to noise at low concentrations but result in faster detector saturation at higher concentrations. It is therefore necessary to be aware of the required detection limit and define the highest concentration to be detected. Despite the wide dynamic range, it is still recommendable to set the highest standard not more than 2 to 3 orders of magnitude above the target limits of detection or quantitation. or l.o.q. If, e.g., an l.o.q of 10 ng/L is intended, the highest standard should be 1,000 ng/L. The reason are mainly carry-over effects in tubing, containers and even in the cell for measurement. The highest standard and the automatic or manual gain setting of the instrument will define sensitivity and s.d. at concentrations close to the l.o.d. After the instrument parameters are set with the highest standard, the system is run several times with the blank solution until the blank shows systematic scatter. From the s.d. of the blank and the sensitivity measured from the lowest standard, the detection limit can be estimated as described already. The calibration curve is run afterward and the homogeneity of the variances of individual standards are determined. Before running the real samples, the stability of the blank is verified. Afterward, as usual, samples, spiked sample, QA samples, etc. are run. The r.s.d. at higher concentrations (>10 times l.o.q.) should be in the range of 2% or below. One run will take about 1 min. Care must be taken that the fluorescence intensity returns to its original blank level at the end of the run.

The calibration curve close to the detection limit is featured in Figure 2.43. The calibration curve was selected such that the l.o.d. and the l.o.q. could be determined from a 10-point calibration curve following the method of DIN 32654 [78]. The calibration curve is defined between 0.010 µg/L and 0.100 µg/L. The values obtained were 0.0005 µg/L l.o.d. and 0.0017 µg/L l.o.q. calculated from the confidence bands of the calibration curve.

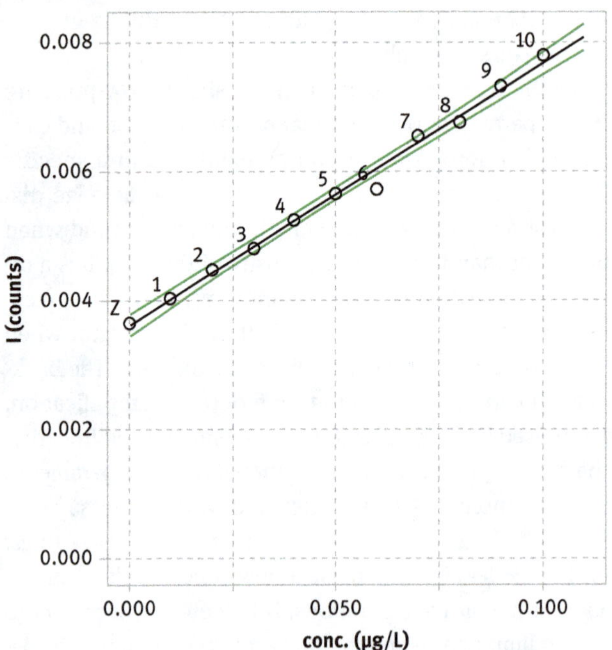

Figure 2.43: Mercury determination under standard laboratory conditions using AFS.

2.1.9 Quantitative detection of atoms: optical emission spectrometry (OES)

2.1.9.1 Basic considerations of emission spectra of atoms

Our most prominent and important light source is the sun. Radiation from the sun is generated at about 5,500 K in the, so-called photosphere. The spectrum emitted by the sun ranges from 100 to 3,000 nm. The UV range (100–380 nm) contributes with only 7% to the total radiation, whereas the visible range (380–780 nm) and the infrared range make up for almost half of the emitted total energy each. The short UV-C range (100–280 nm) hardly passes the atmosphere at all. This temperature translates into excitation of electrons which, upon return into lower energy levels, emit radiation of characteristic wavelength. No surprise that spectroscopy started with the splitting of the sun light (see Chapter 1). Temperature is a measure for the mean average kinetic energy of molecules and elements. In absorption processes we have seen (Section 2.1.6) that we are using an electron volt scale for the interaction of photons with electrons. In the atomic emission process the stimulation of electrons require this energy via thermal excitation. The equation relating electron temperature (2.19) requires thermal equilibrium of the system.
Equation 2.19:

$$E_e = 3/2 k_b \cdot T_e \tag{2.19}$$

E_e is the electron energy in Joule; T_e is the electron temperature in Kelvin; 3/2 stands for three degrees of freedom of movement; and k_b is Boltzmann´s constant, 1.380649 e^{-23} J/K.

The electron energy of 1 eV, as an example, can be translated into an electron temperature of 7,736 K. It becomes obvious that thermal excitation to ionic states requires high temperatures.

The process of thermal excitation is based on collisional excitation by electrons or ions. The population of the individual energy niveaux is defined by the Boltzmann statistics (eq. (2.20)).

Equation 2.20: Boltzmann statistics

$$n_e/n_0 = g_e/g_0 \cdot e^{(-E_e + E_0)/k_b T} \tag{2.20}$$

n_e and n_0 are the average number of electrons in the excited and in the ground state. g_e and g_0 are degeneracies of the excited and ground state, E_e and E_0 are the energy levels of the excited state and the ground state, k_b is the Boltzmann constant, and T is the electron temperature. The number of electrons in the excited state is depending exponentially on the temperature of the system in equilibrium. The higher the temperature, the higher is the number of electrons in the excited state and hence the intensity emitted by the heat source. Requirement for the evaluable emission line is the free atom, however. The initial energy must be high enough to split molecular bonds and to generate free atoms or free ions.

Atomic fluorescence, described above, is an emission technique as well. Electron excitation in this case is caused by photonic energy. We came across emission as well in the section on AAS. Flames are hot enough to split many of the molecular bonds and to excite some of the elements (mainly alkaline and earth alkaline elements) so that their emission spectra can be evaluated. This makes quantitative elemental determinations in the range of µg/L possible. Analysis or determinations of most metals and non-metals of the periodic system requires temperatures much higher than the 2,000–3,000 K obtainable with a flame atomizer.

Thermal control of the processes of splitting molecular bonds and exciting electrons in the free atoms generated is difficult. Therefore, the excitation sources used in optical emission spectroscopy (OES) are, and have been, optimized for highest possible temperature. This assures minimization of molecular matrix effects and provides a maximum of available emission lines with intense photon flux. As an unintended side effect, the wavelength range used in elemental OES is overloaded with emission lines. Hence spectral overlap becomes the most serious interference in high temperature OES. The figures of merit "sensitivity" and "specificity" are antagonists. The obvious corrective is spectral resolution. Although the Etendue (eq. (2.5)) will suffer from narrow optical bandwidth, the true limit of quantitation in complex samples is often defined by the resolving power of the spectrometer.

Though excitation sources such as glow discharge, electrical arc and electrical spark still play some role for specific applications, mainly metallurgy, the emission source most widely used is the inductively coupled plasma (see Section 2.1.6.2). It is often used under hottest possible electron temperatures up to 10,000 °K though it can be tuned down in energy and temperature to reduce ionization or simplify spectra. An attractive alternative to the expensive argon torch may be the microwave-induced plasma based on a nitrogen or air plasma. The latter provides temperatures lower by several thousand Kelvin and may be suited for specific applications. The following chapter will concentrate on ICP-OES.

The applicable spectrum ranges usually from 167 nm (Al) to 780 nm (Rb). This range can be covered with the optical systems described in Sections 2.1.4.2/2.1.4.3. The short wavelength range requires freedom from oxygen and water vapor in the photometer and monochromator parts of the instrument. Oxygen absorbs about 75% of the radiation below 190 nm and almost 100% below 175 nm. Some of the instruments are limited to operate above 188 nm, sacrificing only few analytical lines. Important lines below 167 nm are rare: bromine at 163 nm and chlorine at 135 nm would require specific simplified optical systems for adequate light throughput. Only few instruments offer this range. The wavelength carrying most of the elemental analytical information is between 188 nm and 450 nm. Although this range would cover the applications required by most laboratories, only very few instruments will cover this wavelength range only, although, looking spectroscopically at optics and detectors, a limitation of the wavelength range would make a lot of sense. A more economic approach for very good resolution and simplified detectors would be possible this way.

The energies for excitation in this wavelength range are between 4 and 10 eV (from long to short wavelength). The excitation from the ionic state (ionization + excitation) is in the range from 10 eV to slightly below 20 eV. Few lines are available above 450 nm. These are alkaline and earth alkaline elements with low ionization and excitation potentials in the range of 5 eV and below. Several instruments split the wavelength range and thus make simultaneous detection of the bulk of elements below 400 nm and other selected elements above 400 nm possible. Technically and economically a good compromise!

2.1.9.2 Spectral issues

Other than in AAS, the number of lines is extremely high. Depending on the type of sample, ten thousand emission lines may populate a spectral range of 200 nm. These emission lines are narrow but, depending on the number of emitting atoms (which relate to the element concentration in the sample for measurement), the intensity in the wings of the line may still cause spectral overlap with the analyte. On the other hand, high temperature OES often offers many more intense lines for evaluation compared with AAS.

Spectral lines have the shape of a Voigt profile. Line widths have been discussed in Section 2.1.6.1. The natural width of lines in an inductively coupled plasma depend on the type of element, its volatility, the selected plasma conditions, the matrix load in the plasma, etc. Typical natural half widths (line width at 50% peak height) are 1.3 pm for As, wavelength 194 nm, Fe, 2 pm, wavelength 260 nm, Ti, 3 pm, wavelength 335 nm, Na, 10 pm, wavelength 590 pm. In fact, lower energy transitions result in broader wavelengths. However, volatility of the elements under consideration plays an important role as well. The Li line at 460 nm, for example, is three times wider compared to the Na line at 590 nm.

It would be ideal to resolve even the narrowest lines with the spectrometer. The resolution of the monochromator/polychromator is usually not better than $R \sim 50{,}000$ (see Sections 2.1.4.2/2.1.4.3) so that narrow lines at, say 200 nm, cannot be resolved with the calculated resolution of 4 pm. All the described effects limit the interference-free separation of lines in complex analyte mixtures.

2.1.9.3 Separation of analyte and matrix lines

Just as described in AAS, the accurate background (in emission intensity) would be the reading of the sample at the selected analyte wavelength without analyte. This is mostly impossible in real samples. Background corrections are therefore (with exception of Zeeman AAS, Section 2.1.6.2) offline readings, comparable to continuum source high-resolution AAS:

- The vicinity of the analyte line is measured and the result of the intensity of the readings are corrected with the background readings (see Figure 2.31).

- The calibrated result obtained using two or more different lines of the target analyte element are compared. Deviation outside of the repeatability of the measurement hint toward spectral overlap and can be further evaluated.

Other than in AAS, the background of an emission reading is quasi never zero in intensity. The analytical information, the peak to be evaluated, sits on a base of background intensity. Three main cases will be discussed in the following:

1. The base may be flat as a function of the wavelength indicating a non-specific uplift of the counts (Figure 2.44a). In this case the analytical peak can be easily quantitated by subtracting the counts to the left and to the right of the peak.
2. The uplift can be skew (Figure 2.44b) indicating an interfering intensity on one side of the peak. An average of the background on the left and on the right side of the peak is usually a good approximation of the interfering intensity underneath the peak.
3. The interference is more complex, and the analytical peak manifests itself only by a shoulder of the total intensity (Figure 2.44c). A careful study of the signal is required in this case. The best correction may be a single correction point on one side of the peak.

Understandably, resolution and concentration of the interferent in the sample play an important role in the magnitude of the potential error. Other parameters, such as plasma conditions (mainly temperature!) may contribute to the effect as well. Sophisticated methods for automatic selection of background correction points have been included into the software packages of instrument manufacturers, e.g., [79], to support the user.

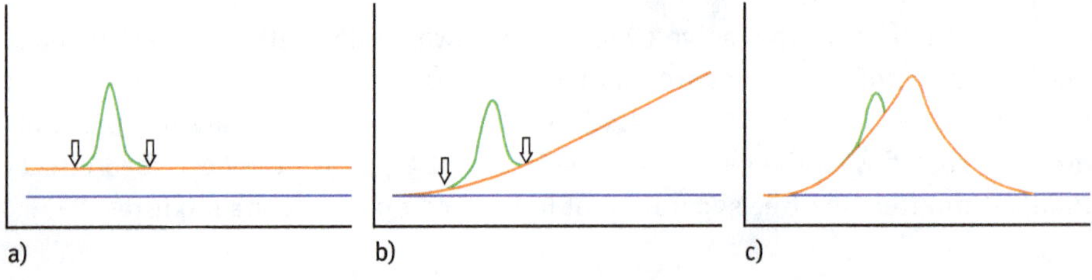

Figure 2.44: Background pattern in high-resolution optical emission. Blue line: baseline without background, orange line: background from matrix causing continuum (a) and from peaking background (b, c). Abscissa: wavelength; ordinate: intensity.

2.1.9.4 Signal to noise from a spectral viewpoint

The issue of noise has been discussed several times in this textbook. We had a look at dark noise of the detector, statistical noise of the detector (shot noise) or emission noise of other bright sources. In optical emission spectroscopy the heat producing source emits an intense spectrum with specific lines and continuum background. In the case of ICP as a source, the emission originates predominantly from ionized Argon. In addition, lines from nitrogen, oxygen, hydrogen contribute to the background spectrum. Lines are scarce below 300 nm and rich between 300 and 600 nm. The continuum background increases from 200 to about 450 nm and from there decreases again. The background below 300 nm is mainly due to recombination between argon ions and electrons. Bremsstrahlung dominates the continuum in the visible portion of the spectrum. Understandably, the intensity of the spectrum depends on the operating conditions of the plasma (energy, gas flow, protection from air, feed of solvent into the plasma). The fluctuations (time-dependent standard deviation of the emission intensity) depend on the stability of these parameters and often define the stability of the background intensity and hence the background noise. A typical background spectrum of an argon plasma is featured in Figure 2.45.

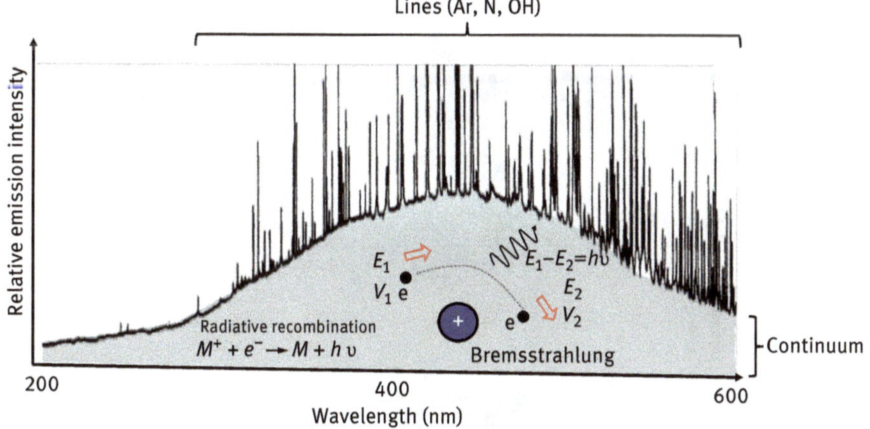

Figure 2.45: Typical background emission spectrum of an argon plasma. Figure 2.45. taken from literature [37].

For estimation of the importance of the background emission for Figures of Merit, the background equivalent concentration (BEC) has been defined as an important analytical parameter. BEC defines the concentration of an analyte which generates an emission intensity equal to the background intensity of the plasma at this very wavelength.

Equation 2.21: Background equivalent concentration (BEC)

$$\text{BEC} = I_{\text{bg}}/I_{\text{a}} \cdot c_{\text{a}} \tag{2.21}$$

I_{bg} is the intensity reading of the background at the analyte wavelength.
I_a is the net intensity of the analyte, c_a is the concentration of the analyte.

The background emission of the plasma is much higher than the smallest concentration of the analyte which can be distinguished from the background. As a rule of the thumb, the noise of the emission background is less than 1%. Hence, the detection limit of the analyte line is not worse but probably up to an order of magnitude better than BEC.

2.1.9.5 Collecting photons

Collecting photons is relatively easy, if the atom carrying volume is clearly defined in space and temperature (as it is the case in graphite furnace AAS or UV spectroscopy, for example, by cuvettes). In Arcs or sparks or in a plasma plume the situation is more complex but as well more flexible. A typical ICP-Plasma is shaped like a candle flame or a plume (see Figure 2.20). The physicochemical effects taking place in a plasma are shown in figure 2.46.

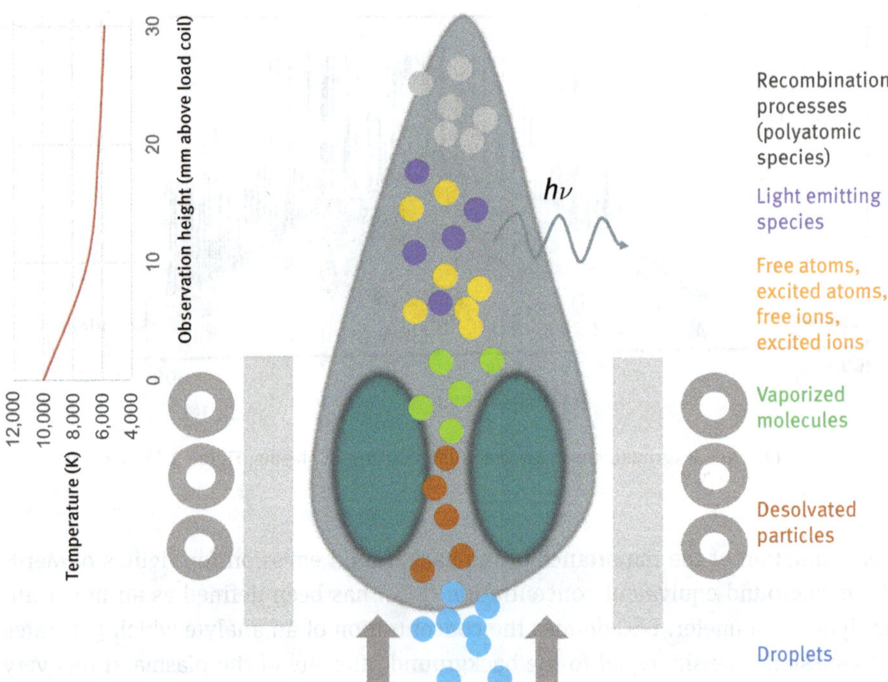

Figure 2.46: Temperature zones in an ICP. Figure 2.46. taken from literature [37].

The bulge of the plume sits inside the load coils. The argon plasma is a toroid, about the size of the coils. The center channel is open for the feed of the sample gas and the microdroplets from the nebulizer. The electron temperature inside the toroid increases to maximum at the top load coil. The temperature in the sample channel is significantly lower as solvent is vaporized, salts are split, molecules and elements are vaporized. The zone of free atoms, ions and excited species starts directly above the load coil and is about 20 to 30 mm long. The temperature in this zone is relatively homogeneous across the plume but atoms start to recombine again into stable molecules. Compared with the axial size of the plasma the diameter of the emission zone is 10 times less, in the range of 2 mm.

The obvious logical way is to look into a plasma from the side and adjust photon sampling to the spot of highest temperature or lowest interferences, in case of complex samples. This was the status up to about 1990 when the observation of the plasma from the front side (in axial direction) was introduced. Viewing the significantly longer emission zone from the front makes the collection of many more photons possible. The background increases as well but still significantly better detection limits could be obtained this way. The number of possible interferences also increases, as the temperature in the viewing zone, and hence the spectrum, is no longer homogeneous. An orthogonal argon sheath gas helps to blow away the part of the plume which is colder and richer in interfering species. The importance for the practical analytical tasks lies in the accessibility of detection limits (e.g., in the field of environmental samples) by ICP-OES, which could only be reached by ICP-MS or GF-AAS before. The limits of detection can be lowered by an average factor of about 5 with axial viewing. The improvement factor depends strongly on the analyte element and the line selected. It ranges from 2 for Ag to 15 for Li. Generally, alkaline and earth alkaline elements, emitting at long wavelengths, benefit most from the axial viewing.

2.1.9.6 Sample and plasma

Sample introduction systems including nebulizer and spray chamber have been discussed in Section 2.1.6. In this paragraph we will look at the influence of sample (matrix) on the plasma. A part of the effects on nebulizer and atomizer are like in flame AAS (Figure 2.38). Nebulizer and atomizer of an ICP are in general more sensitive to matrix than in simple flames. Standard ICP-OES nebulizers feed about 1 mL/min of aqueous or organic solvent into the mixing chamber. This primary aerosol contains small droplets, suited for introduction into the plasma and bigger droplets which would cause instability to the plasma. The primary aerosol from the nebulizer is entering a spray chamber where the aerosol expands. Solvent starts to evaporate, and droplets are coagulating. In a turbulent Ar atmosphere, the secondary aerosol collides with walls and impactors and only particles small enough, and quite homogeneous in size, are transported to the plasma torch. Usually less than 5% of the initial aerosol is introduced to the plasma. This volume is usually tolerated by the plasma. A moderate rate of 1 mg of solvent per second corresponds to 0.06 mmol/s of water. This quantity is evaporated and split up. This process consumes about 20 W/s. Acids and dissolved salts cause an additional cooling influence on the plasma temperature. A typical analytical plasma operates at about 1,500 W with an efficiency of 75%. It becomes obvious that just the solvent water consumes about 2% of the plasma energy upon introduction. Additional energy is consumed by dissolved solids in the solution for measurement. These may be volatile salts, such as chlorides, or salts which undergo a transformation to stable compounds such as nitrates which may change into oxides. Several possible effects of the sample on its way to atoms and ions must therefore be considered:

1. Cooling of the plasma due to solvent and matrix
2. Change of the solvent uptake due to viscosity changes by matrix
3. Change of the droplet size and distribution by matrix (organic solvent, acid, salt)
4. Change of the location of maximal atom or ion density in the plasma
5. Change of the electron density in the plasma due to matrix

Items 1 and 4: Unlike in flame AAS, where solvents may slightly cool or, in case of some organic solvents, may add additional energy into the atomizer, the effect of solvent on the plasma is always a reduction of temperature. This will often have an influence on the intensity of ion lines (reduction) and atom lines (increase) at a given observation height. The decrease in intensity is also due to reduced excitation of electrons. The interference will be effective in radial and in axial observation. The background intensity of the plasma may slightly change as well.

Items 2 and 3: Physical effects on the amount of usable aerosol fed into the plasma are similar to those in flame AAS. The aerosol mass is often reduced by matrix, but it increases if solvents with low surface tension are aspirated. The effect on the plasma background emission is usually more pronounced than in flame AAS.

Item 5: The electron density influences the fraction of the ions and atoms of an element in the observation zone. This will influence the sensitivity of the line under investigation. The effect is in principal comparable with ionization in flame AAS.

All these effects have an influence on sensitivity and quantitation limits, to a lesser portion on noise. They can be calibrated for, if solvent and matrix of the standards is matched with the sample composition. This, however, is not always easy if the sample matrix changes from sample to sample. This is the case for many sample types, for wastewater or digested environmental matter, as examples. Often calibration with another element helps to avoid erroneous results in this case (see interelement correction, Section 2.1.8.8). If, however, the interference is specific for the elemental lines or if it causes additional spectral overlap, neither the method of additions (Section 1.2.6) nor inter element correction (Section 2.1.8.8) can correct for possible errors.

2.1.9.7 Simultaneous versus sequential determination of lines

There are many reasons to detect, evaluate and report the entire spectral range of OES simultaneous, there are significant arguments against it. It is important to separate the determining factors in terms of figures of merit (FOM). We will start with a short repetition of optical systems in OES:

The classical spectrograph was based on the Rawland circle (Paschen-Runge optics). The chemically active foil was bent and positioned on the circle. Illumination of the detector was compromised between lines of different intensity. The measurement was simultaneous, the reading, however, was an integral of all photons which arrived during the time of illumination. Temporally dynamic effects could not be resolved. Later the chemical detector was replaced by photo multiplier tubes at selected places on the circle. Dynamic signals with high temporal resolution could be determined. The PMTs could be controlled individually which guaranteed a wide dynamic range. Measurement to the left and right in the vicinity of the spectral line were not possible. In modern Paschen-Runge spectrographs small CCD detectors replace the PMT detector and allow quantitation of the background in the vicinity of the selected lines.

The classical spectrometer was based on a movable flat grating with high line density and long focal length (Czerny-Turner or Ebert monochromators). The photons of the selected lines were determined sequentially. The time to move from line to line depends on the spectral distance of the lines and on the technical layout of the monochromator drive. It requires few seconds up to a minute. The classical detector was the photo multiplier tube. Measurement in the vicinity of the spectral line required a movement (scanning) of the monochromator. Thus, measurement of background is possible but with a minimal time distance between analyte and background reading. In modern instruments with classical optics, small CCD detectors replace the PMT and allow simultaneous read out of the vicinity of the line under investigation.

Echelle spectrometers can be designed for simultaneous or sequential operation. In spectrographs, the two dispersive elements (usually a grating and a prism, or two gratings) are fixed. The entire spectrum or a part of the spectrum is focused on a CCD, segmented CCD, CID or CMOS detector. The measurement is simultaneous. Analyte intensity and background correction is determined simultaneously. Optimization of the time for photon collection of strong and weak lines may be a critical issue. Illumination of the entire Echellogram onto one detector is a compromise between detector size, spectral resolution, and read out time, or detector and instrument cost.

In the sequential version, the Echelle spectrometers operate like a classical spectrometer (see above). Both dispersive elements are moving simultaneously and thus wavelengths can be addressed sequentially with high speed. A small CCD detector with fast readout allows simultaneous information of signal and background intensity with individually optimized integration time.

If we compare modern instrumentation with Active Pixel Sensor detectors the following rules of thumb usually hold true:

Speed:

- Simultaneous detection is obviously faster than sequential detection as the time for hopping from one line to the next is not required.

- The Paschen-Runge setup does not have to compromise integration time or read out speed for multiple line detection. The number of possible lines, however, is limited.
- The Echelle type offers a wide simultaneous range and high speed. The type and cost of the detector contribute strongly to analytical quality and cost of instrument.
- Sequential detection is slower, as several detection cycles are following each other.
- If only few elements (less than 10 per sample) are measured, the difference is small.
- Sample change and analyte feed to steady state conditions require about 30 s for all types of instruments. On top of this pre-measurement time, the additional time for wavelength change is only crucial if many elements per sample are determined.

Resolution:
- Paschen-Runge systems and classical flat-grating monochromators are compromising resolution with instrument size.
- Echelle spectrometers must optimize coverage of the total wavelength range versus detector cost and/or resolution.
- The sequential Echelle approach offers best resolution and small instrument footprint.

Signal to noise:
- Resolution and signal to noise in complex sample matrix are closely related in ICP-OES.
- Sequential instruments are technically predestined for best detection limits.
- Settings, individually optimized for each element such as integration time, and variation of the plasma observation zone contribute to optimal signal to noise.

Background correction:
- Knowledge about the line and its immediate neighborhood is essential for correct intensity reading.
- Today most scanning instruments provide simultaneous reading of line and background. Simultaneous instruments provide the additional advantage of simultaneous reading of several lines of the same element, or simultaneous reading of nearby lines for inter-element correction, an advantage of the spectrograph.

Working range and straylight:
- The advantage is on the side of sequential measurement. Individual optimization of the amplification conditions and selection of only a small portion of the spectrum provides optimal signal dynamics.

Chemistry:
- Preparing of standards, matched with the expected main matrix components of the sample becomes more and more complex the more elements are added to a standard.
- Chemical reactions between the elements, adsorption, and precipitation may hamper proper calibration.
- This effect is independent of the measurement mode, but it may limit the number of elements which can be calibrated by one measurement cycle.

2.1.9.8 Calibration in OES spectrometry

Optical emission spectrometry usually covers a wide linear dynamic range of 5 orders of magnitude or more. This tempts the analyst to use only few or even only one calibration solutions for setting the calibration curve. Statistically the same rules as discussed for methods with shorter linear dynamic range apply: the error in preparation of the individual standard and the total number of repetitions for the calibration solutions will finally define the confidence bands of the calibration function. The concentration of the calibration should match the order of magnitude of the expected value in the sample. This is not only motivated by statistical considerations but also issues of carry-over, adsorption phenomena, etc.

Although linear calibration functions are used almost exclusively in OES, non-linear functions, the method of bracketing, the methods of addition, and addition calibration and other suitable calibration methods should be used if beneficial for the specific analytical situation.

Trueness and accuracy of the reading depend on the influence of matrix on the sensitivity of the unknown sample (multiplicative effect relative to the calibration curve) or on positive or negative additive bias relative to the baseline (additive effect relative to calibration).

Whereas the first effect can be balanced by the method of addition, the latter cannot.

Other methods of error compensation are required in the latter cases to facilitate higher accuracy.

One prominent method suggested from early ICP-OES work is the Inter Element Correction IEC [82]. If the line of a known element overlaps the analyte line within the resolution of the spectrometer completely, the measured intensity I_{tot} consists of the analyte intensity I_a and the intensity of the interferent I_1 at the target wavelength λ_1. The interferent usually offers many lines where no interference with the analyte is observed. The intensity ratio of the two wavelengths I_1/I_2 can be measured with a reference solution which contains the interferent only. This quotient is used to correct for the interference on the analyte line. We name it m_{IEC}.

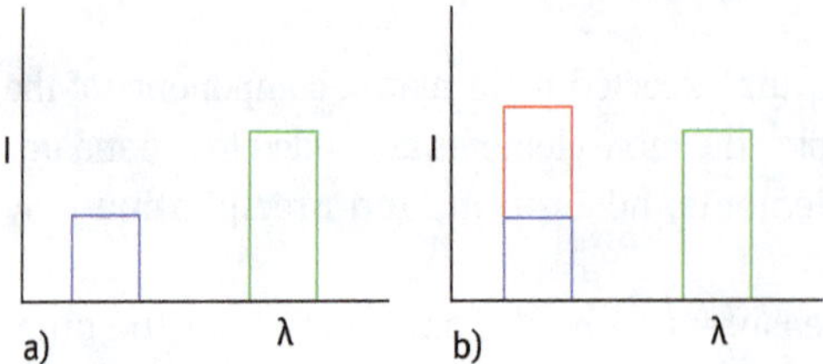

Figure 2.47: Inter Element Correction a) The intensity ratio between the interfering element at the analyte wavelength (blue) and the secondary wavelength (green) is determined as $m_{IEC} = 0.5$ ($I_1/I_2 = 6{:}12$). b) The intensity of the analyte wavelength in the sample is 14 Intensity units. The intensity of the interferent at the secondary wavelength is 12 intensity units. The red bar indicates the analyte intensity.

The corrected analyte signal is calculated according to eq. (2.22), numerical values appearing there refer to the example from Figure 2.47.

Equation 2.22:

$$I_a = I_{tot} - (I_2 \cdot m_{IEC}) = 14 - (12 \cdot 0.5) = 8 \tag{2.22}$$

Important interelement factors of wavelength pairs of elements are known and listed in instrument software packages. The method is considered as robust, however, several requirements must be fulfilled to obtain a true background correction:

- The contributions of analyte and interferent at the analyte wavelength need to reside complete within the spectral window selected.
- The secondary line of the interferent should be identical or very similar concerning excitation conditions. If not, matrix may influence the two lines of the interferent differently and m_{IEC} in matrix might be different to the calibrated m_{IEC}.
- In multi-element determinations IEC may become a complex process, for preparation of the calibration solutions as well.

In the case of IEC, the single analyte concentration is effectively replaced by a vector containing two concentrations. This vector is then obtained using two intensity measurements at different wavelengths. This basic idea can now be extended in several aspects:

- More than two intensities can be used
- Concentrations of several elements with potentially detrimental effects can be added to the vector mentioned above
- The number of intensities can exceed the size of this vector, i.e., we can work with redundant information

Combining these three basic ideas, we end up with multivariate methods, where the explicit eq. (2.22) is replaced by a different procedure which yields the concentration vector from many observed intensities. In this procedure, a linear model, relating the vector of concentrations to intensities (i.e., a sensitivity matrix) is used. The concentration vector is then varied until it best matches the observed intensities.

In contrast to IEC, this method also works in situations where each intensity is depending on more than one concentration. Making use of all available – and at first sight even redundant – information it results in improved analytical quality. Multivariant correction methods gained more and more usage in software packages since computing power became readily available [83]. The exact algorithm of these methods, named, for example, Myers-Tracy compensation, is in-depth knowledge of the manufacturers and not published in detail. A closely related technique is based on a specific version of Kalman filtering [84]. Although these correction algorithms are effective, they are not able to reduce the demand for highest possible spectral resolution in OES.

Internal standardization is another routine procedure for the reduction of matrix effects in OES [85]. The method corrects for non-spectral effects. An element which should not be present in the sample to be determined is added to blank, calibration solutions and unknown sample. The intensity of a suitable line of this internal standard is monitored between calibration solution and sample and the factor between the intensity of the standard is used to correct the analyte concentration. The method works well if the interference concerns efficiency of nebulization, but it is prone to errors if the excitation conditions in the plasma are changing due to the matrix. In this case the internal standard works properly if the excitation conditions of the analyte line and the line of the internal standard are equal. Just like the method of standard addition, the internal standard ought to be evaluated in the linear part of the calibration curve.

2.1.9.9 Method development in optical emission spectrometry

In low temperature sources like flames, few elements and few lines are suitable for OES. These are sensitive, and determinations are straightforward. The line of interest, e.g., K at 766.5 nm, is selected via software. A maximum concentration, representing the expected highest concentration in the sample, plus an added standard if the standard addition method is intended, is aspirated. The multiplier gain is set, and the line may be refocused for maximum intensity, if required. The setting of detector amplification conditions makes sure that there is no photon saturation in the detector, and the signal to noise is optimal for the expected working range. Subsequently blank, calibration solutions, samples, quality control samples, etc. are run just like in AAS (see Section 2.1.6).

Systems working with electric arc, spark or glow discharge atomization and excitation are usually limited to solid samples. The type of sample is usually known and is rather constant in its composition. The parameters which are optimized are the electrical conditions of the discharge used for excitation and the selection of the analytical line providing lowest probability of spectral overlap and best signal to noise for element and sample under investigation. The fundamental requirement for a successful method development is the availability of a reference material which represents the sample as closely as possible. In the following paragraphs we will concentrate on the ICP plasma source to describe an approach to method development in optical emission spectrometry.

ICP-OES

Most elements offer several analytically usable lines. For convenience, the sensitivity of the lines is defined in BEC (see eq. (2.21)). This is the concentration of an element which provides element intensity as big as the background on the selected line. The smaller the BEC value, the better is the sensitivity of the line. The BEC has a similar function as the characteristic concentration or characteristic mass in AAS. BEC is related to the limit of detection (see Section 1.2.5) It is a useful tool for line selection. As a rule of thumb, the background noise is in the range of 0.5% to 1% of the background intensity. The expected detection limit (based on 3 standard deviations of the blank) is 0.015 to 0.03 times the BEC value.

Assume 0.02 BEC is the detection limit, then 0.05 BEC is the limit of quantitation (10% r.s.d. criterion) and BEC is also the concentration where the repeatability of the determination starts to be close to optimum. Depending on the expected concentration range, lines with a lower or higher BEC concentration can be selected. Although the linear working range of ICP-OES is about 5 orders of magnitude, it often makes sense to select a less sensitive line if concentrations to be determined are higher.

The emission line of the analyte is usually narrower than the resolution of the spectrometer. Thus, the maximum intensity may not be resolved and be located between two detector pixels. Provided, the intensity distribution of the analyte follows a Gaussian function –which is usually the case – the maximum intensity (peak intensity) can be calculated from the adjacent pixel intensities. Another way to evaluate data is, like in AAS, the integration of a peak from a set zero level before and after the peak. It should be kept in mind, that pixels with higher intensity provide a better signal/noise than those close to the baseline. The best signal to noise for undisturbed lines is therefore usually obtained if only the pixel of highest intensity is evaluated.

Most important for the line selection, however, is the spectral environment of the selected line. Matrix will usually influence different lines of the analyte completely differently. The fundamental idea of background correction is to quantify element non-specific intensity directly at the point of the selected line without the emitted light of the analyte (see background correction in AAS, Section 2.1.6.2). It is, however, usually impossible to obtain or prepare a reference without analyte and/or with exactly the same matrix composition as the sample. Therefore, non-specific emission must be detected by the shape of the spectrum and must be corrected by setting the right measurement points for quantification. The procedure is related to high-resolution AAS (Section 2.1.6.2.) and has been discussed above. The pattern of signals with partial spectral overlap may look very different. There may be a shoulder, a rising or falling baseline, or other deviations from the Gaussian shape. Usually, a significant spectral interference becomes visible from the resolved spectrum. In all these cases, sophisticated methods of setting background correction points or using mathematical algorithms help to minimize or eliminate the interference. Probably the most important method for validation of the result is the comparison of the calculated analyte concentration between analytical lines. This is often but not always possible; serious limitations must be tolerated if elements with significantly less sensitive secondary lines are determined.

A widely used method for correction of spectral interferences in OES and MS is the inter element correction, IEC described above. This requires knowledge of the sample matrix and consultation of wavelength tables. It should be pointed out, that IEC will work best when the spectrally distorted line and the reference line, used for IEC, is measured simultaneously to avoid scatter by temporal change of excitation conditions.

Optimization of sample flux

Matrix has an influence on the efficiency of nebulization and the long-term stability of aerosol generation. ICP-OES is often using small aspiration rates to save sample and to keep the load of matrix in the plasma small. Small flows are limiting the concentration of dissolved solids. Nebulizers and mixing chambers of various designs are commercially available which have specific fields of use: multi-purpose design, low sample consumption, tolerance against higher concentration of dissolved solids, micro injection volume [80]. Not all combinations of nebulizer and mixing chamber are available und useful. In general, self-aspirating nebulizers are not rugged against high sample load. The higher the concentration of dissolved solid in the sample, the higher usually the flow of sample and sampling gas. Mixing chambers for heavy samples are bulkier than those for, e.g., clean water analysis. High concentrations of salt may require humification of the sampling gas to avoid crystallization from nebulizer to plasma torch. The variety of nebulizers, mixing chambers and plasma torches in the laboratory is usually limited. The main type of samples expected should have been a selection criterion for the laboratory. Once in use, the technical setup should be known with respect to the tolerable amount of solvent and matrix. If the sample exceeds save and rugged limits it must be diluted beyond the critical point. This will influence l.o.d. and l.o.q. but will not necessarily deteriorate linearly. Sensitivity effects due to transport can be easily calibrated. The method of additions will usually compensate the effect.

Matrix will also affect the plasma. This will be discussed below. However, the effect of crystallization at the tip of the sample tubing in the torch or eventual blockage inside the sample tubing may occur. Torches with bigger diameter of the sample tube will usually help. In ICP-OES salt concentrations higher than 1% in the test sample are usually avoided.

Plasma and sample

The temperature and the operation conditions of the plasma are essential for emission intensity, signal to noise and possible chemical interferences. The hotter the plasma, the higher usually the emission intensity of ion lines and the lower the intensity of atom lines. Higher temperatures decrease the probability of recombination of atoms to molecules in general, and in the cooler parts of the plume in particular. Mass of solvent, type of solvent, concentration, and type of dissolved solids in the solution for measurement influence the plasma temperature. This obviously has an influence on the background and the BEC. These effects are usually stronger when the plasma is observed in axial direction. Radial direction allows compensation of some of these effects by optimization of the observation height (like the effects in flame AAS, Section 2.1.6). For method development, the first important decision is the choice of atom and ion lines. The second one is radial or axial observation. The latter one is useful only if the required detection limits are low and the samples are comparatively clean. As usual, a reference sample representing the worst-case matrix should be prepared to study the effect of the matrix on the intensity of the analytical lines, BEC, signal to noise, ruggedness. If axial plasma observation is not really required, small changes in the plasma observation zone in radial observation are recommendable. A standard without matrix should be optimized against the worst sample. Aim of the optimization is not best signal to noise but minimal sensitivity difference between reference and sample.

The effects described are influenced by nebulization efficiency and plasma temperature as well. Distinguishing may not be necessary in routine work. However, the relative plasma temperature can be easily estimated by comparing the relative intensities of a suitable atom and ion line. This method [81] became popular to obtain a better understanding of plasma performance, specifically with respect to ruggedness. The lines often used are the magnesium atom line at 285.21 nm (Mg II), and the ion line at 280.27 nm (Mg I). They are spectrally close to each other. BEC of the atom line is about 0.05 mg/L, of the ion line 0.01 mg/L. The ion/atom ratio of the lines is temperature dependent. If the Mg I/Mg II intensity ration is > 10, the plasma temperature is high, and the plasma is called rugged. For classical spectrometers like Czerny-Turner or Paschen-Runge, the spectral distance between the two lines is small enough such that spectral effects are negligible. This may be different in Echelle spectrometers as lines are often measured in different spectral orders even if they are nearby. In this case the factor 10 mentioned above must be corrected by a spectrometer factor.

2.1.9.10 Application examples

EPA Method 6010D (SW-846) “Inductively Coupled Plasma – Atomic Emission Spectrometry” [86] describes the determination of elements in groundwaters, soils, sludges, sediments, and industrial waste using ICP-OES. Jianfeng Cui and Timothy Traynor, application scientists of Thermo Fisher application laboratories published an application note which describes the procedure for modern ICP-OES equipment [87].

About 100 mL of the liquid samples is digested with 3 mL of concentrated nitric acid on a hot plate and reduced 1:20 in volume by boiling. The concentrate is refluxed with 10 mL of 1:1 hydrochloric acid and brought back to the original volume with ultrapure water. There is no sample dilution involved.

One gram of the dry sediment or soil is boiled on a hot plate with 10 mL 1:1 diluted $HNO_{3\ conc.}$, reduced in volume and again treated with 5 mL of the acid mix. All NO_x fumes generated by the oxidation reaction should disappear at the end of the oxidation. The final reduced nitric acid concentrates are treated with H_2O_2 (2 mL of water and 3 mL of H_2O_2, 30%) until no more reaction takes place. Remaining solid residue is filtered. To the remaining concentrated solution 10 mL of HCl 1:1 is added, and the solution is again refluxed for 15 min. This final solution is diluted to 100 mL. The dilution factor of the solid is 100. The methods described above are time and reagent consuming. At the boiling temperature of close to 100 °C they represent a chemically aggressive extraction of possible solid residue.

A more attractive alternative is microwave assisted pressure digestion. To 0.5 g of the dried soil or sediments, 9 mL of concentrated nitric acid and 3 mL of 30% H_2O_2 are added. The first vigorous reaction is awaited before the pressure vessel is closed. The microwave system is heated under program and feed-back control to 175 °C in 5 min and kept for another 10 min at this temperature. The sample is cooled down and depressurized. Possible residues are separated by centrifugation or filtering. The digest is subsequently diluted to an appropriate volume which again results in a dilution factor of 50 to 100. For some elements the addition of HCl may improve the element recovery.

In the paper described above, microwave digestion never resulted in lower recovery than the hot-plate procedure indicating at least equal effectiveness with much higher efficiency.

Many elements, in total 34, are covering a wide range. Some are ubiquitous in the environment, such as Ca, Mg, Na, Si; others are expected at very low concentrations, such as Ag, Cd, Co, Ni. Important for the limits of quantitation required, are the regulations of the environmental boards of the countries, e.g., EPA or the European regulations. The threshold levels required in water range from the low µg/L range to the high mg/L range. In soils and sludges, the values are much higher; however, the dilution factor during sample preparation results in similar concentrations that need to be detected in the samples. The issue has been discussed in the application example for flame AAS in detail already (Section 2.1.6.5). The wide range to be covered is handled by simultaneous plasma observation in axial direction, for element traces, and in radial direction for ppm concentrations. The plasma is operated at 1,250 W with a total gas consumption of about 14 L Ar/min.

The instrument used can run all elements in a single cycle which reduces the measurement time for one individual blank/sample/standard to less than 2 min. The time for sampling and flushing between samples is reduced by making use of the volume-based injection of 2 mL of sample into a constant flow of blank solution. This method has been described in the AAS section already.

A glass concentric (Meinhard-type) nebulizer is used in combination with a cyclonic mixing chamber. About 2 mL of sample is injected into the carrier flow. The measurement time after stabilization of the signal is 7 s, both for radial and for axial viewing.

Due to different expected concentrations the calibration solutions are mixed from single element standards. The calibration curves are linear and are defined by blank, low standard and high standard. The high standard is usually above the highest expected element concentration. Yttrium is added as an internal standard. Yttrium offers several lines suited for correction of non-spectral interferences of the elements. Suitable lines are individually allocated to the respective elements of interest.

Elements with elevated concentrations in the permille range are usually known to the laboratories in the field of environmental testing. The concentrations found in an explatory survey analysis provides a lot of information on possible spectral overlap. The number of elements that might interfere is not insignificant though. The method therefore consists of several individual element standards (16!) which are mixed as a reference for inter element correction (IEC, see above). The respective lines are assigned to analyte lines. This is a complex procedure, if performed manually. Modern software resident in the software packages of leading manufacturers, are supporting the procedure, however. The same intelligent software support is provided for the selection of the analytical lines which are optimized based on probability of spectral overlap, and the required limit of quantitation.

The method must handle complex samples with various matrix composition, a wide dynamic concentration range between elements, and various non-spectral and spectral interferences. An individual careful method development for liquid samples and dissolved solids is required to master possible unwanted effects. A number of samples for validation are required to assure analytical quality. These are periodic checks for the validity of the calibration, checks of the solutions for IEC and reference material samples for accuracy control. The limits of detection reported range around 1 μg/L for elements determined with axial observation and in the range between 1 and 10 μg/L for radial observation. These are more than sufficient to meet the regulatory threshold values with good repeatability. The recoveries of certified values in reference materials [87] were good. The example shows that ICP-OES nowadays can be used for the determination of the element concentrations relevant in most important environmental samples without the need of special equipment such as ultrasonic nebulization. In the earlier days of ICP-OES, some of the elements would have required other techniques (GFAAS, ICP-MS) to meet the FOMs.

Method development is demanding. Many parameters, foremost possible spectral interferences, must be controlled. Software inherent intelligence helps to set up the method. The fast, economic method with impressive figures of merit justifies the effort of the optimization process.

A second application example concerns the determination of minor and trace elements in edible oils [88]. A similar approach would be used for the determination of elements in crude oils or refined oils. Fewer elements than in the previous application are usually determined. Challenges in method development concern sample preparation and plasma stability. As the digestion of oil in acids is time consuming, and may be even dangerous, the aim is a direct injection of the sample following a dilution with a suitable organic reagent. Compared to aqueous solutions, organic solvents have a strong negative temperature effect on the plasma. Organic solutions therefore require rugged plasma conditions.

The aim of the oil analysis is consumer safety (traces of potentially toxic elements in oil), product quality (influence of contaminants on taste and/or wear of the oil), and possible detrimental influences of contaminants on the production or refinement process. Each of the contaminants must be determined with a regulated or specified limit of quantitation. Consumer safety, for example, asks for detection limits of elements such as As, Cd, Pb, Ni, and Sn in the low µg/L range. Fe, Cu, or Ni must be monitored at slightly higher threshold values to assure product quality with respect to wear, and Ca, Mg, Na Zn, P in elevated trace concentrations may influence the production process negatively. The sample preparation and the resulting dilution factor of the sample must obviously match the lowest required limits of quantitation.

Oil and fat are usually dissolved in Kerosene, or n-butanol [89]. The authors of the paper [88] decided on xylene as this solvent allows adequate dissolution with low dilution factor. Thus, the final dilution of liquid oil samples is only 2; the dilution of samples solid at room temperature is 5. The latter samples are gently heated to the liquid state of matter, diluted, and homogenized. Calibration solutions are treated like the samples. Oil-soluble Conostan standards are prepared following the analytical standard [89]. The trace elements are calibrated with 4 standards between 0.1 and 1 mg/kg. Elements at elevated concentrations range from 0.5 to 10 mg/kg. Oil-soluble yttrium is added as an internal standard. The sample is fed with 0.8 mL/min. As the analyte lines are run sequentially (see below), individually optimized conditions for the flow of sample, the sample gas flow, and auxiliary gas flow, and the generator power can be set as well. This allows, for example, the measurement of atom lines of the alkaline and earth alkaline elements under reduced plasma temperatures. The high carbon content of the samples is compensated by addition of small amounts of oxygen to the plasma gas.

The ICP-OES system used is scanning the individually selected wavelengths. The spectral environment at both sides of the wavelength is measured simultaneously together with the analyte wavelength. Wavelength selection and detection is like the continuum source AAS approach described in Section 2.1.6. The disadvantage of slightly longer analysis times due to the sequential measurement process is compensated by the very good resolution of $R \approx 100{,}000$. The sample introduction is realized with a standard system with concentric nebulizer and cyclonic mixing chamber. All transfer tubing and sealing rings must be organic resistant. Automatic sampling of minimal diluted oil requires specifically designed sampling devices including homogenization of the samples shortly before measurement. A sampler optimized for this type of application was coupled to the spectrometer.

Differences in sample transport and nebulization are corrected by the internal standard. Due to the very high instrument resolution most element lines are measured by direct examination of the spectral vicinity of the analyte. The process is supported by the instrument software with an automated algorithm based on least square methods. Only few elements require additional inter element correction (see Section 2.1.8.8). Among these is Mn. The effect of IEC on the background is shown in Figure 2.48a and 2.48b.

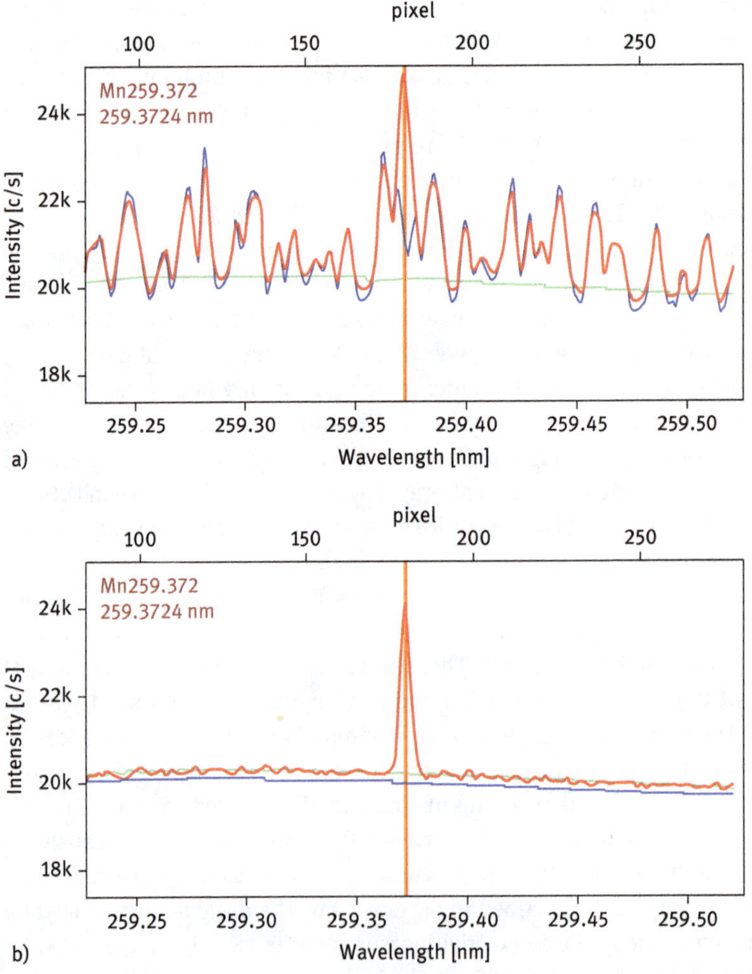

Figure 2.48: Analyte line of Mn at 259.372 nm without (a) and with (b) IEC correction (courtesy of Analytik Jena AG Jena, Germany).

The detection limits obtained are between ≈ 1 µg/kg and ≈ 10 µg/kg. The recoveries of added standards to the samples range from 90% to 110%. Additions on the sample are in the range of 300 µg/kg. The repeatability at this concentration level is generally below 1.5% r.s.d.

In this second example the sequential measurement approach with a very high-resolution spectrometer enables individual optimization of the best plasma conditions. Direct background correction of the spectra ensures high accuracy and excellent detection limits.

2.1.10 UV/VIS molecular spectroscopy

2.1.10.1 General remarks

UV/VIS spectroscopy uses the wavelength range between 180 and 900 nm. Radiation from continuum source type lamps, i.e., D2, tungsten or Xenon flashlight lamps is spectrally separated in scanning or array instruments and is passed through the sample to be analyzed. This is often an organic molecule dissolved in a suitable solvent. It may as well be an ion (typically an anion) dissolved in water. The classical sample containers are quadratic glass or quartz cuvettes with 10 × 10 mm inner dimension, holding approximately 3 mL of sample. Cuvettes with longer measurement path are used to improve measurement sensitivity. Characterization of biomolecules has gained an important share of the applications and instruments are specified specifically for this purpose. Cuvettes or application of microdroplets to optical systems become increasingly important accessories. Often only 1 µL of sample is available. UV/VIS is often used to characterize transparent solids (glasses) with respect to their spectral transparency. Transmission of smooth surfaces or the thickness of layers or films are determined in reflection mode. Powder surfaces can be characterized in transmission mode with additional integration spheres. Gaseous molecules or ions can be measured in special cuvettes or absorption volumes. The information obtained is often qualitative. The spectrum (absorption or transmission versus wavelength) describes the analyzed material so that it can be defined in type and cleanliness. The measurement can be used as well to quantify the concentration of an analyte. In this case the determination is usually run with a fixed wavelength.

UV/VIS spectrometers are a standard equipment in most analytical laboratories. They range from simple low-cost equipment to sophisticated instrument with high-performance optics and lots of accessories.

Basic law for absorption is the Lambert-Beer law described in the AAS section above (eq. (2.10)). The length of the absorption path is usually defined by the cuvette. Like in AAS, quantitative UV measurements have a limited dynamic range due to non-absorbable light. Compared to AAS the phenomenon is intensified, as broad band emission is used for illumination. The aim in UV still is to ensure a linear dynamic range over at least four orders of magnitude up to more than 3 A. Prerequisite is therefore to efficiently separate radiation within the window for measurement from stray radiation. Specific standards, such as Pharmacopeia, require extremely high linearity such that simultaneous spectrometers can usually not meet the strict regulations. High quality quantitative analytical measurements are therefore usually run with scanning spectrometers, sometimes even with double monochromator or premonochromator.

Qualitative spectra scans or simultaneous images of a part of the spectrum are technically simpler and can be performed with both straightforward scanning optical systems and simultaneous array instruments.

2.1.10.2 Technical layout

Molecular spectra are usually broad band compared to elemental spectra (Figure 2.49). In most cases the sum of electron transitions, vibration, twisting, gyration, are recorded during a scan or determined when absorbance at a fixed wavelength is measured. The optical bandwidth, the resolution of the spectrometer, is in most cases less important than it is in elemental spectroscopy. However, there are exceptions which will be exemplified in the application section. For many applications where scans are required, a resolution of around 5 nm is sufficient. The simplest UV spectrometers are therefore equipped with fixed slits.

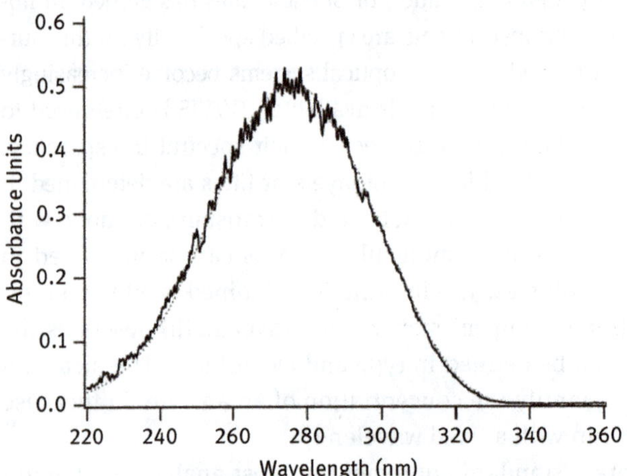

Figure 2.49: Example of a UV spectrum: acetone. Chartier, R, Greenslade, M.. (2012). Initial investigation of the wavelength dependence of optical properties measured with a new multi-pass Aerosol Extinction Differential Optical Absorption Spectrometer (AE-DOAS). Atmospheric Measurement Techniques. 5. 10.5194/amt-5-709-2012.

Monochromator

Regarding selection of the optical bandwidth, UV and AAS spectrometers are similar. Whereas in classical AAS the exact position of the monochromator grating is set by scanning a given source emission profile for its maximum (optical resolution determined by the width of the emission profile), the monochromator in UV must find its exact position with the help of a mechanical or electronic benchmark or by using a filter with exactly defined absorption characteristic. The speed of wavelength scanning is a secondary performance criterion in AAS and a minor parameter for total analysis time. In UV scanning speed and absolute scanning performance is a major performance criterion. Although UV monochromators are very similar in design to AAS types, they are optimized toward other priorities. Monochromator types with a flat reflective grating and focusing mirror have been discussed earlier (see Section 2.1.4.2). Modern photo lithography as method of grating production has gradually gained share against mechanically ruled gratings. Using lasers to imprint a grating structure onto the surface of a photoresist coated mirror has made it much easier to combine the functions of a focusing mirror and a reflective grating into one component. Monochromators became essentially an entrance slit, an astigmatic grating and an exit slit. An example is shown in Figures 2.50 and 2.51. Optical systems become very compact this way. Regarding analytical performance these differences in design do not play an important role. The design of the entire system must be optimized with respect to specification (low end, standard, high end) and with respect to prize to performance in its market segment. For the analytical laboratory figures of merit must be the selection criterion.

For quantitative determinations UV spectrometers should be shot noise limited. Low baseline noise depends on strong illumination from lamps with a flicker smaller than the statistic detector noise. Lowest baseline noise is expected at wider slit width. The demand for a wide linear dynamic range on the other hand is satisfied at high optical resolution (low slit width) providing minimal straylight.

Double beaming

Longer-term fluctuations in photon output of lamps (non-flicker events) must be corrected for. Just as described for AAS, single-beam optical systems provide best light throughput and photon integration time. Optical double-beam systems which measure two beams quasi simultaneously are the most accurate way to compensate for these fluctuations but at the cost of light loss, higher cost of equipment and, depending on the number of detectors, loss of photon counting time. Other than in AAS, where double beaming is exclusively used to compensate photon intensity of the lamps, double beaming in UV is used to compensate effects due to solvents, etc. The second beam may pass through a cuvette of the same type filled with solvent and thus compensate for additional effects. A UV/VIS double-beam spectrometer is sketched in Figure 2.50. Double beaming in UV is way more important than in AAS.

A method for compensating just the radiation source is the split beam technique. The radiation, after passing the monochromator grating and exit slit, is focused on the absorption volume in the sample compartment. Before entering the compartment, it is passing a quartz plate where a small portion of the light is reflected onto a diode detector which acts as a reference to the sample beam passing through the absorption volume. A split beam UV spectrometer is sketched in Figure 2.51.

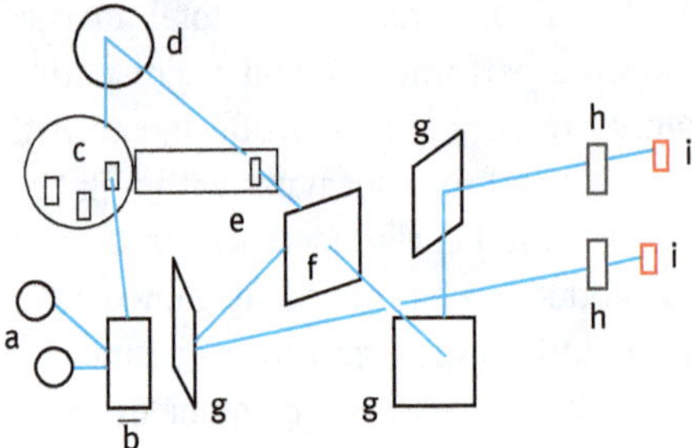

Figure 2.50: AnalytikJena Specord 210 + double-beam UV/VIS spectrometer; schematic of optical setup. a: light sources; b: flipping mirror; c: filter wheel with entrance slits; d: moving, concave, focusing grating; e: exit slit; f: beam splitter; g: mirrors; h: cuvettes for sample beam and reference beam; i: diode detectors.

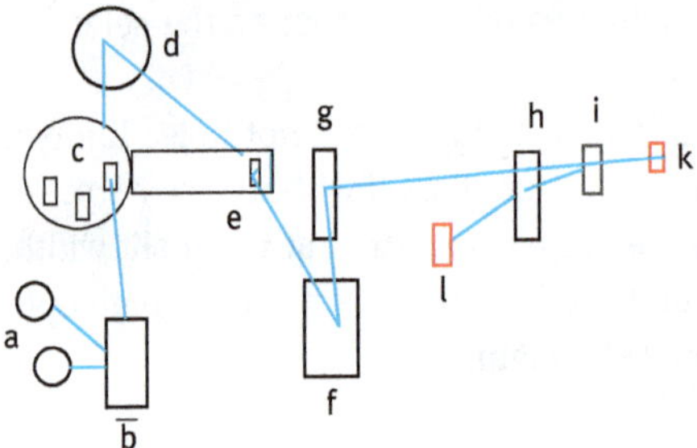

Figure 2.51: Analytik Jena Specord 50 plus split beam spectrometer; schematic of optical setup. a: light sources; b: flipping mirror; c: filter wheel with entrance slits; d: moving, concave, focusing grating; e: exit slit; f: focusing mirror, g: flat mirror; h: quartz plate; i: cuvette; k: detector diode; l: reference detector diode.

Illumination

Scanning instruments are usually equipped with two sources, a D_2 lamp for the wavelength range between below 200 nm up to about 320 nm and a tungsten halogen lamp for the visible part of the spectrum. The lamps can be run continuously or can be modulated. High photon throughput is a selection criterion. However, for special purposes it may be required to reduce the output energy of the source. Filters are used to attenuate light. The entire wavelength can be illuminated with Xenon flashlight lamps. They emit their maximum at about 400 nm. At low wavelength, at 200 nm, their output is usually below 10% of the maximum. Still, they are the devices of choice for simultaneous spectra recording in UV.

Detectors

Photodiodes and other active pixel sensors are an attractive alternative to photo multiplier tubes. The main reason is lower cost and, particularly at wavelengths higher than 1,000 nm, the superior quantum efficiency of these detector types. Depending on the specified analytical performance, which defines the market segment offering, the instruments are equipped with one or several types of detectors ranging from photodiodes and photomultiplier tubes to specifically designed detectors such as PbS and other wavelength optimized types.

Stray light and linearity of calibration curves

Stray light has been discussed extensively in the section about AAS (see Section 2.1.6). In classical AAS, the operational conditions of the line source play a significant role on the amount of light which cannot be absorbed by the analyte (non-absorbable radiation). In UV (just like in continuum source AAS) the radiation is wavelength independent within the optical window set by the monochromator. The conditions of the monochromator window (optical slit width) define the absorption window and have a dominant influence on photon intensity. The absorption profile needs to be constant within the optical bandwidth, i.e., the interaction of analyte and radiation bundle must be equal. In this case radiation of a different energy may only come from imperfections in separation in the monochromator. In particular the reflective grating is a non-ideal dispersive element and will contribute to stray light. Other sources of straylight are reflections at mechanical parts, such as slit, mirror, lens, and cuvette. This straylight is often below 10^{-3} such that analyte absorbance up to 3 A (0.001 T is possible). Both imprints of a flat master grating and holographic concave gratings which focus the selected wavelength directly on the exit slit are relatively low in straylight. If still lower straylight is pursued, the light can be separated by a pre-monochromator, collimated on a first exit slit and sent into the second monochromator of the same type. Double monochromator systems allow extremely low straylight levels and excellent linearity up to 5 A. Yet, high absorbance readings will suffer from increasing noise (low photon counts) and non-linearity of the calibration curve. The latter can usually be corrected by suitable mathematical curve fitting, though it is generally not tolerated in international UV standards.

Standard deviation of the measurement at low photon counts is not only a question of shot noise. Dark noise of the detector in this case is a bigger contribution to the total read out noise. Instruments designed for accurate reading at high absorbance are therefore equipped with high-quality, high-resolution optics, and cooled detectors.

Scanning versus simultaneous instruments

Simultaneous spectra reading has been discussed in the section on optical emission spectrometry. The classical approach of the Rawland Circle is only used in ultra-low wavelength molecular spectroscopy and will not be discussed here. Laboratory instruments for standard applications make use of holographic imaging gratings and photodiode arrays or CCD-type detectors. Unlike in scanning instruments there are no moving parts in this type of instrument. Polychromator and diode array are usually imbedded into a housing which offers high thermal and mechanical stability. An example is displayed in Figure 2.52.

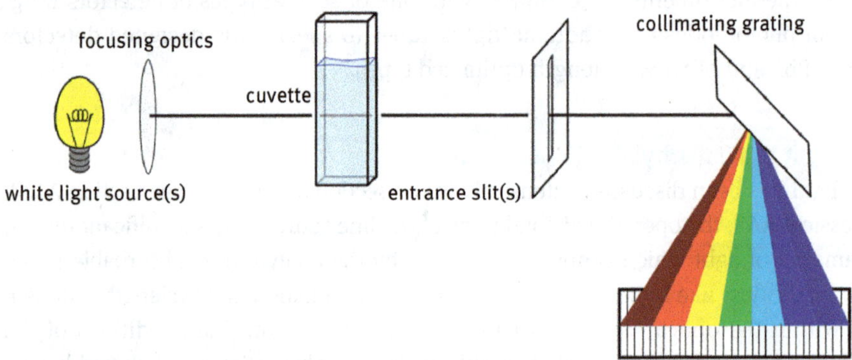

Figure 2.52: Simultaneous UV/VIS spectrometer; Reference: OrgChemist – https://commons.wikimedia.org/w/index.php?curid=66766909; CC BY-SA 4.0 modified.

The spectrometer is behind the sample cell. Xenon flashlight lamps or a D2 and a tungsten-halogen lamp illuminate the entire wavelength range from about 190 to 1,000 nm. Spectrum collection can be fast, usually in the range of a second or even below. Averaging of multiple spectra can be used to improve signal to noise characteristic. Within the short time for spectra collection the change of photon output from the sources is negligible compared to shot noise. Double beaming to compensate for lamp drift is therefore not required. However, due to illumination of the sample with the entire spectrum, stray light will be higher in this spectrometer type. In particular in the short UV range, the stray light requirements of the strictest standards may not always be met. Simultaneous readers are often used for samples which are rapidly changing. Simultaneous readers can easily follow samples from liquid chromatography; multicomponent analysis, dissolution, or kinetics are other typical fields of application.

2.1.10.3 Specifications and instrument performance verification

The range of price, performance, size, and offering in UV is immense. It is most important to specify the type of application and the expected or imperative analytical performance. The latter is often defined by international standards.

Brochure specifications will usually inform about the most important features. This is exemplified with a comment column in Table 2.9.

Instrument performance verification tests are demanded by international standards. Among the strictest ones are pharmaceutical regulations such as the European Pharmacopeia or the United States Pharmacopeia [90, 91, 92].

The agreement between set and measured wavelength is confirmed with chemical solutions showing distinct and sharp peaks between 200 and 700 nm. An example of a holmium oxide solution in perchloric acid is displayed in Figure 2.53. Solutions of rare earth salts are available as standard reference materials for this test. A spectral filter with added holmium can be applied as well as sharp lines emitted by a Xenon lamp.

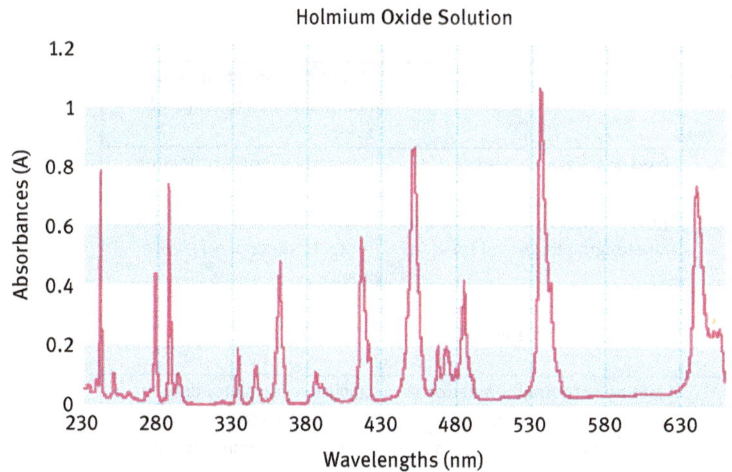

Figure 2.53: Spectrum of a holmium oxide reference solution used for wavelength accuracy checking. Courtesy of Starna Scientific ltd. Ilford, UK.

The requirement of agreement is ± 1 nm between 200 and 400 nm and ± 3 nm between 400 and 700 nm in the European Pharmacopeia and ± 1 nm and ± 2 nm in USP, respectively. The standard deviation between repetitive comparisons shall be <0.5 nm. It is obvious that this test should be run with an instrument with a resolution of better than 1 nm.

Certified filters, lamps and accurately prepared solutions or Certified Reference Materials (CRMs) will certainly exhibit the expected spectrum. The situation becomes more complex if absorbance is tested. In this case solutions of $K_2Cr_2O_7$ with concentrations of 40 mg/L, 60 mg/L, 120 mg/L are measured at four wavelengths from 235 to

Table 2.9: Technical and analytical features for UV spectrometers.

Feature	Specification	Comment
Optical principle	Single/double/split beam	Reference for drift and physicochemical effects
Illumination	Type of lamp	May include wavelength range of individual sources
Detector	Type of detector; often more than one type	May include wavelength range of individual detectors
Wavelength range	e.g., 190–1,000 nm	May be limited by accessory in use
Range of display	e.g.,–6 to +6	Defined by electronics and software, less meaningful
Photometric range (A)	e.g.,–3 to +3	Requires additional information on noise and linearity at high levels
Deviation of 0% line as function of wavelength (corrected)	Offsets of baseline should be in the range of standard deviation of baseline	Short-term offsets
Max scanning speed	nm/min	Shows only speed of drive but not quality as a function of speed
Spectral bandwith	Fixed or several choices	Defined by monochromator and mechanical slits
Trueness of wavelength	Agreement between set/read out wavelength at a known spectral event	Determined with the help of filters or spectral reference lines
Reproducibility of wavelegth	Standard deviation of repetitve readings of wavelength	
Transmission 0-line	Baseline without illumination as function of wavelength	Quality criterion for detector and electronics
Photometric accuracy (A)	Agreement between set/read absorbance compared to a known spectral event	Filters and chemical solutions used for various wavelengths. Includes accuracy of reference
Photometric reproducibility	Standard deviation of repetitve readings of absorbace	
Straylight % T	Wavelength of interest is blocked; rest light determined	Based on chemical standard solutions; requires immaculate standard
s.d of baseline	Determined at a fixed wavelength	No absorbance; should be lower than the deviation of absorbance at 0-line
Long-term stability of baseline	In A units per hour of operation	Shows quality of referencing
Spectral resolution	Peak to dip information of a known spectrum	Requires immaculate solution

350 nm. These shall generate exactly defined absorbance readings between 0.2 and 1.73 A. In combination with 6 repetitive readings for each concentration and wavelength, accuracy, precision range of agreement and linearity can be detected. Although reference materials are offered, they must be immaculate and handling during the test must be perfect. This type of test requires an absolute analytical technique. It seems to be more appropriate to control the accuracy with appropriate filters just like in AAS. Neutral density standards are available by NIST and should be used for this test.

Non absorbable light is the main reason for bending of calibration curves in absorption spectrometry. For straylight testing, solutions which are known to block the light at a certain wavelength range completely are used at a wavelength falling into that wavelength range. In this case the absorbance reading shall be above 2 A which indicates a stray light level below 1%. The test is run with 4 different solutions between 190 and 400 nm. The most critical range for stray light is the far UV range at 190 nm.

Finally, the resolution is determined with compounds which have sharp absorption bands. A simple quantitative test is obtained with a solution of 0.02% toluene in hexane. The mix shows a pronounced valley at $\lambda = 267$ nm and a peak at $\lambda = 269$ nm. The ratio of absorbances between valley and dip are a measure for the resolution of the spectrometer (see Figure 2.54).

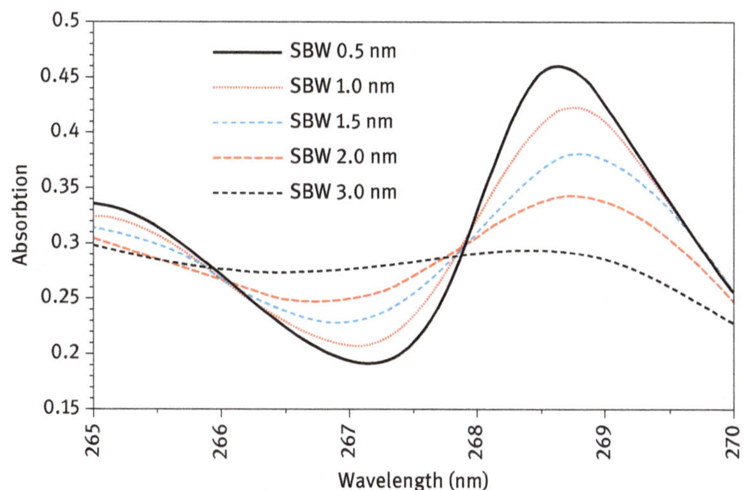

Figure 2.54: Toluene/hexane test for resolution of UV spectrometers. The spectra are displayed as a function of the optical slit width (nm). Courtesy of Starna Scientific ltd., Ilford, UK.

The ratio can be measured as a function of the mechanical slit width. From the ratio, a resolution can be approximated. In case of the scans shown above, the ratio recorded with the smallest bandwidth is approximately 0.46 A/0.19 A = 2.4. This indicates a bandwidth of 0.5 nm. The pharmacopeia asks for a resolution better than

1.8 nm. The blue line (SBW 1.5 nm) provides a ratio of 1.65. Tables indicate that the specification is outperformed. Calculation of the ratio and back-calculation to an optical resolution is usually a part of the instrument quality software.

If one is looking at the specifications in the brochures of the manufacturers, the requirements mentioned above are usually met with standard, middle class instruments. These specifications are usually obtained under best instrumental and laboratory conditions. It is advisable to select an instrument which outperforms the requirements of the standard by a factor of 2–3 to compensate for aging of components under laboratory conditions.

2.1.10.4 Selected application examples

UV-VIS spectroscopy is often used as qualitative determination of the molecular structure. Strong absorption takes place if specific compounds are present which interact with light easily (chromophore). These are usually double bonds. The energy transition of the electrons is a n to π^* or a $\pi \rightarrow \pi^*$ transition. Typical examples are homocyclic compounds, such as benzene, aromatic and heterocyclic compounds, and compounds with C–C double bonds. Issues like conjugation/cumulation, formation of complexes, e.g., charge transfer complexes and transition metal complexes can be studied with the help of UV spectra. An example molecule is featured in Figure 2.55a.

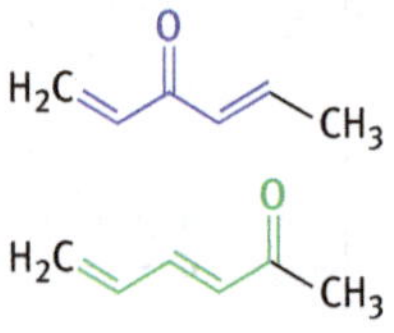

H_2C O CH_3

H_2C O CH_3

Figure 2.55a: Two ketones with the same sum formula but different structure.

Quantitative determinations follow Lambert-Beer´s law (see above). In UV spectrometry the molar extinction coefficient ε is in the range of 10^3 to 10^5 L/mol cm. We solve Lambert-Beer's law for *c* (eq. (2.10)).

Equation 2.23: Lambert-Beer law

$$c = \log I_0/I \cdot \varepsilon^{-1} \cdot d^{-1} \tag{2.23}$$

Three standard deviations of the baseline are defining the minimum concentration to be positively distinguished from zero. I_0/I at the detection limit of many instruments is in the range of 0.001. *d* is usually about 1 cm. An estimate of the detection limit is therefore in the range of about 10^{-7} mol/L. For a molecule like $K_2Cr_2O_7$, as an example, this would translate to a detection limit of slightly below 30 µg/L. The standard for photometric detection states 2–50 µg/L.

Element compounds such as CrO_4^{2-} are colored. Species can often be transferred into colored complexes with suitable agents. The most well-known complex is dithizone.

Anions, such as halides, NO_2^- and NO_3^-, ammonia, phosphates, sulfates and sulfur compounds, phosphates and silicates are determined with the help of colored complexes.

In case of organic compounds, photon activity results from double bonds, such as unsaturated hydrocarbons, alkenes, alkynes, fatty acids, oxy-, -azo, -nitro compounds, sulfur containing compounds and sulfates. Sugars, amino acids, peptides, and proteins as well as organic metal compounds show good UV activity.

All compounds mentioned above show characteristic spectra which make quantitative detection possible. However, as described in earlier sections, the compounds are usually surrounded by solvent and matrix. They will usually influence the spectrum.

Absorption and concomitants

Sensitive UV/VIS absorption is induced by the chromophore, a structure of delocalized electrons, as described above. The corresponding bonds or free electrons in multiple bonds, oxygen-, nitrogen-, sulfur-, phosphorus groups show characteristic absorption maxima with strongly different ε-values. These are usually in the short wavelength range around 200 nm up to 300 nm. Some of them are beyond the usual measurement range of standard instruments, such as benzene at 184 nm or pyridine at 174 nm. However, many groups show secondary maxima at longer wavelengths but usually with a significantly lower ε. The maxima can be related to electron transitions. The strongest maximum at short wavelength is related to a transition of π-electrons from the binding to the anti-binding molecular orbital.

Vast classifying spectra tables are available [93, 94]. Nonetheless, a few general guidelines shall be given here:

Hydrocarbon chains with conjugated double bonds reveal the maximum toward longer wavelength and growing ε with increasing chain length. The same number of double bonds in a ring absorbs at shorter wavelengths with similar ε.

Nitrogen substitution into a ring (benzene to pyridine) shifts the wavelength toward UV but increases ε. In double rings (quinoline versus naphthalin) the shift is small. The position of the foreign atom may have a strong influence on the sensitivity though (quinoline versus isoquinoline).

Hydrocarbon chains with atoms of main group 6 show UV activity at short wavelengths with moderate sensitivity (ether, thioether). The foreign atom S shows higher activity than O (the electrons are in orbitals of higher energy in the ground state).

Terminal end groups such as amines, thiols, halogens, carboxyl are active with relatively low sensitivity. An exception are sulfur compounds and nitro compounds. These compounds are nominated auxochrome groups, i.e., functional groups which change the absorption characteristic to higher wavelength (bathochrome effect), shorter wavelength (hypsochrome effect) higher sensitivity (hyperchrome effect), or lower sensitivity (hypochrome effect).

From the observations mentioned above it becomes obvious that solvents will influence both absorption maximum and molar extinction coefficient. Polar solvents, such as water or acids interact predominantly with n-orbitals of the chromophore and have only a minor influence on the π orbitals. The interaction occurs mainly via hydrogen bridges. The energy of n and π orbitals is lowered to a different extent. The energy of π to π^* transitions become lower with increasingly polar solvents; ε and wavelength increase. The energy difference between n and π^* transitions increase with increasing polarity of the solvent. The peak moves to shorter wavelength, and ε becomes smaller.

Absorption as a function of technical parameters

All parameters which influence the activity of solvent and/or chromophore may have an influence on absorbance and exact position of the peak. These are mainly the purity of reagents and the temperature during measurement. The cuvette material must be suitable for the intended spectral range (quartz must be used below 300 nm). The cuvettes in the measurement and reference channels must be identical within tight tolerances. The solvent obviously must not show significant absorbance in the wavelength range of quantitative determination.

Showcase applications

The quantitative determination of inorganic ions such as phosphate and nitrite are important determinations for environmental control. The ions do not show strong activity and must be stimulated by ligands. It may therefore be argued that these quantitative determinations belong to Chapter 3 of this compendium.

Phosphate concentrations in water must be controlled. Plants and fish require phosphate for nutrition whereas surplus results in excessive growth of plants, in particular algae, which disturbs the equilibrium of plants and finally leads to a deficit of oxygen in waters. The concentrations of interest are in the range of few milligrams per liter. Small concentrations of phosphorus in drinking water, on the other hand, are required to keep the pipe system stable. These low concentrations are in the range of micrograms per liter.

The sequence of phosphate determination is the generation of a stable compound between phosphate and ammonium heptamolybdate in acidic medium. The complex is then reduced by ascorbic acid. A dark blue complex (molybdenum blue) is formed which is highly active for quantitative UV determinations. Antimony ions are stabilizing the color. The absorbance is determined at 885 nm. The method is an international standard method [95] for the determination of orthophosphate, orthophosphate after solvent extraction, hydrolysable phosphate plus orthophosphate, and total phosphorus after decomposition. The method is optimized for phosphorus concentrations in the range of 5 to 800 µg/L. The total procedure is chemically laborious as, in addition to the reference solutions and the samples, six reagents must be prepared and mixed. Therefore, reagent sets have been developed to simplify the procedure. This standard procedure is in worldwide use [96].

Elevated nitrate and nitrite concentrations in waters are predominantly the outcome of strong fertilization of soil. Inorganic nitrogen salts are the most important nutrient for plants. In sound environment nitrogen-containing salts are removed as molecular nitrogen following a denitrification process via nitrite and nitrate. This process may be inhibited by too high pH value of water, so that soil, groundwater, and drinking water may contain high levels of nitrate. The threshold level of nitrate in drinking water in Europe is 50 mg/L, the maximum concentration of nitrite in drinking water is 0.5 mg/L [97]. Nitrite and nitrate are determined sequentially using the same basic measurement process.

Just as in the example described above, nitrites must be coupled to a colored complex which is suitable for UV determinations at quantitative levels in the range of above 10 µg/L. Nitrate diazotizes with primary amines in acidic medium. The azo salt is coupled with nucleophilic aromatic compounds forming a red azo dye shown in figure 2.55b. The reagents used are sulfanilic acid amide for formation of the diazo salt, and naphthyl ethylene diamine for formation of the azo complex.

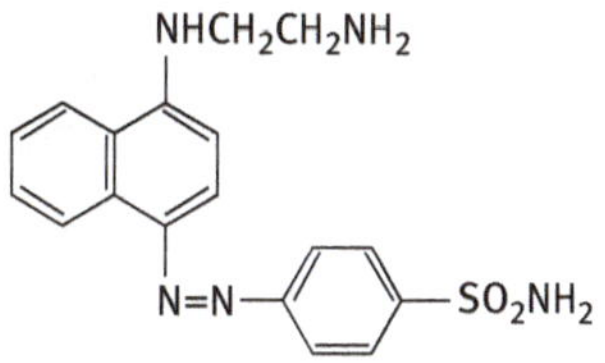

$NHCH_2CH_2NH_2$

N=N–

SO_2NH_2

Figure 2.55b: Azo dye obtained from nitrite by diazotization and azo coupling reaction.

The reaction is firstly used to determine nitrite at low concentrations in a 5 cm cuvette. The nitrate in the test water is subsequently quantitatively reduced to nitrite in a column, filled with copper coated cadmium chips. If the nitrate concentrations are high, a 1 cm cuvette is used for quantification. The method is standardized in the ISO norm 6777 [98]. The limit of quantification is in the range of 0.1 mg/L.

High-resolution molecular absorption spectrometry allows quantitative measurements of anions at absorption maxima of bands structured by rotational transitions. Research applications with the high-resolution continuum source AAS instrument described in Section 2.1.6, focused strongly on ion determination, mainly F^-, Cl^-, SO_4^{2-}, PO_4^{3-}. An example of such type of absorbance structure is sketched in Figure 2.56 [99].

The closed absorption volume graphite furnace in this case is combined with molecular absorption spectrometry. The anion must be present in form of a thermally stable compound with low solubility. Halogenide salts of this type are, e.g., Ga, Ca, Sr, and Ba. In the case described in [99], the sample is prepared and handled like in GF-AAS applications. The reagent forming agent is added as modifier to the sample, often acting as stabilizing modifier as well. In the example described, the strongest narrowband absorption is observed at 606.440 nm (the highest peak in the picture shown above). This is a favorable spectral range where interferences by other concomitants in

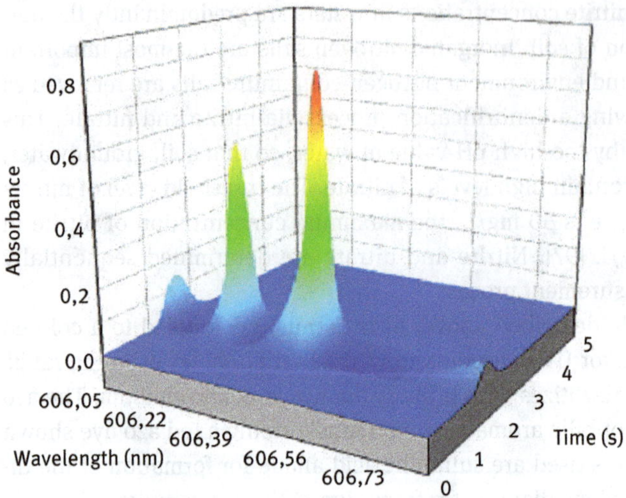

Figure 2.56: Absorbance structure of the CaF molecule at 606.440 nm; from [99].

the sample are rather unlikely. The detection limit reported was 0.16 mg/L. The upper limit of the working curve was 25 mg/L The working range of the method is sensitive enough to determine tea infusions showing between 21 mg/L F^- and 56 mg/L F^-.

Indirect determinations using the time dependence of chemical reactions

Important substances, e.g., in food chemistry or medicine, can be quantitatively determined by measuring the speed of decay of a substance. The decay for the reaction can be calibrated. The decay thus allows to calculate the concentration of the analyte species. The test substances integrating the analyte are mostly enzymes. The concentration of urea, for example can be quantitatively determined by measuring the decay of its main absorption band at 290 nm. Reagent in this case is urease which transforms urea into ammonia. The enzyme is added to the test solution at a certain time t_0. After a delay of 10 to 20 min the absorbance is measured at equidistant times t_1, t_2, t_3. The measurement sequence is an integral part of the instrument software. The ΔA is averaged. With the help of the calibrated factor the original urea concentration can be determined. The detection limit is in the range of 0.15 mg/L urea. Enzymatic tests are widely used in food chemistry and for medical applications.

Biochemical applications

Quantitative determinations of DNA and proteins became a primary field of application for UV/VIS spectroscopy. The instrument manufacturers usually offer specifically optimized systems for this purpose. The bases adenine, guanine, cytosine, and thymine in DNA, and uracil in RNA are UV active with a maximum at 260 nm. DNA and

RNA solutions are usually contaminated by proteins. Proteins are structured long chained amino acids consisting of thousands of units. These absorb maximally at 280 nm. The absorbance ratio at 260 nm and 280 nm is used as criterion to estimate the purity of the DNA/RNA solution. A DNA spectrum is sketched in Figure 2.57.

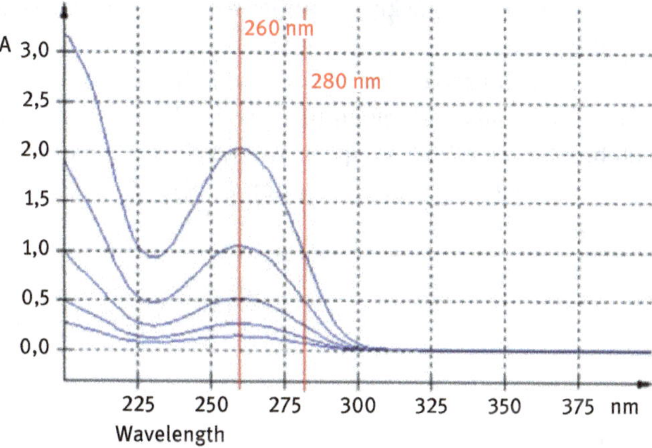

Figure 2.57: Absorption of DNA reference solutions (courtesy of Analytik Jena AG Jena, Germany).

As pointed out already, proteins are determined directly at 280 nm. Colored reagents, such as tri-phenyl-methane, a red compound with an absorption maximum at 470 nm, are used to improve the sensitivity of the determination. The compound forms complexes with cationic and non-polar side chains of the protein in acidified medium. The absorption maximum of the complex is shifted to 595 nm with a strong increase in ε upon complexation. The difference in ε during complexation is a highly sensitive measure for the protein concentration in solution. Detection limits of only a few μg/mL of protein become possible.

Disintegration of DNA

The strings of the DNA double Helix are connected via hydrogen bonds between adenine-thymine and guanine-cytosine. The bonds are very weak. The G-C bond is stronger than the A-T bond. Another stabilizing effect is based on dipole interactions of elongated helices. Longer strings are therefore energetically favorable, and helices with higher G-C content are more stable than DNA with lower G-C content.

At higher temperature the DNA is unwinding. The hydrogen bonds break down. This phenomenon is known as melting. The melting temperature is defined as the point where half of the hydrogen bonds are disintegrated. With the help of this temperature the G-C content in the DNA can be determined using the empirical eq. (2.24):

Equation 2.24:

$$c_{(GC)} = (T_m/^\circ C - 69.4) \cdot 2.44. \tag{2.24}$$

c is the G-C concentration in %, T_m is the melting temperature of the genome under investigation. The melting point of the human genome is about 86 °C. With the help of the melting point, different DNA can be quickly quantified and categorized. This way bacteria can be rapidly classified.

Disintegration of the DNA results in separation of the double helix into two independent strings. This effect is accompanied by an increase in UV absorption at 260 nm of up to 40% (hyperchromicity). A DNA sample is heated slowly and exactly controlled from temperatures below 69 °C. The absorption at 260 nm is plotted against the temperature (see Figure 2.58).

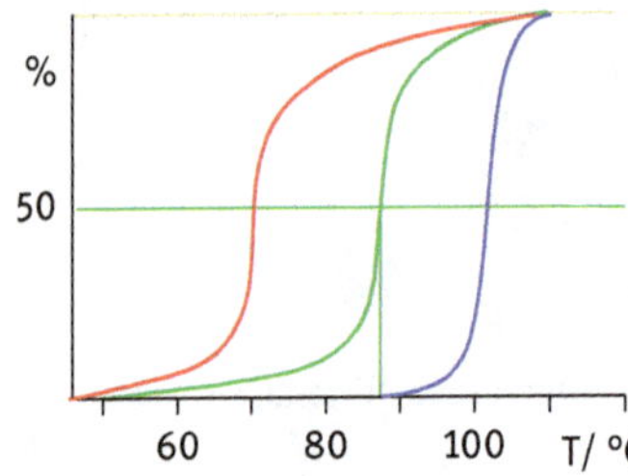

Figure 2.58: Determination of the DNA melting point: sketch of a melting curve of a DNA composed of A-T and GC (green), of A-T (red) and of GC (blue).

The point of inflection of the curve defines the melting temperature and approximates the G-C content of the sample.

As DNA is usually available at very low masses and as the temperature must be measured very accurately, this type of analysis requires microcuvettes and a precisely controlled heating/cooling system of the cuvette. Systems which use volumes in the range of a µL have become an important accessory, dedicated instruments have been designed for this purpose. The temperature is usually set by Peltier systems with a temperature resolution of 0.1 °C, which is another important accessory for modern UV spectroscopy.

Color and surface

Fundamental applications of UV spectroscopy concern themselves with the exact definition of color and surface structure. Although the applications are rather qualitative than quantitative, they are used widely for classifying definition of these parameters. Radiation passing through glasses (transmission spectra) is handled alike radiation passing through cuvettes. Spectra are recorded in the usual way and standard equations are valid. It should be mentioned that the determined absorption maximum is complementary to the information of the human eye. A yellow glass absorbs predominantly in the blue wavelength range at about 480 nm. Numerous national and international standards such es ASTM E-308 [100] use transmission spectra to clearly define the psychophysical color stimulus to the human eye.

Every surface, even transparent glass or other transparent layers is reflecting light, depending on the angle of the incoming light and on the wavelength. Reflected light can be used for photometric measurements in various ways. Depending on the structural properties of the surface, the reflection can be directed or diffuse. If the incident beam is clearly defined in direction and the surface is mirror-like, the qualitative information taken from the spectrum is comparable to the situation mentioned above. The important quantity is wavelength-dependent transmission. As the spectra are usually not fine structured, absolute wavelength accuracy is more important than spectral resolution. If the reflecting layer is underneath a transparent coating or layer, the incident beam will generate two reflected beams. The same happens if transparent layers adjoin air. The reflections are sketched in Figure 2.59.

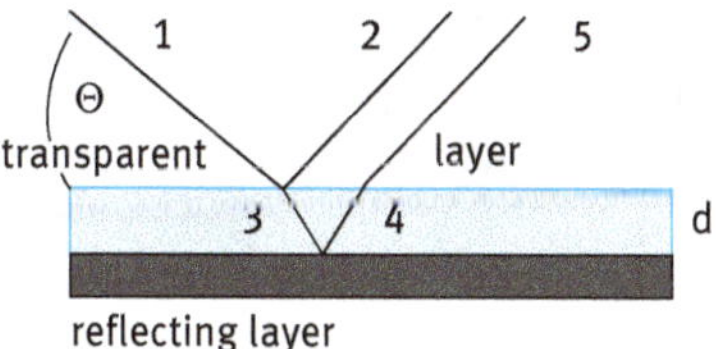

Figure 2.59: Reflections at and underneath a transparent layer: the incident beam (1) is partially reflected (2) and partially entering the transparent layer with thickness *d*. The beam is bent (3) reflected (4) and bent again to a second beam (5).

The reflected beams interfere and generate an interference spectrum. The interference spectrum is dependent on the thickness of the sample (*d*) index of refraction of the sample (*n*) and reflection angle (Θ). The spectrum provides information on the number (*m*) and wavelength (λ) of interference maxima. Θ can be precisely set by a suitable accessory of the UV system. *m* and *λ* are extracted from the spectrum. If *n* is known, *d* can be calculated from the spectrum and vice versa.

If rough surfaces are scanned, the angle of the incident beam cannot be exactly defined, and/or if the source is emitting disperse, the radiation must be harmonized and normalized. This is accomplished with a so-called integration sphere (Ulbricht-sphere). It is a hollow sphere where a source is directly in front of an entrance port (slit/hole). The light is diffusely reflected inside the sphere at layers of $BaSO_4$ or optical PTFE. Usually, the exit port is perpendicular to the entrance port such that the light is homogenized before it is leaving the sphere. All ports are much smaller than the size of the Ulbricht sphere. With the help of this device the luminous flux of sources can be quantified and compared. Sources can be normalized this way to compare photon detectors (Figure 2.60).

If rough surfaces are spectrally analyzed, the sample is illuminated by the source under a defined angle and the reflected radiation is collected and scattered

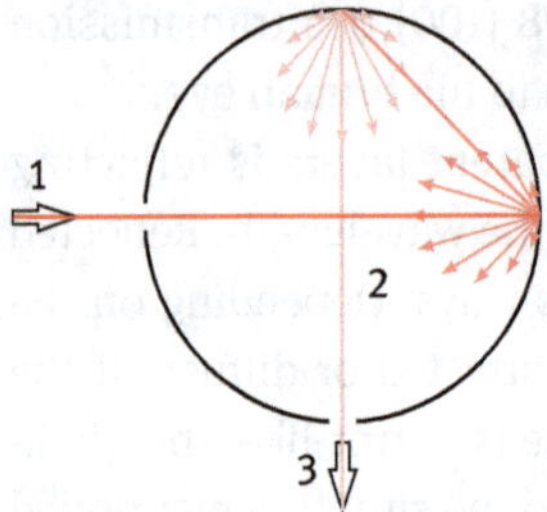

Figure 2.60: Ulbricht sphere. Light from the source is entering the sphere and is multiplied inside the sphere (2). The harmonized light exits the sphere in orthogonal direction (3).

inside the sphere. The detector is evaluating the harmonized radiation in transmission mode. If reflection properties are investigated, the sample is illuminated with light harmonized inside the sphere. An application example describes, e.g., the determination of the color (whiteness) of teeth [101].

The sample is illuminated with light from the sphere. The spectrum in the visible range (380–780 nm) is recorded in reflection mode within about a minute. A specially designed software for color evaluation helps to define the color of the teeth compared to "ideal white" defined by standard.

2.1.11 NIR and IR molecular spectroscopy

2.1.11.1 General remarks

The wavelength range next to the visible part of the electromagnetic spectrum is called near infrared (N-IR) followed by the mid-infrared range (M-IR) and Far Infrared range (F-IR). The N-IR is directly adjacent to the range visible to the human eye (780 nm). Technically, there is a slight overlap to UV/VIS spectroscopy. It spans about 600 nm (up to 1,400 nm) in the higher energy part (called IR-A) and 1,600 nm (up to 2,500 nm or 2.5 µm) in the lower energy range (IR-B). The mid-IR range covers the classical IR applications in the range between 2.5 and 25 µm and the F-IR range is between 25 and 1,000 µm. Often the wave number $\tilde{\nu}$ is used instead of the wavelength (eq. (2.25)). In spectroscopy it is traditionally based on the unit cm [102].

Equation 2.25:

$$\tilde{\nu} = \nu/c = 1/\lambda \tag{2.25}$$

where ν is the frequency and c is the speed of light.

The wavenumber is indicated in cm^{-1}. Using this terminology, the M-IR, for example, ranges from 4,000 wavenumbers ($\lambda = 0.00025$ cm) to 400 wavenumbers ($\lambda = 0.0025$ cm).

The splitting of the IR range in at least three sections, as described above, has technical and fundamental reasons. The latter are motivated by different interactions of radiation with molecules: the low-energy F-IR radiation activates mainly rotation of complete molecules. At increasing wave numbers (shorter wavelength) the movement of atoms or group of atoms relative to each other is activated. N-IR activates mainly harmonics of oscillations providing plenty of information mainly for -CH, -NH and -OH bonds. Dividing the total IR range into segments is technically important. Illumination of the sample, detection of radiation and supporting spectra software are expensive and should be dedicated to the targeted types of application. The universal instrument solution would be expensive. The respective information can be found in the relevant technical sections at the beginning of Chapter 2.

IR spectroscopy belongs to the very early methods of analytical instrumentation. For several decades IR was technically implemented by classical monochromator instruments before interferometers became the exclusive solution in the mid-IR range.

The spectra obtained in the M-IR range are diverse. Radiation interacts by stimulation of a relative movement of atoms in the molecule. These may happen in different directions. They are classified in symmetrical and asymmetrical stretching, bending, rocking, wagging, twisting (see Figure 2.61). In addition atoms and molecules are stimulated to rotate. This results in an overlay of multiple absorbed wavelengths on top of the simpler oscillation spectrum.

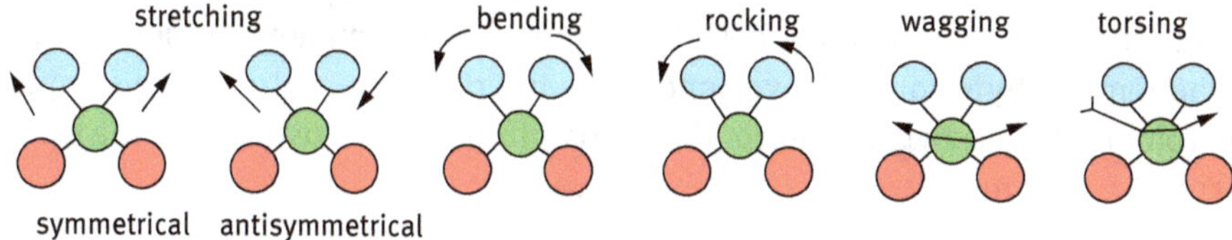

Figure 2.61: Stimulation of atomic movement in molecules by IR radiation.

As described for the shorter wavelength UV spectroscopy, the interaction can be explained with the help of quantum mechanics. Molecules tend to adopt the state of lowest energy if not stimulated by energy. This ground state is characterized by an optimal distance between the atoms. Stimulation by an electromagnetic wave may result in a short-term transformation to a new discrete energy state (see Figure 2.62).

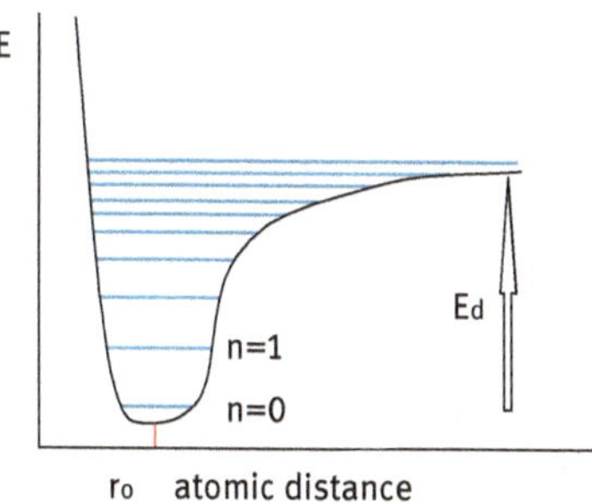

Figure 2.62: Quantum well of a simple molecule consisting of 2 atoms (e.g., HCl); energy (ordinate) versus atomic distance. E_d is the dissociation energy.

As a first approximation the next higher energy state in vibration (v) and rotation (J) is stimulated. $\Delta v = 1$; $\Delta J = 1$. Excitations of even higher vibrational and rotational states are possible but require obviously higher energy (shorter wavelength) and are much less probable.

The classical model of interaction of molecules with an electromagnetic field is based on a movement of electrical charges. Molecules like HCl are electrical dipoles which can be easily stimulated by electromagnetic waves with matching energy. Symmetrical molecules such as H_2 or N_2 do not show noteworthy IR activity. The quantum mechanical model, however, allows interaction of those molecules with very low probability.

In essence, and important for analytical purposes, the IR spectrum obtained often allows the identification of the molecule in the sample. A quantitative determination requires calibration with a suitable reference solution and quantitation of absorbance or transmission of an IR band or of a series of bands.

2.1.11.2 Basic instrumental setup

Figure 2.63 shows the fundamental components of a Fourier-transform spectrometer.

The source is imaged with focusing mirrors (b) through an assembly of filters (c) into an interferometer. A reference beam (i) (pink) is guided into the same interferometer and collected onto the detector. It provides information on the position and the temporally resolved movement of the interferometer mirror. The sample beam (blue) is focused via the mirror (b, bottom) through the cuvette with the sample (f) to the detector (d).

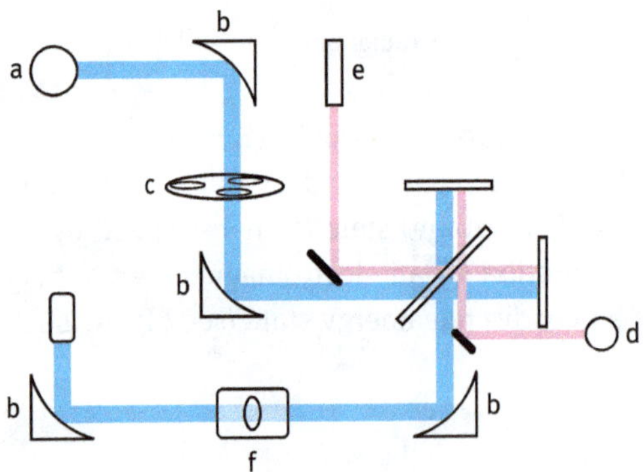

Figure 2.63: Basic setup of a FT-IR spectrometer.

2.1.11.3 Presentation of the sample

Sample preparation is an essential part for IR spectroscopy as well as for the other methods of instrumental analytics. The following paragraph mostly concerns the M-IR range. N-IR follows different rules and will be discussed later. The samples determined with M-IR spectroscopy can be solid, liquid, and gaseous. Solids can be handled as pressed pellets (usually embedded in KBr), embedded in polymer material as a film, or as suspension. Liquids are handled dissolved or as the pure liquid substance (for structural questions only). Gases are usually measured in cuvettes which can be evacuated and filled with the analyte gas. The unique challenge for this technique is the housing of the volume of interaction with radiation. Glass is not permeable to IR radiation. Additionally, almost all solvents show strong interaction with IR radiation and are therefore hardly usable for specific sections of the spectrum.

Alkali halogenides are widely permeable for IR radiation and can be used both as cuvettes and as carrier material. They can be formed into stable thin-walled cuvettes. However, the material is soluble in polar solvents such as water or alcohol. Even low water contents in the sample or in solvents may quickly destroy the cuvettes. A variety of materials are available with different optical permeability and different stability toward polar solvents, see literature [122]. These are potentially stable against higher water concentrations but may be usable in sections of the electromagnetic spectrum only, or they may be expensive.

KBr and NaCl offer the widest spectral range in IR, from 0.25 μm (40,000 cm^{-1}) to 5 μm (2,000 cm^{-1}).The range of NaCl is limited toward shorter wavelengths. The same is true for AgBr, which is transparent from about 20,000 cm^{-1}. Its big advantage is the insolubility in water; its disadvantage is mechanical sensitiveness and cost. ZnS is water insoluble as well and mechanically stable. It is useful for longer wavelengths starting at about 10,000 cm^{-1}. CaF_2 and SiO_2 are both transparent from the visible range of the spectrum. SiO_2 looses its permeability at about 2,500 cm^{-1}. Both materials are rugged but CaF_2 cuvettes are expensive.

Almost all solvents show strong absorption bands in the M-IR range. Water, for example shows close to zero transmission in the range of 3,400 wavenumbers (3.1 μm) and about 25% transmission in the range of 1,600 wavenumbers (7.1 μm). The IR spectrum of water is sketched in Figure 2.64.

Polar solvents in particular, but even almost unpolar solvents exhibit absorption bands in the important wavelength range between 500 and 3,500 cm^{-1} (2.9 μm and 20 μm). The analyte bands for identification and even more for quantification must not overlap strong bands of the solvent. A sketch of important absorption bands of some widely used solvents is listed below.

The concentration of the analyte substance in the solvent should be high, i.e., in the range of at least 5%, for good signal to noise ratios. Non-polar solvents, if suitable for the required dilution, are preferential as their absorbance bands are significantly weaker. All solvents may interact with the analyte and may induce interferences which may even hamper the validity of Beer's law. Alcohol and water are usually avoided as they absorb very strongly and deteriorate the cuvette material. Other solvents must be dried to avoid these effects.

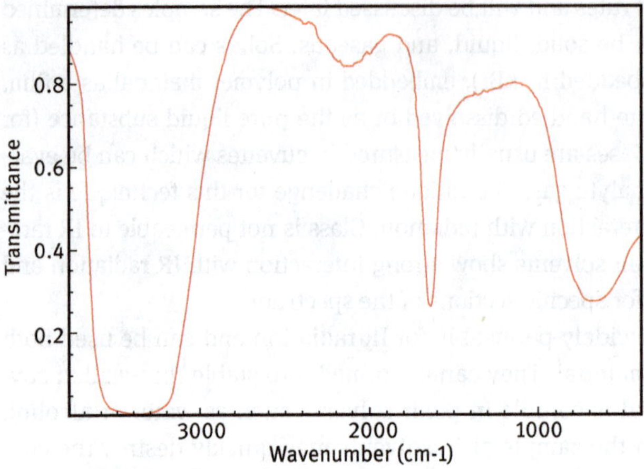

Figure 2.64: IR spectrum of water; source: NIST Chemistry WebBook. NIST Chemistry WebBook; National Institute of Standards and Technology, Gaithersburg, MD, USA.

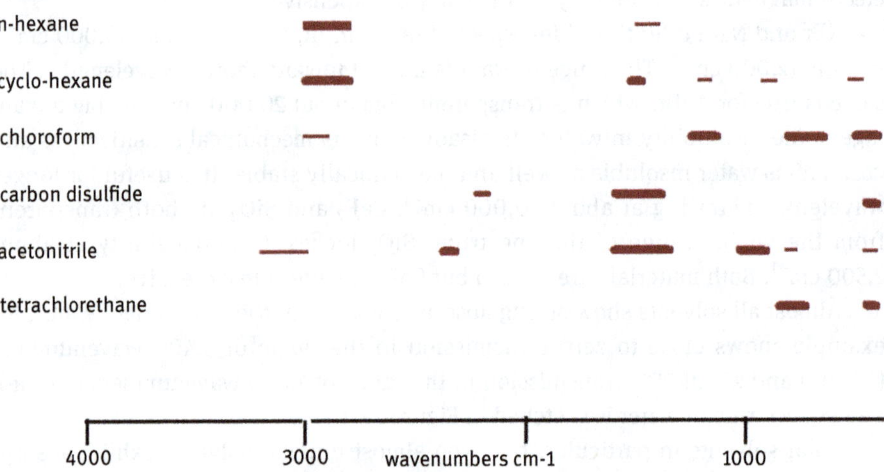

Figure 2.65: Solvents and their main IR-active absorption. Red bars indicate close to 100% absorption (wide bars) or about 50% absorption (narrow bars). Data from Raman/IR Atlas organischer Moleküle, Verlag Chemie GmbH, Weinheim (1974).

If polar solvents must be avoided and non-polar solvents are not suitable for dissolution, non-polar liquids such as long chained paraffins can be used to generate a suspension of fine powdered substance with solvent. The procedure for preparation of the suspension is like that of the slurry technique in AAS. The analyte must be finely powdered (optimal are particle sizes of less than 10 µm). It is homogenized with the paraffin (usually a mix of paraffins is used which is available under the name "Nujol") so that a viscous suspension is formed which can be imbedded into a slim IR cuvette. Note that even paraffins have relatively strong absorption bands in the IR spectrum. Nujol, e.g., absorbs strongly at about 2,900 cm^{-1} (3.45 μm). Just as discussed in the section on slurries, calibration is the name of the game, if quantitative determinations of slurries are pursued.

IR is particularly suited to handle solid samples. From the early days the pressed pellet is a key application for this technology. Requirements are finely ground powders of the analyte substance, a salt which does not significantly absorb in the wavelength range of interest, a perfect homogenization of analyte and carrier and a reference of known concentration if analyte quantitation is aimed at. Quantitative determinations are rather rare, however.

The carrier material should form transparent pellets. The most widely used material is KBr or other alkaline or silver halogenides. Polyethylene may be used as well. After mixing carrier and sample, the fine powder is pressed to relatively small pellets, roughly a centimeter in diameter and one mm thick. The analyte/matrix ratio is usually in the range of 1/500. About a mg of analyte is pressed with roughly half a gram of matrix at pressures of 1,000 MPa (10,000 bar). Smaller masses of analyte in the range of a few μg can be pressed with specific tools.

Just like in the case of solvents, chemical interactions between matrix and analyte should be excluded to avoid misinterpretation of spectra for structure information. Alkaline halogenides are hygroscopic. To avoid strong bands of moisture (OH-bands) preparation and storage of pellets needs dry environmental conditions. A big difference between the index of refraction of matrix and analyte would also falsify the obtained spectrum. The matrix must obviously be selected with respect to the wavelength range under investigation.

If polymers are analyzed, they may be soluble in a volatile solvent. This can be removed in moderate vacuum or on slightly heated surfaces, leaving back a thin but defined polymer film on a cuvette. The polymer may be suited for drying at elevated temperatures without solvent as well.

2.1.11.4 IT for infrared spectroscopy

Unsurprisingly, the complexity of information in IR is high. Numerous transitions generate an enormous number of spectral bands with various intensities. Matching an unknown analyte spectrum with filed data is generally cumbersome. An example figure is displayed in Figure 2.66. Sophisticated software programs and spectra data bases are therefore an essential part of an analytical system. Software packages are available from instrument companies and from vendors specialized on software only. Open-source software is also available.

The most prominent module is an electronic functional group dictionary. It recognizes spectral bands and assigns it to functional groups from a dictionary. Based on matches found, a list of suggestions is presented to the user. Besides automatic

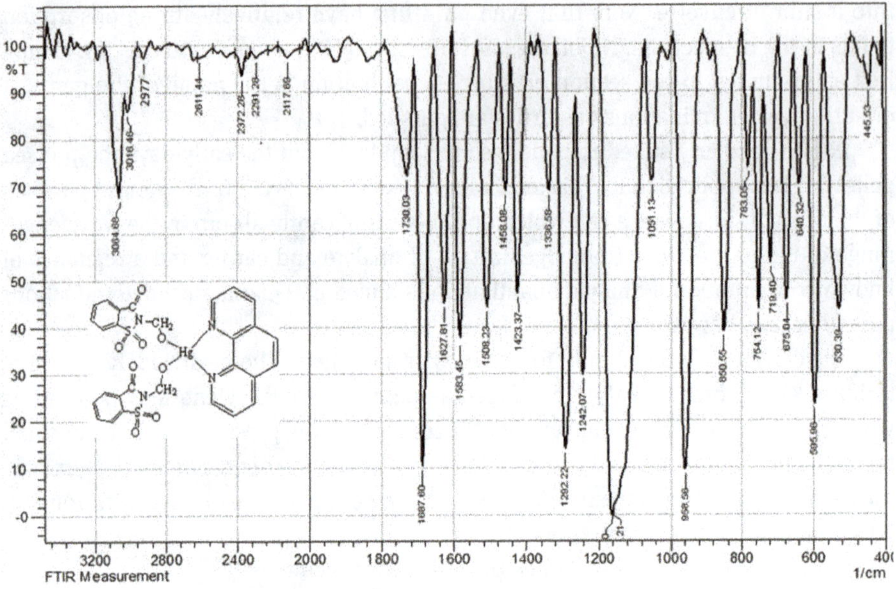

Figure 2.66: Complex IR spectrum. Source [103]; researchgate.net.

assignment the user will find a wealth of spectral information – of several hundred functional groups – for manual selection in the dictionary.

As discussed already, the chemical activity of the analyte compounds, and their interaction with solvent, may generate subtle changes to relative size (transmission) and exact wavelength position. The dictionaries use standardized data which may have to be modified according to the users' requirements. Modifications, however, are following rules which are software inherent as well. If the user is changing the data dictionary according to his requirement, he will receive guidance by the software inherent algorithms again. The presentation of automatic or manual assignment, two- and three-dimensional presentation of spectra, and design of reporting are important functions of software packages as well. The same is true for statistical modules for data interpretation and quantitative analysis.

2.1.11.5 FTIR coupled with microscopy

Coupling of microscopy and FT-IR spectroscopy is a powerful method to obtain chemical information about the sample. The infrared detector may detect light at a single point, a linear array, or a focal plane array. The sample under investigation is illuminated by IR radiation. With the help of microscopic magnification micro samples can be investigated and parts of samples, e.g., body tissue can be scanned with utmost precision. Light from the source is limited by a variable aperture. The sample can be analyzed in transmission or reflection mode. The optical system is based on reflecting mirrors as quartz or glass lenses are not transparent for IR radiation (remember the section on cuvettes!). Usually a Cassegrain-type optical system is used where the beam from the source is folded back by a condenser mirror to the focusing mirror and from there focused onto the sample (see Figure 2.67). If the light is transmitted, the beam is guided through a second Cassegrain optical branch to the detector. In reflection mode, the part of the monochromator which is focusing the light onto the sample acts as collector for the reflected light which is guided on the detector with a semi-permeable mirror. The resolution of infrared microscopy (the microscopic magnification) is limited by the diffraction of the wavelength of the IR light. Most IR microscopes in practical use have a maximal spatial resolution of the wavelength of the light source, i.e., 5–30 µm.

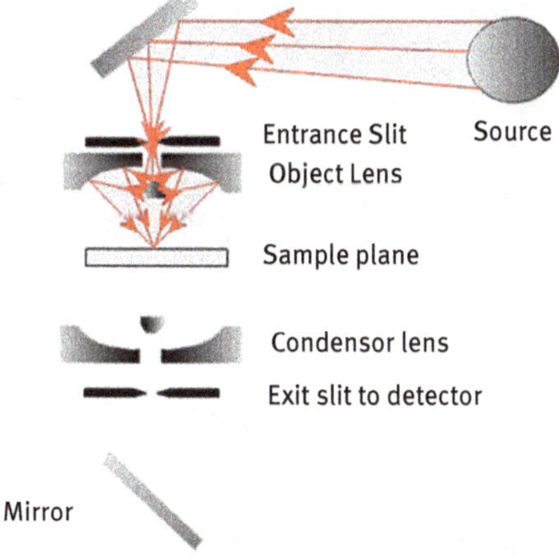

Figure 2.67: Cassegrain optics for IR microscopy. Source: Chemgapedia modified.

If the light from the sample is detected with an array detector, a two-dimensional image of the transmitted or reflected IR radiation is obtained. Additionally, this technology can be combined with a fast controlled movement of the interferometer mirror (step scan spectrometry). This will add an additional dimension to the data, namely the exact time at a certain wavelength. Time-dependent processes can be measured using the latter technique. Typical applications for the FTIR technique coupled to microscopy are:

- Medical diagnostic tool to distinguish sound body tissue from marred body tissue
- Analysis of microsamples in forensic applications (e.g., drug crystals)
- Contamination control in semiconductor chips
- Material science in general (polymer foils, emulsions, etc.)

2.1.11.6 N-IR spectroscopy

The near infrared ranges from 760 to 2,500 nm (13,000 and 4,000 cm^{-1}). It overlaps with the UV-VIS technology in the lower range and with the classical FT-IR instrumentation toward longer wavelength. Both types of instruments (UV instruments and IR instruments) are frequently equipped to handle the N-IR range. The wavelength range does not need specific technical components required in the UV range or starting from the mid-IR range. N-IR instrumentation can therefore be technically comparatively simple with respect to light source, optics, and detection. Flexible components such as wave guides can be used. There are numerous instruments on the market which are specified just for these wavelengths. Instruments which extend the wavelength range from UV-VIS into the N-IR range are usually equipped with classical monochromators (e.g., Czerny-Turner). IR instruments extending into the shorter wavelength range and instruments just covering the N-IR range are often based on interferometry. Other, even simpler monochromator setups are possible as well.

Radiation with energies higher than the mid-IR range is stimulating harmonic components of the fundamental movements discussed above. Higher energies facilitate a better penetration of the waves into sample which makes possible analysis through thicker sample layers. The combination of vibration and rotations, however, lead to rather broad bands in the spectrum. OH-, NH-, and CH-groups show strong absorption bands. Water is absorbing very strongly in the range of 1,900 nm and 1,440 nm. The latter peak is often used for the qualitative determination of water in all types of inorganic and organic samples. Other typical applications are based on carboxy groups, all types of hydroxy groups and CH- bonds. Proteins, albumen, amino groups, fibers, etc. are quality controlled by NIR. The width of the bands and the intricacy of spectral overlap require chemometric methods for spectra evaluation, interpretation, and quantitation. NIR became a widely used analytical method with, and due to the help of these mathematical tools. It is used in many fields of routine controls in food industry and chemistry, pharmacy, waste separation, etc.

2.1.11.7 F-IR spectroscopy

Far-infrared (F-IR) spectroscopy covers the spectral range beyond 25 μm (400 cm^{-1}) up to 1,000 μm (10 cm^{-1}). It requires specific radiation sources and detectors. Lamps used in the shorter wavelength are often based on thermal effects. As the temperature related to F-IR wavelengths is extremely low, this way of illuminating the sample would no longer work out. Gigahertz and Megahertz radiation is produced electronically but the F-IR range (the range of Terahertz frequency) lacks suitable radiation sources. Photon mixing [105] is a relatively recent way to produce high-intensity radiation in this wavelength range. Standard sources, such as a mercury pressure lamp emit radiation at these long wavelengths, but with low intensity. The same challenges can be found for detectors of F-IR radiation. A widely used detector is the so-called Golay detector, based on movements of a blackened membrane due to the F-IR radiation. The cyclic vibration is matched with the speed of a chopper and read out [106]. Other types of detectors are bolometers. F-IR is mainly used for structural investigations of ions, hydrogen bonds, lattice vibrations of crystals, etc. It has become a very valuable tool for structural investigations of biochemical samples and body tissue. It is used as body scanner for security purposes. Its field of quantitative measurements is very limited though and further discussion would be out of scope of this book.

2.1.11.8 ATR infrared spektroskopie

Light can be totally reflected at surfaces such as prisms and crystals. Thus, light should not be present behind the reflecting surface but it can in fact be detected. This disappearing (evanescent wave) is evolving in the medium behind the layer where total reflection takes place, and it decays there exponentially. The effect is sketched in Figure 2.68.

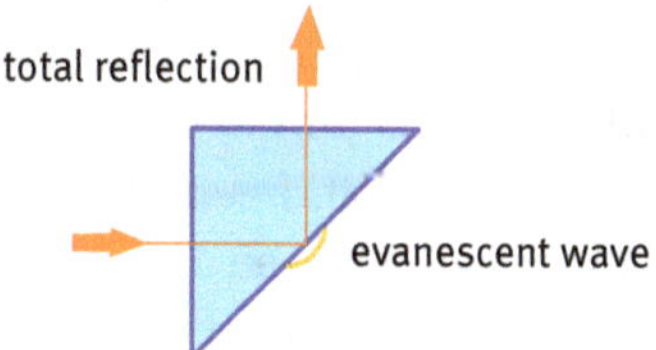

Figure 2.68: Evanescent radiation behind a prism where total reflection takes place.

Evanescent radiation must be picked up very close to the surface where total reflection takes place, usually in the range of ¼ of the radiation wavelength used for illumination. Other factors defining this distance are the angle of illumination/reflection, the refractive index between surface, and the surrounding medium. If light at a wavelength of 2,000 nm is used, the reflected light may be picked up at 500 nm = 0.5 µm. This evanescent wave can be picked up and guided into a sample which is otherwise impermeable by light, e.g., polymers, lacquer, or solutions for measurement. It can be used to obtain important analytical information. The technique is known as attenuated total reflection-IR spectrometry. The technique requires an optical element where multiple reflections are possible. This is named internal reflection element. It may be a fiber-optic light guide or a crystal.

2.1.11.9 An example of quantitative determination

Amoxicillin is a penicillin compound widely used as an antibacterial drug. Analytical certification of the drug is important to assure and secure the efficiency of the antibiotic. Different methods for the determination of amoxicillin are available, among them chromatographic methods like HPLC. These methods, however, are based on extraction with potentially toxic solvents. In addition, they are not focused on quantification in capsules. The reported IR-based method [104] uses the genuine tablet as the sample. One tablet usually contains 500–1,000 mg amoxicillin trihydrate as the active ingredient. The ingredient is, among others, mainly embedded in matrix containing sugar compounds, stearate, and surface-active agent. The matrix should be known to avoid spectral interferences. Pure amoxicillin is required to prepare the standards for calibration. About 2 mg of the standard and of the tablet are powdered, homogenized, and pressed with KBr to obtain translucent standard and analyte tablets of 150 mg. These are inserted into the holder and run in an FT-IR spectrometer.

The spectral region included in the analysis was from 4,000 to 400 cm^{-1} (the mid-infrared region). The analysis was held in transmittance, and the spectrum was obtained with the aid of "IR Solution" software (Shimadzu, Kyoto, Japan). The same procedure was performed with amoxicillin in capsules. Finally, a comparison of the spectra obtained capsule was performed to verify the similarity between them.

In addition, three independent pellets were prepared, containing each adjuvant present in the pharmaceutical dosage form (croscarmellose sodium, sodium lauryl sulfate, and magnesium stearate), in a concentration of 1.0 mg/pellet. For this, 1.0 mg of each component was homogenized with 149.0 mg of KBr. Each mixture was compressed by a mechanical press for 10 min to obtain translucent pellets. The spectral region included in the analysis was from 4,000 cm^{-1} (2.5 µm) to 400 cm^{-1} (25 µm), the mid-infrared region. From the spectrum the best region, free from matrix interferences is selected. The authors identified the range between 1,815 and 1,736 cm^{-1} to be suited for the quantitative analysis. This contains a specific carbonyl band of the amoxicillin molecule. The peak (height of the band) is used for quantitation.

The pure substance is used as a standard and a 5-point calibration curve from 0.5 to 1.5 mg per KBr pellet is constructed. The content in the sample is based on the calibration curve assuming linearity. The suitability of the method is validated by control of linearity, precision, limits of detection and quantification. Linearity is tested in the mass range mentioned above by linear regression of the calibration curve. Precision was obtained from repetitive measurements of six individually prepared pellets of the same analyte concentration using the second high standard. It was found to be roughly 4% r.s.d. The recovery is assured by a mix of the genuine tablet with the standard as % recovery. This is quasi a method of addition.

The analyses were performed in the spectral range of 1,815–1,736 cm^{-1}, and the samples were analyzed as potassium bromide pellets.

2.1.12 Raman spectroscopy

2.1.12.1 General remarks

Raman spectroscopy has become one of the fastest growing techniques in analytical instrumentation. This is mainly due to the rapid growth of information demand in bio-analytics. Raman spectroscopy is closely related to IR spectroscopy, but it provides complementary information. Thus, matrix with high IR activity which hampers the quantitation of certain molecules in IR spectroscopy may be easily detectable with the Raman technology. Water, a molecule with very high IR activity, is almost omnipresent in biological samples and a strong interferent for IR applications. Water, on the other hand is mostly Raman inactive. Biological applications are therefore easier using the Raman technology. We have discussed (see Section 2.1.11) that strong IR interaction require a shift of an electrical dipole moment while vibrations are activated by the radiation. Although interaction of IR radiation with homonuclear molecules is not completely absent, its intensity is extremely low. In the case of Raman spectroscopy, changes in the polarization of the electron shell of the entire molecule provides the measurable modification to the incident radiation which is used for analytical purposes. The Raman effect is virtually independent of the wavelength of the incident radiation. Usually, the visible wavelength range and the NIR range are used for illumination. Although the Raman effect was predicted about 100 years ago and experimentally confirmed [107] around 90 years ago, the use as analytical method was very limited and thus scarce because of the extreme weakness of the Raman-scattered radiation. With the straight road of success of lasers, very intense monochromatic radiation became accessible at tolerable cost. The instrumental method was quasi re-invented about 50 years ago.

Laser radiation of, say, 1,064 nm (Nd-YAG-laser radiation) or 532 nm (Nd-YAG frequency doubled), light of a He-Ne-lase at about 633 nm or radiation of a Ga-As diode laser in the range of about 780 nm is directed on the sample. Lasers provide a parallel light beam. Focused illumination is therefore easy. Most of the light passes the sample without interaction, but less than 1 per mil is scattered in all directions. This Rayleigh-Scattering (see Section 2.1.6.2 AAS) is due to elastic scattering of photons at molecules. The wavelength of this scattered radiation is identical to the incident laser wavelength. A minimal fraction of the radiation (in the range of 1/100) is inelastic scattering. This is the so-called Raman-scattering. This radiation carries the information about the interacting molecule. Bonding electrons of the molecule as well as the core structure are moving periodically. Unlike in IR spectroscopy, symmetrical bonds such as benzene or the C = C double bond are strong bands in Raman spectroscopy. If we observe the spectrum of a Raman-active substance, we can observe the following spectrum (Figure 2.69a).

The spectrum is explained with the following processes (see Figures 2.69b):

1. The molecule is interacting with photons of the source and is excited to a defined higher energy state. The excited molecule emits radiation and returns to the original energy state. Adsorbed and emitted energy are identical, the interaction is elastic: Rayleigh scattering.
2. The molecule is interacting with photons of the source and is exited to a defined higher energy state. The excited molecule emits radiation at lower energy (longer wavelength, smaller wavenumbers) and remains in an exited state of

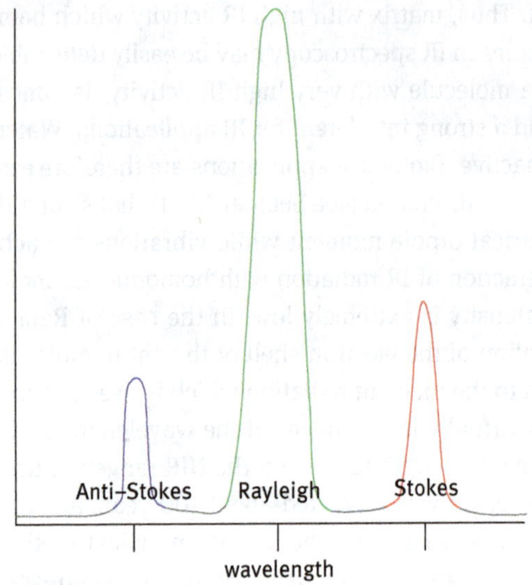

Figure 2.69a: Typical Raman spectrum, intensity versus wavelength, with only one Stokes and one anti-Stokes band. Rayleigh scattering in the center. If the Raman spectrum is displayed using wave numbers, which is often the selected diagram, Stokes and Anti-Stokes lines exchange their position relative to the Rayleigh line. The incident radiation generates a strong signal due to Rayleigh scattering. The Raman scattering generates at least two peaks located symmetrically left and right of the incident intensity. These peaks represent lower energy (longer wavelength, lower wavenumbers) and higher energy (shorter wavelength, higher wavenumbers).

molecular vibrations. The emitted radiation is shifted by a wavenumber which is characteristic for the analyte molecule. The peak in the spectrum is called Stokes-scattered radiation.

3. The molecule is interacting with photons of the source. It is already excited (photonically or thermally) to the excited state described under 2. It is relaxing to the ground state by emitting photons of a slightly higher energy (shorter wavelength/higher wavenumber) compared to the stimulating radiation. The resulting peak is shifted symmetrically to the Stokes scattering off the Rayleigh scattering peak and is called anti-Stokes scattering. This peak is significantly smaller than the Stokes scattering. Stokes and anti-Stokes bands are emission phenomena!

The spectra are often presented in wavenumbers (cm^{-1}) where the Rayleigh scattering peak is set to zero. Lower energy Raman scattering (Stokes peaks) is defined with minus, higher energy scattering (anti-Stokes bands) is defined with plus (increase in wavenumbers). The typical wavelength shift from the incident radiation is usually in the range of more than 50 cm^{-1} corresponding to roughly 10 nm on the wavelength scale.

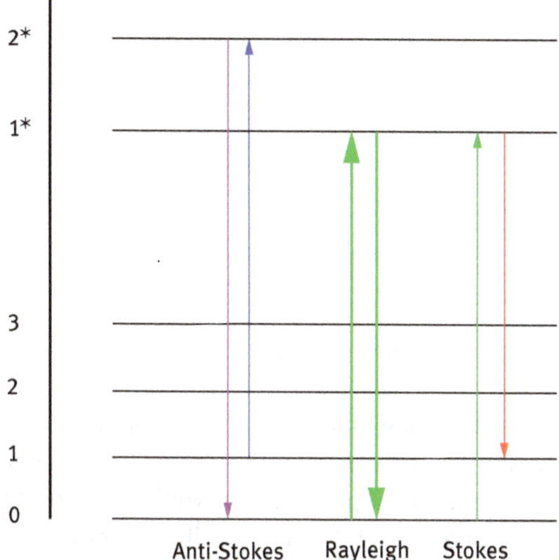

Figure 2.69b: Scattering processes in Raman spectroscopy. The initial radiation leads to a virtual energy state 1* (broad green arrow upward) and a strong scattering emission (broad green arrow downward) and to a weak Raman scattering to an elevated energy mode of the molecule (1) (red arrow downward). Weak excitation from an elevated energy mode (1) to a virtual elevated energy state 2* (blue arrow) results in higher energy anti-Stokes Raman scattering to the ground level (violet arrow).

2.1.12.2 Technical layout

The incident radiation for Raman scattering must be strong. Laser light has almost completely replaced other sources of stimulating light. The source is intense and emits a parallel monochromatic beam. A frequently used laser is the Nd-YAG-laser with a primary wavelength at 1064 nm (9398 cm^{-1}). The laser is high in energy, moderate in cost and very stable in emission. The relatively long wavelength, however, generates substantial heat in the sample which may impose a potential problem for some samples (see the discussion on temperature effects on the spectrum below.) The Rayleigh scattering peak is way more intense than the analytically useful bands. It is therefore important to minimize the incident peak with the help of suitable filters. The remaining radiation is usually separated with a Fourier-Transform spectrometer like that of complementary IR-spectrometry. The wavelength range of the spectrometer, however, is in the NIR or even visible part of the spectrum as the major lines of interest display bands in this range. Its ultimate layout obviously depends on the wavelength of the source used for illumination.

The setup of a Raman spectrometer is sketched in Figure 2.70. The arrangement of light source, optics and detector is like that of an FT-IR spectrometer. Limitation of the wavelength allows simplification of the instrument size, the optical components (e.g., the use of glass fibers) and the detector. The Raman spectrometer often resembles the N-IR spectrometer more than the M-IR instruments. It may be

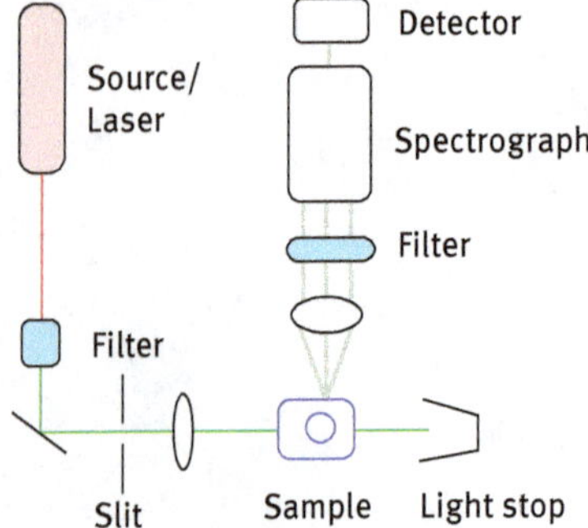

Figure 2.70: Sketch of the optical arrangement of a Raman Fourier spectrometer.

portable and even independent of standard voltage mains. As also required in fluorescence, separation of incident radiation from the scattered light or fluorescent light is of utmost importance. Straylight inside the spectrometer would deteriorate the detection capability drastically. Even more so as the Raman light is only one millionth or less of the incident radiation. The scattered light is therefore collected under an angle of 90° or completely turned backwards from the incident direction. A second most important component is the filter used to block the Rayleigh scattered radiation. A wide range of very efficient compounds with a pass/block ratio of $1/10^6$ is meanwhile available at tolerable cost.

Like in IR, solid, liquid, or gaseous samples can be analyzed. The requirement for the cuvette material is less stringent compared to the UV or M-IR wavelength range. Due to the still limited sensitivity of the Raman process, measurement in gas samples is less frequent, however.

Despite the very intense laser light sources, quantitative analysis with conventional Raman spectrometry is far from trace or ultra-trace ranges. Detection limits are usually in the upper ppm range. The reason is that from the incident photons between 10^{-6} and 10^{-8} photons only result in Raman scattered intensity. Additionally, strong illumination stimulates fluorescence which is often stronger than the emitted Raman radiation.

When the stimulating wavelength approaches the wavelength of an analyte molecular transition, the signal intensity is enhanced. In case of a wavelength match this may be an enhancement of up to 6 orders of magnitude. This resonance Raman effect is used to bring Raman spectroscopy into the range of trace determinations.

The probably most important finding for the development of ultra-sensitive Raman determinations was the observation that analyte molecules in close contact to specific (mostly noble metal) surfaces show strongly enhanced signals. These findings date back about 50 years [108] and were systematically investigated and used for analytical purposes since about 30 years [109]. One mechanism of this effect is related to plasmons on metallic surfaces. The electrons on the metallic surfaces interact and modify the radiation sent to or emitted from the molecule. Depending on the geometry of the surface and the radiation frequency and incidence angle, this can lead to a massive amplification of the field strengths at the molecule. Thus the molecule is effectively coupled much more strongly to the radiation field. For Raman emission this amplification increases absorption and emission efficiency. It enters quadratically into the strength of the effect, while normal fluorescence is only amplified linearly. At the same time the effect is potentially reduced by a decrease in quantum efficiency due to the metal surface.

If the analyte molecule is directly adsorbed to the metallic surface, it acts no longer as a free molecule but as a molecule-metal-surface complex with modified electronic structure and possible electronic transitions between orbitals. The chemical SERS effect is the other mechanism discussed for the observed strong enhancement of the Raman scattering.

The sensitivity of Raman scattering in the vicinity of small metallic particles with large surface may increase by 6 to 14 orders of magnitude, making the technique useful for ultra-trace analysis. It has been reported [110] that, under favorable conditions, even single molecules can be detected. One additional important effect of SERS is the extinction of fluorescence by metallic surfaces close to the active molecule and thus the minimization of unwanted stray radiation (see above).

To make reproducible SERS possible, the analyte must be brought into direct contact with micro-structured surfaces, electrodes with nano-structured metal surface, or solutions containing the activating agent in nano-particulate suspensions. These suspensions of metal particles – mainly suspended gold or silver nano particles – are commercially available. It is obvious that matrix molecules, solvent, temperature and many more factors may change the interaction of the analyte molecule with the activating surface. SERS is therefore a powerful tool for analytical Raman applications but, in most cases, reliable results require sophisticated experimental protocols and calibration strategies of highly sophisticated and specialized laboratories.

2.1.12.3 Application

Raman spectroscopy is used, amongst others, in the fields of pharmacy, food-chemistry, semiconductor industry, paint industry. As a typical example, the quantitative determination of designer drugs is described by F. Stahlkopf [111].

The determination of the level of active agent is, among other things, important to state the litigability of the possession of a drug. Designer drugs are intoxicants which are chemically modified such that they are not listed in the drug register yet. Cocaine and amphetamines are used as model samples in the dissertation mentioned above.

The powdered drugs and mixes of the drugs with blend are analyzed as solid powder. Sample preparation requires a thorough homogenization of the sample. The sample can be handled on a microscope. Small quantities are focused with a microscope on a sample carrier, bigger quantities can be filled in a transparent cup or sheet and positioned in an accurately defined distance from the laser source.

Raman spectrometers are often equipped with a permanent standard, which allows instrument performance verification prior to the measurement. This is often a polystyrene sample disk. Peak positions intensity and reproducibility of the measurement are determined prior to the actual measurement.

If a class of substances is to be identified or quantitatively determined, the position of the Raman peaks are recorded and compared to spectra bank data, if available. The analyst generates a genuine data base. This can be used for identification of the type of drug (the substance class) and designed derivates. Amphetamine sulfate and cocaine hydrochloride have been used for quantitative analysis in the example described above.

For quantitative analysis reference samples must be prepared which span the concentration range of interest. The sample concentrations should be above the lowest reference and below the highest reference sample. In the case of drugs, mixtures with the blend substances, such as lactose, caffeine, and phenacetine, must be prepared and used in various combinations to assure the accuracy of the measurement. The freedom from cross interferences from the blend substances must be assured. With this base of information, various real samples have been determined and successfully characterized. The results were compared to classical gas chromatographic methods.

It must be pointed out that the figures of merit of the direct spectroscopic approach do not meet the classical methods using chromatographic separation and mass spectroscopy as detector. However, sample preparation is significantly easier, and an approximate result is obtainable rapidly and with portable instrumentation. As discussed already, the specification of the figures of merit for the specific situation or the purpose of analysis is extremely important. Speed and cost must be balanced against precision and accuracy. The more complex the interpretation of the spectra and the more subtle the quantitation of spectra the more important become chemometric methods. This approach is characteristic for molecular spectroscopy in general and for Raman spectroscopy in particular. Partial least squares analysis and principal component analysis are often standard procedures inherent in the instrument software packages.

2.1.13 Molecular fluorescence

2.1.13.1 Fundamental considerations

Molecules possess multiple states of excitation next to the energetically lowest state. Most of the transitions take place without the emission of visible light. The relaxation is thermal. Relaxation often results in a split of chemical bonds. Molecules may emit a characteristic spectrum upon transition to a lower state of energy as well. The process of molecular light emission is named luminescence. Fluorescence is a part of the observed phenomena, widely used in analytical spectroscopy.

The excitation to the light emitting state may be photonic, thermal, or by chemical reaction. Fluorescence phenomena have been discussed in the section on atomic fluorescence spectroscopy (AFS, Section 2.1.7). In AFS the stimulating wave and the fluorescent wave are usually resonant. Molecular fluorescence is mostly detected at a longer wavelength compared to the incident radiation (Stokes shift). A simplified absorption and fluorescence process is sketched in Figure 2.71. Just like in molecular absorption spectroscopy the number of possible transitions is much higher than in atomic spectroscopy. This leads to dense spectra with small energy difference between the individual lines. Those are usually not resolved but evaluated as a broad band peak. Unlike in absorption where the incident radiation does not change the absorbance, the fluorescence is proportional to the incident radiation. Other parameters defining the emitted fluorescence are – just like in absorbance – the molar spectral absorbance, the radiation beam length (width of the cuvette), the concentration of the analyte and the fluorescence quantum efficiency, the ratio between emitted photons absorbed from the incident radiation (eq. (2.26)):

Equation 2.26:

$$I_F \quad I_I \cdot \varepsilon \cdot Q \cdot d \cdot c \tag{2.26}$$

I_F is the intensity of the fluorescence, I_I the intensity of the incident radiation, ε is the molar spectral absorbance, Q is the ratio between emitted and absorbed photons (the quantum efficiency), d is the width of the transmitted layer, and c is the analyte concentration. The proportionality includes the efficiency of the instrument used for measurement. It is determined experimentally. The quantum efficiency ($0 \leq Q \leq 1$) is defined by a quotient of fluorescence and all the other possible relaxation processes. They are proportional to the rate constants of these processes.

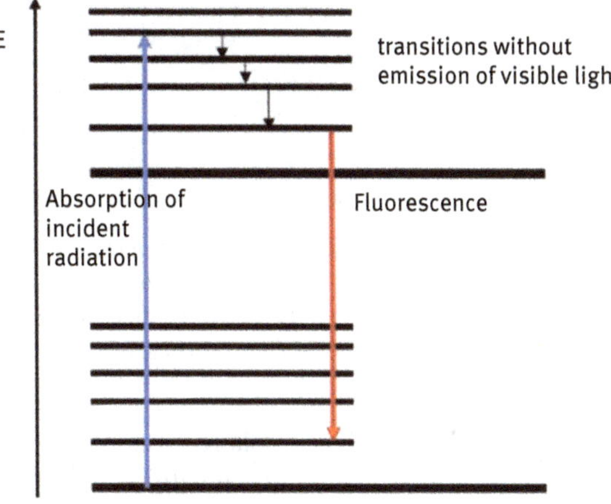

Figure 2.71: Simplified sketch of the absorption and fluorescence process in molecules.

Excitation with energy rich radiation often results in destruction of the molecule. Molecular fluorescence spectroscopy is therefore predominantly applied if excitation at longer wavelengths (usually above 300 nm) results in strong fluorescence at longer wavelengths. The electronic systems of molecules which show strong fluorescence are usually systems with many double bonds, more specific aromatic compounds. As relaxation occurs predominantly without fluorescence, if free rotations and other movement within the molecule are possible, rigid aromatic ring compounds will provide the most intense fluorescence.

2.1.13.2 Technical layout

Molecular fluorescence spectroscopy is often very strongly application focused. If specific molecules are traced, a very specific wavelength may be used for excitation and fluorescence at a specific wavelength range collected. This will obviously influence the radiation source, and the wavelength selection for source input, and fluorescence output. Universal molecular fluorescence spectrometers are relatively scarce in the market. Considering the restricted wavelength range, an instrument for universal use in molecular fluorescence may be characterized by Figure 2.72.

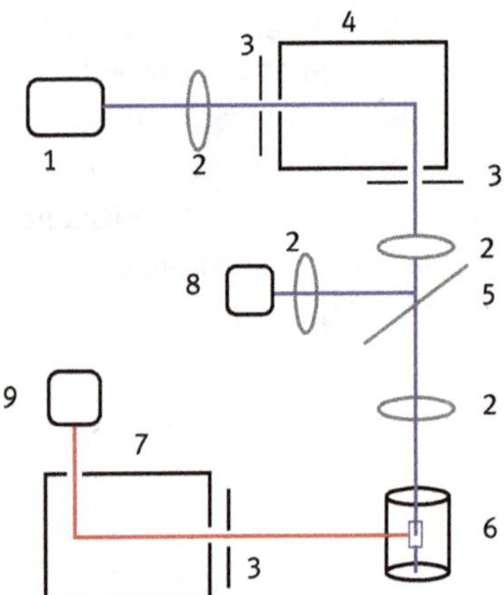

Figure 2.72: Wide range molecular fluorescence spectrometer.
1, Light source, usually halogen lamp; 2, beam focusing elements; 3 slit; 4, monochromator for exciting radiation; 5 beam separation with semipermeable mirror or rotating mirror; 6, sample chamber with sample; 7, monochromator for luminescent light; 8, reference detector measuring the intensity of the exciting radiation; 9, detector for quantitation of luminescence.

One major application of molecular fluorescence spectroscopy is real time PCR (see Chapter 3). The amplification process of a target molecule is tracked with the help of a fluorescent dye. The intensity of the fluorescent dye indicates the amount of nucleic acid produced simultaneous to the amplification. In this case the spectrometer must illuminate a standard microtiter plate of, say, 48 positions, and read out the fluorescence of the individual positions. A possible setup is sketched in Figure 2.73. An array of 48 light sources holds 48 light emitting diodes (LED). The narrow band emission of the diodes is guided to the sample containers in the plate with a suitable optics. This may be a fixed optical system as displayed in the sketch or it may be a suite of optical fibers. A second, third, fourth array of diodes may be used to excite other dyes at different wavelengths. The sample is illuminated for a few milliseconds with variable lengths of time to control and match the amount of fluorescent light which obviously increases strongly during amplification. The fluorescence is again imaged well by well onto a CCD with an optical system. Suitable filters limit the wavelength range in order to minimize stray light.

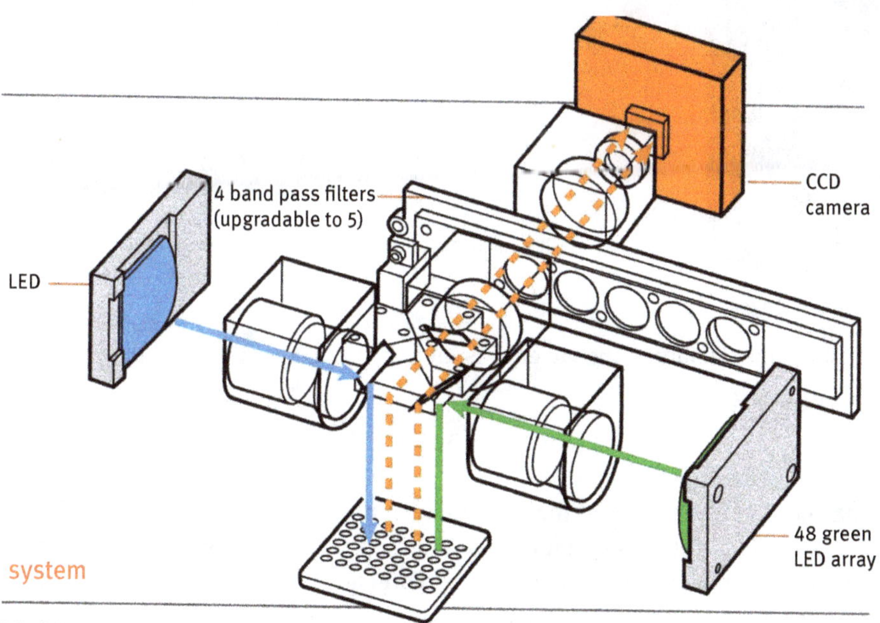

Figure 2.73: Fluorescence spectrometer for real time PCR Techne Prime. Courtesy of Cole Parmer, Staffordshire, UK.

In addition to the setup for universal applications and the multiple readout of many sample cups in parallel, a broad range of systems designed for dedicated applications are available. As molecular fluorescence focuses often on bioanalytical applications, minimal sample mass/volume became an important specification. Like in UV/VIS spectroscopy, this can be realized with cuvettes which hold only microliter volumes. An example is displayed in Figure 2.74.

Other than in UV/VIS applications where volumes of 1 µL became possible, fluorescence usually is limited to volumes of at least 50 µL. The demand on excellent

Figure 2.74: Ultra-micro cuvette for fluorescence spectroscopy. Courtesy of Hellma GmbH, Müllheim, Germany.

transmission stability of the cuvettes is high as it directly influences the analytical result.

2.1.13.3 Measurement correction

It has been emphasized that the magnitude of absorption is intrinsically independent of the light flux of the source. The very opposite is fluorescence, where emission is proportional to illumination. Therefore, additional parameters need to be controlled to minimize systematic errors. The effects on emission intensity may be instrument or sample induced. In the first case systematic errors are scarce as the effects can usually be calibrated or corrected by technical means. Sample-induced errors need to be corrected with methods already discussed in earlier sections. Instrumental effects are predominantly caused by variations of the lamp intensity. Excitation takes place in a spectral window which is wider than UV and AAS. Double beaming sketched in Figure 2.72 makes use of the same spectral window as excitation and corrects for the changes in lamp intensity with high speed. Changes in the transmissivity of optical components, such as filters, as well as quantum efficiency of the detector as a function of wavelength will influence the absolute intensity of the detected fluorescence. This will influence detection limit and reproducibility but not the analytical result.

Interferences due to the sample composition are frequent:

- Fluorescence of the chromophore may be time-dependent as the molecule can be gradually decomposed by the incident radiation. The effect may be influenced by the surrounding matrix including the solvent.
- Rayleigh or Raman scattering of the incident light are processes taking place in parallel to fluorescence. Their magnitude depends on matrix composition.
- The intensity of the incident radiation is weakened by absorption of matrix molecules and chromophore. The effect depends on the concentration of these molecules.
- Partial absorption of the emitted fluorescence by matrix molecules or the chromophore itself may be distinct and may lead to complete quenching of the fluorescence. The effect depends on the molecular structure of the matrix, the concentration of concomitants and the concentration of the fluorophore.

In all these cases systematic errors can be minimized by proper calibration. As usual, sample and calibration solutions should be as similar as possible in composition.

2.1.13.4 Phosphorescence and chemiluminescence

The main difference between phosphorescence and fluorescence is the time between excitation and decay of the emission. Fluorescence is fast. It happens within less than a millisecond. Phosphorescence is a slow process which last for seconds up to hours. The electrons are excited from ground state S_0 into the excited state S_1 as shown above. The transition is probable as S_1 is a singlet state. From there the electrons may relax rapidly with a fast probable transition (fluorescence) or they may move via a non-probable transition into a slightly lower energy state (triplet state). The process is called intersystem crossing. From there the transitions to the ground state are not probable any more. The result is a long-lasting weak process. Phosphorescence of organic compounds takes place predominantly in solids. It is seldom used for quantitative analysis.

Chemical reactions between a compound A and a reactant B may result in an energetically elevated intermediate compound AB^*. Most reactions from AB^* to a lower energy end product AB will generate molecular movement and will express itself as heat. Some compounds, however, will release a small part of the energy as photons. The relaxation of the excited electrons may be fast (fluorescence, probable transition) or slow (phosphorescence, improbable transition). The emitted light can be detected and quantitated like fluorescence excited by photons. In the case of chemiluminescence, the light source is quasi replaced by a suited reactant A for the detection of analyte B. The excitation to the energetically higher intermediate may be stimulated with electrical charges as well. In this case the term "electrochemiluminescence" is used. The method is frequently used for the detection of hazardous compounds such as NO_X, gaseous sulfur compounds or ozone in gases (air) or liquids. The detection limits for NO, for example, are as low as ppb or even below

[112]. Chemiluminescence is a process which would better fit into Chapter 3 of the text. The detection, however, is like described above.

2.1.13.5 An application example

Beer is produced in high quality according to "Reinheitsgebot" since the fourteenth century. A clearly defined number of ingredients (water, malt, yeast, hop) made production in simple breweries possible. There was no need for analytical control, but the processes were clearly defined using strict recipes. Deviations from the taste, such as bitterness, could be detected organoleptically only in the final product. Industrial beer production needs to control several parameters which influence the taste and quality of the product. Bitterness, among others, is a key characteristic of beer products. Bitterness is mainly defined by the amount, the quality, the form of hop, and by the temperature and time of the boiling and fermentation processes. Although highly automated, disruptions within the processes may lead to a product with unwanted properties. Tracking of the parameters during production are therefore necessary. The methods applied should be simple concerning sample preparation, should deliver fast results and ought to be economic. The number of reagents applied should be minimal. Bitterness is mainly defined by so-called iso-acids (IAS). They are formed during the boiling process, by extraction from the added hop. They define taste, help to preserve the beer, and are involved in the formation of a stable froth. Iso-acids are compounds such as the α-lupulinic acid, α-hop bitter acid, and α-humulone.

The standardized methods for the determination of bitterness are complex. The stipulated method is based on extraction of the acids from beer with iso-octane. Extractant and sample are shaken thoroughly. The phases are separated and the iso acids in the organic phase are quantitated with UV spectrometry. The method is time-consuming; the process of extraction is complex, and the method requires a relatively large amount of chemicals which need to be depolluted. A second standard is based on separation of the iso acids by high-performance liquid chromatography (HPLC) followed by UV detection. The cost of instrumentation and the complexity of the determination is too high for small laboratories in small breweries.

A simple method has therefore been developed by Wilke and coworkers [113, 114]. Iso-acids cannot be activated to intrinsic intense fluorescence. Beer is a complex matrix. Dissolved CO_2 makes volumetric measurement difficult. The color of beer and the matrix mix results in strong absorption in the wavelength range used for excitation and detection of the fluorescent light. If treated with an activating reagent, strong and specific fluorescence can be detected, however. The activating reagent found by the authors are lanthanoid ions, in specific Eu and Dy. The fluorescence intensity allows a strong dilution of the matrix by at least 1:20, and up to 1:500. The authors found a strong decrease of the fluorescence signal of the complex at IAS concentrations below 5 mg/L. An additional activating reagent, trioctylphosphine oxide, helps to overcome this effect which is attributed to inhibition of complexation. With addition of these two reagents, the sample preparation as well as the detection of the fluorescence becomes very simple:

- Beer is degassed from CO_2 in an ultrasonic bath.
- 1 mL of the treated sample is diluted 1:250.
- 100 µL of a 20 mmol dysprosium solution in 1 normal HCl, and 50 µL of a 0.1% trioctylphosphine oxide solution are added to the diluted sample. The final mix has an acidity of pH 1 and pH 3.
- The mix is pipetted into a suitable cuvette.
- The sample is excited at 285 nm, fluorescence is detected at 575 nm.
- The correlation between fluorescence intensity and IAS concentration is calibrated with a suitable mix of iso-acids.

The authors correlated the result of a wide number of beer samples with the result of HPLC determination (the gold standard method) and found very good agreement. The correlation of fluorescence and the measured IAS concentration is sketched in Figure 2.75. The usual bitterness of beer ranges from 10 to 40 mg/L IAS. The method is capable of quantifying these values within tight tolerances of less than 1 mg/L IAS.

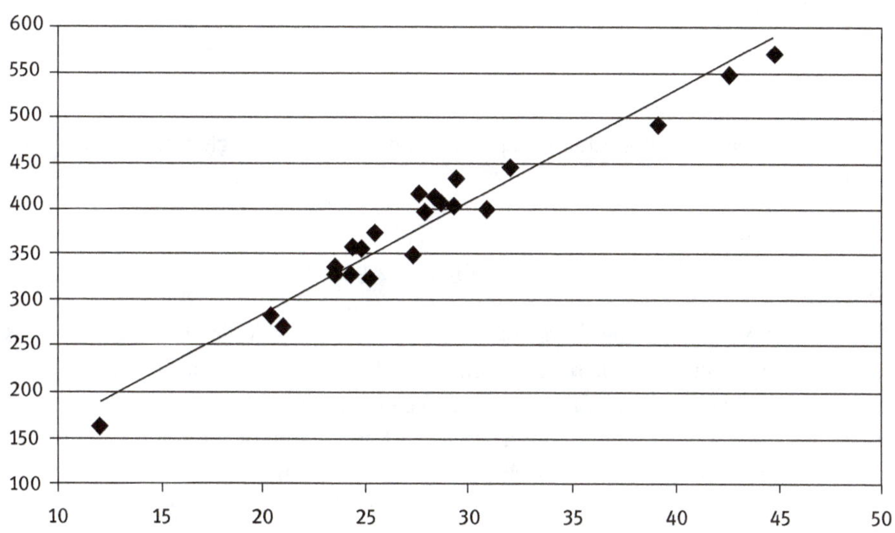

Figure 2.75: Bitterness of beer expressed as IAS concentration in mg/L (abscissa) and fluorescence intensity in arbitrary units (ordinate). Taken from [113].

The simplicity of sample preparation and measurement allows to use a simple experimental setup. A spectrometer can be simplified from the universal setup (Figure 2.72) as follows:

- The source (1) can be a simple, commercially available halogen lamp.
- Monochromators (4) and (11) are replaced by suitable filters.

- Detectors (4) and (9) are inexpensive photodiode detectors
- The cuvette (6) can be easily replaced by a flow through system

The easy and cost-efficient measurement can be run offline, or it could be automated to a completely online procedure. Although prototypes of a dedicated system have proven their accuracy and precision, the method unfortunately was never commercialized. Beer bitterness is still detected with the standardized methods sacrificing efficiency in laboratory cost and simplicity.

2.2 Mass spectrometry

2.2.1 Fundamental considerations

Ions are charge bearing atoms or molecules. The mass range starts from atomic mass 1, a hydrogen cation or anion, up to high masses of bio-organic molecules. Massive particles are following Newton's second law (eq. (2.27)).

Equation 2.27:

$$F = m \cdot a \tag{2.27}$$

Force equals mass m times acceleration a.

Charged particles are interacting with magnetic and/or electric fields. The force is the Lorentz Force (eq. (2.28)).

Equation 2.28:

$$F = q\,(E + v \times B) \tag{2.28}$$

F is the acting force. q is the charge of the particle. E is the electric field strength. $v\ x\ B$ is the vector product of current velocity v and magnetic flux density B. $v\ x\ B$ is a vector which is perpendicular to v and B. The force on an ion in an electric field is $q \cdot E$ and $q \cdot v\ x\ B$ in a magnetic field. Inertia allows spatial or temporal separation.

Mass spectroscopy separates charged particles in an electromagnetic field by speed or direction of flight. This happens via the following sequence of fundamental steps:

- In a first step the analyte atoms or molecules must be volatilized such that they are separated from their matrix.
- The second step is ionization: cations or anions are generated by electron loss or charging of molecules or atoms with an additional electron.
- The ions are accelerated into a high vacuum where collisions with other particles and among themselves are minimized. Usually, the analyte is flying fast toward a detector. However, they may be trapped in a defined space for measurement for a longer time as well.

- The ions are separated according to their speed or direction of flight in electromagnetic fields, or according to their electromagnetic behavior (frequency of rotation or oscillation).
- Ions with a particular mass/charge ratio arrive at a position, where they are detected with a suitable electronic device.

2.2.2 Instrument setup

Ion source for mass spectrometry

Mass spectrometry is selective for ions. The generation of ions which can be absorbed into a high vacuum of a mass spectrometer is often a challenge. Details will be discussed in the respective sessions of the individual techniques. Elemental ions are generated under ambient pressure in a plasma source, mostly an inductively coupled argon plasma, in an electrical arc or spark or by excitation with laser light. In organic mass spectrometry, the substance is volatilized and ionized by a beam of electrons interacting with the analyte molecules. Sophisticated volatilization methods have been developed to volatilize and ionize molecules originating from gas or liquid chromatography.

Optics for mass spectrometry

In mass spectrometry the analyte to be quantitated and detected is handled directly. This is a fundamental difference to optical spectroscopy where photons emitted or absorbed by the analyte are the measure. Lightweight or heavier charged particles are passed through the analytical instrument from their point of generation (ion generation) to the detector. The ions are flying in a high vacuum chamber. The means to shape, focus, speed up or slow down, and separate the particles are predominantly electromagnetic. The systems are called ion optics, referring to the photon optics of the techniques already described. It would exceed the scope of this textbook to derive the fundamental equations of the different spectrometer types from basic physics. The principles will be explained qualitatively only. The authors decided to dig deeper into the quadrupole mass analyzer. Firstly, because few users have a real understanding how the principle is working. Secondly, because this complex function principle became most successful in analytical instrumentation.

The chronologically oldest realization of mass spectrometry is a static magnetic or electric field or the combination of both. Various geometric configurations are used and have been used. They are named according to their inventors, e.g., Mattauch-Herzog, Bainbridge-Jordan, Takeshita, Nier-Johnson. Each design (which is meanwhile often modified from the original configuration) consists of a bent electric sector followed by a magnetic sector with different angle of deflection. The spatial setup and the applied electric and magnetic fields focus the ions with a certain mass/

charge ratio to an exact spot in space where they can be detected (see Figure 2.76). Just like described in optical spectroscopy, the sector field mass spectrometer is scanning from mass/charge to mass/charge.

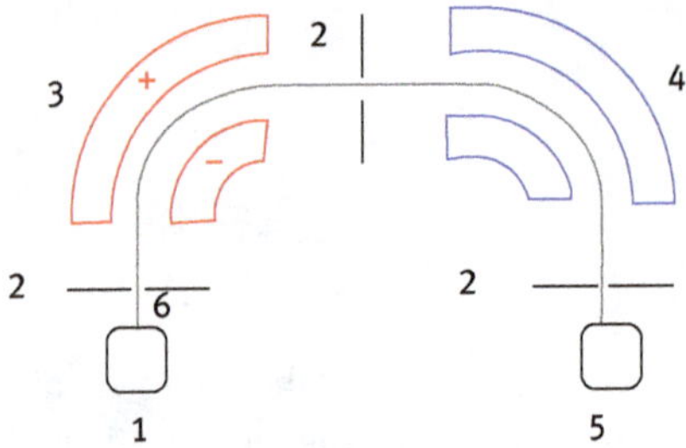

Figure 2.76: Sketch of the Nier-Johnson double focusing mass spectrometer: The ion beam (6) is generated in the ion source (1) is focused in the slit (2) and is entering the circular-shaped electric field (3), is refocused (2) into the magnetic field (4) and is entering the detector (5) via the beam slit (2).

The separation of ions with a different charge to mass ratio is achieved using a completely different mechanisms in a quadrupole MS. Whereas in sector field instruments circular motions and static electromagnetic fields are used, stability of a linear motion through the assembly is the discriminating principle. An electrical quadrupole is a (mostly) quadratic configuration of two pairs of charged electrical rods at distance a. The voltage between the positively and negatively charged electrodes consists of a static component U_0 and a superimposed radio-frequency part with amplitude V_0 and frequency $f = \omega/s2\pi$. The resulting voltage V_{res} between the electrodes is time-dependent and a sum of the two components (eq. (2.29)):

Equation 2.29: Resulting voltage in a quadrupole field with AC and DC component:

$$V_{res}(t) = U_0 + V_0 \cos(\omega \cdot t) \tag{2.29}$$

The arrangement of the electrodes, the resulting electrical field, and an ion with charge q is sketched in Figure 2.77. It shows a plane orthogonal to the central axis of the setup. This axis is also the flight direction of the ions.

An ion moving at position (x,y) close to the axis experiences an electrostatic force. The x- and y-components of this force depend on its position; for a "perfect" quadrupole the x-component only depends on x and the y-component only on y, i.e., the motions in x- and y-directions are entirely decoupled. The setup using cylindrical rods depicted above provides a good approximation to such a perfect quadrupole; even better realizations can be obtained using electrodes with hyperbolic shapes on the side facing the path of the ions.

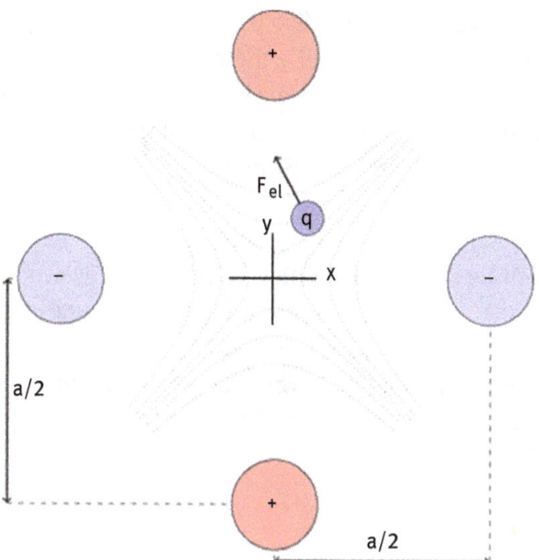

Figure 2.77: Cross section of a quadrupole arrangement with coordinate axes x,y. Ion of charge q between the electrodes. The dashed lines indicate positions with equal electrical potential. The electric field at the position of the ion (q) is perpendicular to the dashed lines.

Applying the laws of mechanics this leads to the two (decoupled) eqs. (2.30).

Equations 2.30 a, b: Differential equations describing motion of ions in x- and y-directions. C is an instrumental factor depending on the geometry of the setup:

$$\ddot{x} = -Cq/m\, V_{\text{res}}(t)\, x, \qquad \ddot{y} = +Cq/m\, V_{\text{res}}(t)\, y \tag{2.30}$$

The first equation describes the stable (linear) motion of an ion in a harmonic oscillator potential with strength modulated by the radio-frequency term. Without the superimposed radio-frequency, this would be an oscillatory motion with an intrinsic ion-specific frequency f_0, which will reappear below. The second equation describes the unstable (linear) motion of an ion in an inverted oscillator potential, again modified by the radio-frequency potential. See Figure 2.78.

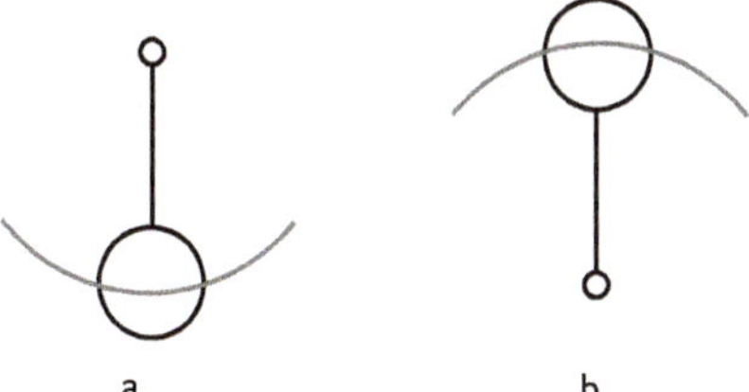

Figure 2.78: Harmonic oscillator (a) and inverted harmonic oscillator (b).

With the help of the RF field it is now possibly to stabilize the motion in the previously unstable y-direction. However, this will also modify the motion in the stable direction. As a result, one ends up with two conditions on the RF field, which include q/m. These conditions are expressed in terms of two dimensionless parameters:

- A ratio V_0/U_0 of the amplitudes of the RF and the static field
- A ratio $(f_0/f)^2$ of the intrinsic and the modulating frequency

More technically, eqs. (2.30) can both be related to Mathieu equations [115], which include two parameters that can be related to the two ratios above. Thus, one can relate stable movement conditions in the quadrupole to the existence of bounded solutions to the Mathieu equation. There are diagrams which depict the parameter regions where stability hold. Such a diagram, which shows these regions for the stable and unstable directions, is displayed in Figure 2.79.

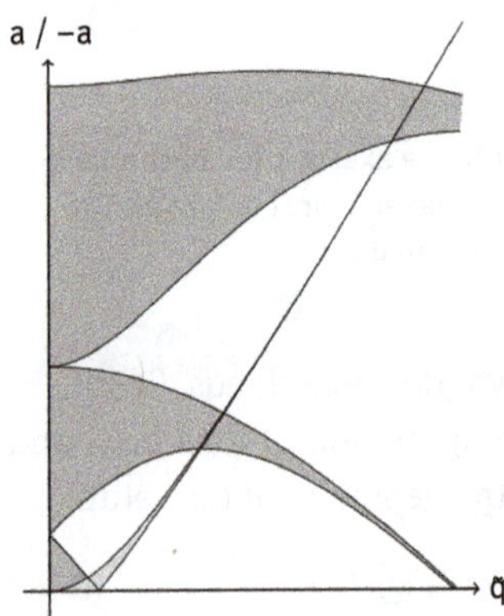

Figure 2.79: Regions of stability (dark gray for the stable, light gray for the instable axis). Note the narrow stability region for the instable direction. This is both expected and technically useful (reasonable q/m resolution).

If the parameters are outside these ranges, the motions described by the equation will typically move away from the axes, and the ions will either not make it through the exit stop, hit the electrodes, or leave the area enclosed by them altogether. In principle, there are several regions of stability for both axes. However, as the areas related the instable direction has the shape of a narrow "spike," the region chosen to operate the quadrupole is the small triangle-shaped area starting at the origin.

The coordinates of the points delimiting this area are scaled by q/m. If ions with different q/m-ratios are present, we get the following picture (Figure 2.80). To be able to separate these ions, one thus wants to scan (i.e., tune f and V_0) along a curve such as the red line in the previous picture.

The parameters in one of the triangles correspond to a q/m-ratio for one of the ion-types present as analyte ion. As long as the parameters are inside this triangle, only the corresponding ion type will move in stable oscillations along the center line of the quadrupole.

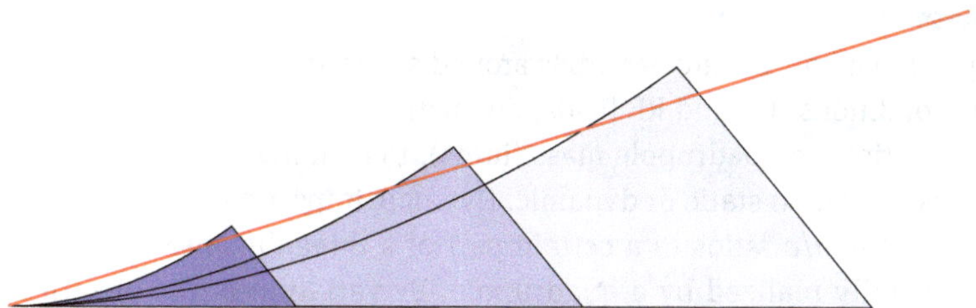

Figure 2.80: Stability regions for three ion species with different q/m ratio. The red line represents value for the parameters selected by the instrument during a scan.

Remarkable properties:

- To first order, the speed of the ions orthogonal to the x-y-plane does not matter, if they spend enough time in the trap to perform oscillations with potentially growing amplitude.
- This q/m filter does not use a combination of electric and magnetic fields, but electric fields only.

  Upper and lower limits for the q/m values compatible with a given quadrupole setting result from stability conditions and not (directly) from stops which, e.g., cut off trajectories with curvature radii above/below a certain value.

Time of flight (TOF)

Ions are started from a point zero toward a detector within a defined linear electric field. They accelerate according to their inertia. The accelerated ions move along a drift tube toward a detector. Light weight particles are faster than the heavier ones and reach the detector within a shorter time. They can be separated by time in the Time-of-flight mass spectrometer. As the ions are starting at the same time and reach the detector with a time difference of split nanoseconds they can be looked upon as simultaneous measurement device. The time resolution is similar to a polychromator in optical spectroscopy. Often the movement is not linear but curved in an electric field to increase resolution and remove matrix ions. The optics to change direction is based on an electromagnetic field and is named reflectron. It reverses the direction of flight or changes it significantly. The field strength is rapidly adjustable.

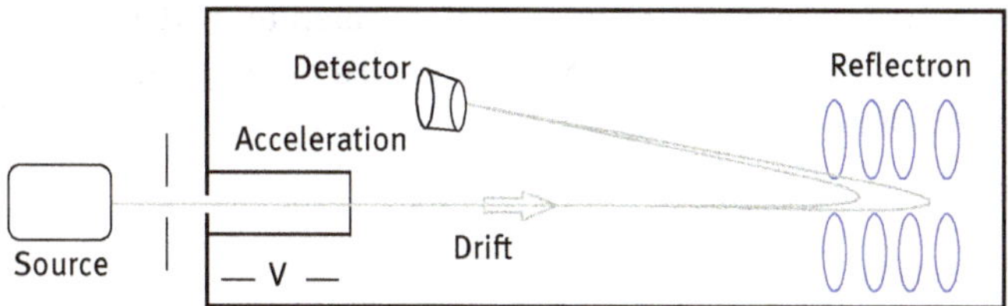

Figure 2.81: Sketch of a time-of-flight mass spectrometer with reflectron.

Ion-trap type mass filters and analyzers

Ions can be trapped in a box, on circular orbits around an electromagnetic field, if the electromagnetic conditions, DC and RF fields, are manipulated in a carefully timed manner (see the section on quadrupole mass filters). Depending on the operating principle the traps are called static or dynamic. Dynamic traps are used to concentrate ions of selectable m/e ratios in a certain part of a three-dimensional space. The trapping is usually realized by a quadrupole. By variation of the electrical conditions a defined (selected) mass charge ratio can be guided to the detector or a secondary mass selective device.

In the static situation the ions are circulating around a device generating a static magnetic or electrical field. Depending on the m/q ratio the ions are moving with a characteristic frequency (cyclotron frequency). The moving ions contain electromagnetic information which can be received and translated into mass/charge information. As an example of a static trap, the very popular Orbitrap [116] system is briefly described. A spindle-shaped electrode is surrounded by a similar shaped chamber. The ions are accelerated and enter the trap at one side of the electrode. They are circling around the electrode but, due to their inertia, are oscillating as well along the spindle (see Figure 2.82).

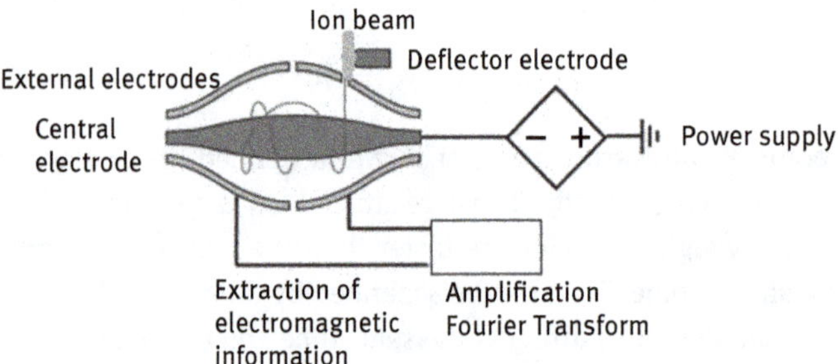

Figure 2.82: Sketch of an Orbitrap: Figure modified from literature [117].

The oscillation is depending on m/q and generates an additional electromagnetic field which can be detected by metal elements. This time information is translated into a mass information by Fourier transformation. The Orbitrap system is a FT mass spectrometer with extremely high mass resolution.

Other ion-optical components which are needed to accelerate or decelerate ions, to change the direction of flight, to reflect ions to increase the flight distance, etc. are based on electrically charged nets, rings, etc.

Detectors for mass spectrometry

Mass spectrometers detect ions. These are generated in a suitable source (see section above). The ions are separated and detected as intensity of an ion current. The detectors in mass spectrometers convert the ion current into a current of electrons. Depending on the type of mass/charge separation according to space or according to time, detectors must follow different specifications.

Just like in optical spectroscopy the first detectors were photo plates. Advantages and disadvantages are essentially identical to those discussed in optical spectroscopy.

Array detectors in mass spectrometry are different from those in optical spectrometry. Known as channel electron multiplier or micro channel plates they work like a PMT tube with many very small individual channels. The opening for the ions is the cathode, the end of the thin channel is the anode. The channels are covered with a doped coating which releases electrons when hit by an ion. The electrons move toward the anode and hit the wall again, releasing a cascade of electrons on their way. As shown in figure 2.83, the channels are shaped in a way that electrons can move rapidly while the cations are largely blocked from moving back to the cathode.

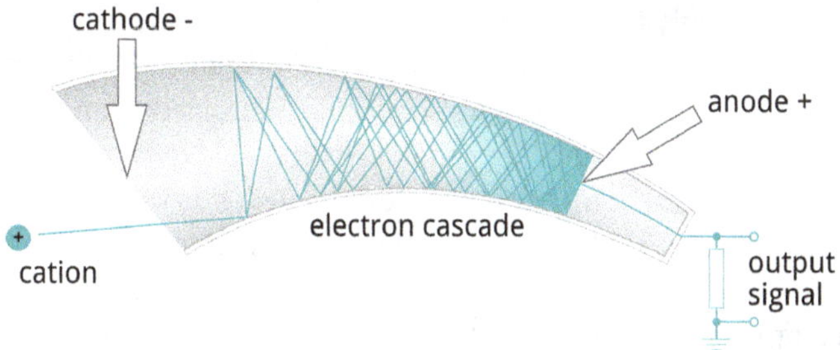

Figure 2.83: Channel electron multiplier. Source: Chemistry Libre Texts (modified).

If mass separation is realized by the time of flight only a single detector is necessary to detect the ions. The smaller the mass/charge ratio is, the higher is the speed of the ions and the shorter is the time difference between ions of similar mass. Detectors must therefore be very fast.

A fast detector type is the scintillation counter. The principle is not so different from the one described above already. The cations hit a cathode with a high negative potential. Electrons are released upon impact. They are accelerated toward a fluorescing screen where electrons are converted into photons and measured on a photomultiplier tube.

A slower detector type is the Faraday cup. The ions coming from the flight tube are guided into a small box acting as the Faraday cup. They hit a plate which is connected via a large resistor to ground. Charges collected there are compensated by currents to ground via this resistor. The voltage drop across the resistor is proportional to the current, i.e. ions per time hitting the plate.

Photomultiplier tubes are often used for mass spectrometers as well. They must be equipped with a dynode, converting ions into electrons just as described above. The photomultiplier tubes used for mass spectrometry are usually significantly more sophisticated and hence more expensive than the ones used in optical spectroscopy.

Other analytical systems which separate and detect ions, like chromatographic methods, are using the same type of detectors as described above.

2.2.3 Figures of merit in mass spectrometry

Signal to noise

The high vacuum in the detection chamber ensures a very clean environment for the detector Other than in optical spectroscopy the baseline noise on the detector is close to the detector dark noise, provided that the mass filter is very efficient in separating the mass to be determined. On the other hand, the detector is a powerful amplifier. Dark noise levels of 1 count per second (cps) are obtainable if only electronic noise is present. The sensitivity of 1 ppm of an element ion introduced to an ICP-MS is in the range of $1{,}000 \times 10^6$ counts/s. It becomes obvious that the signal to noise of mass spectrometers is usually superior to every other instrumental analytical technique. The detectors used in ICP-MS are generally vulnerable to large loads of incoming ions as the thin detector tubes are quickly overloaded. It is as well difficult to remove the ions from the tubes. The species of interest in mass spectrometry are cations, i.e., elements or molecules with a positive charge. Photons, neutral particles, electrons, anions or cations with a mass which cannot be separated by the mass filter are a burden to the detector. Requirement for low noise is therefore the minimization or absence of unwanted particles on the detector. These will increase the blank noise due to their own statistical noise. They may as well have an influence on the sensitivity of the detector and cause memory effects. Protection of the detector against overcharge is essential for signal to noise, memory, and detector lifetime. Essential technical parameters influencing FOM are mass resolution and protection of the detector by ion optical components.

Separation of cations

Separation of cations from neutral species, mainly photons, atoms, and recombined molecules as well as from negatively charged species is put into effect with the help of baffles and electrical fields which stop the linear movement of particles in the vacuum chamber of the spectrometer. The ions of interest may be turned by 90° from their original flight direction, are diverted around baffles to a parallel beam relative to their original flight direction, etc. before the bulk of the sample is entering the mass filter.

Mass resolution

We have seen that the principles of mass selection are fundamentally different in different types of mass spectrometers. They may be based on time separation, movement on different orbits or generation of stable flight conditions. All these principles provide a typical mass separation due to the technical components and mechanisms in control of the physical parameters.

Mass resolution R is a measure to quantify separation of two neighboring peaks. It is usually defined as (eq. (2.31)) the ratio between the selected mass m and the width of the peak at 5% of its maximum height.

Equation 2.31:

$$R = m/\Delta m \tag{2.31}$$

If, say, the mass under consideration is 50 and the resolution is 500, the 5% peak level (Figure 2.84) has a width of 0.1 mass units. The specificity is obviously very strongly dependent on resolution. The situation is comparable to optical spectroscopy.

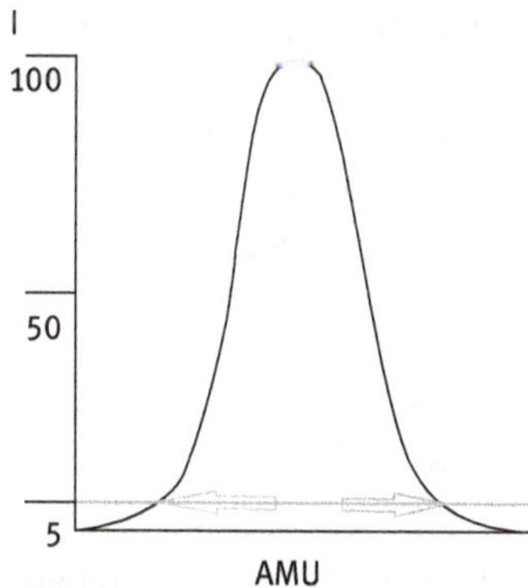

Figure 2.84: Mass peak with 5% criterion.

Besides a normalized definition which allows to check the resolution of a mass spectrometer, the analytically more important information concerns the separation of two ions with very similar mass. If both masses, m_1 and m_2 are exactly known, the virtually identical 5% criterion asks for a dip in between the peaks to the 10% level (see Figure 2.85) and eq. (2.32):

Equation 2.32:

$$R = {}^1/_2(m_1 + m_2)\,/\,(m_2 - m_1) \tag{2.32}$$

In our example from above, masses of 50 AMU and 50.1 AMU will be separated with a dip of 10% in between the two peaks. Unfortunately, the mass of interest and the unwanted mass are very seldom present in comparable concentrations. Resolution, in real life often becomes the most important factor for the limit of quantitation. The origin of spectral interferences will be discussed in the respective sections.

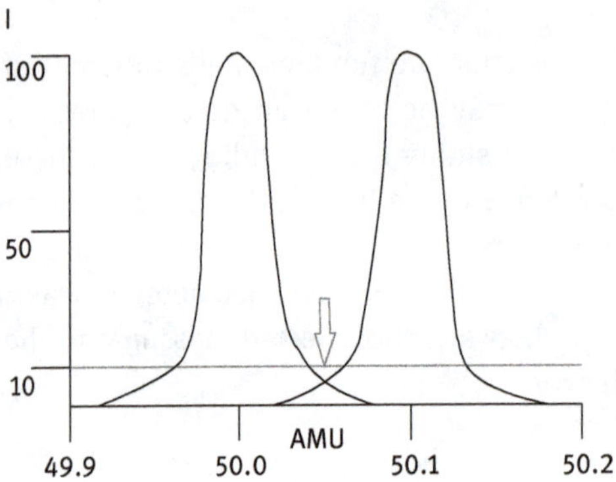

Figure 2.85: Separation of two adjacent peaks in mass spectrometry.

It has been shown that separation of masses depends on movement of a mass (quantified by impulse, energy) and interaction of the charged particle with the electromagnetic field of the instrument's ion optics. Both fundamental variables have an inherent scatter which result in a slightly different time-space behavior of the particles.

In most mass spectrometers the ions are sucked into the vacuum with a constant flow. The particles of interest are imbedded or surrounded by a carrier matrix. The difference in speed is relatively small and may be due to small variations in the ion source, the vacuum, turbulences in the inlet system into the mass spectrometer, etc. In a time-of-flight setup, the ions of interest must be stopped in a suitable compartment and must be accelerated to the inlet of the time-of-flight analyzer such that they all enter at a defined time with the same speed. The scatter in resolution originates from the scatter in penetration time into the flight tube and the scatter in speed.

The resolution of the ion optics depends on the stability of the electromagnetic conditions. As the settings must be fast in order to rapidly scan through the mass spectrum, the scatter will increase with increasing scanning speed, resulting in a lower mass unit resolution.

In a time-of-flight arrangement, the length of the flight path, the speed of the accelerated ions and the time resolution of the detector will define the mass resolution. It is obvious that heavier masses will challenge the speed of the detector electronics much less than the light elements. For elongation of the flight distance and still compact instrument dimensions, the ions are often reflected to pass through the flight tube twice (see Figure 2.81).

A magnetic sector field spectrometer discriminates by speed of the ions at the end of an acceleration segment and using an electromagnetic field. After passing the magnetic sector, the ions are at a different positions in space. They can be detected by a suitable chemical detector (a photographic plate), or by electronic detectors located at the expected positions, defined by the mass to charge ratio of the ions. These latter instruments are called multi collector simultaneous mass spectrometers. In case of a single detector the acceleration conditions and/or the magnetic field strength must be scanned to collect different m/z ratios. While setting of the magnetic field is accurate and precise, the scanning process of the magnetic field is comparatively slow. The dispersion of the speed of the accelerated ions is usually compensated with an electric filter as a kind of pre-or post-filter to the mass spectrometer. The filter consists of bent plates with a potential difference. The ions are forced to a bent propagation. Only ions with a (doubled) kinetic energy $m{\cdot}v^2$, which meet the radial requirement r of the plates with an electrical charge $q{\cdot}E$ will reach the detector (eq. (2.33)). Double focusing mass spectrometers provide excellent resolution of up to 100,000, however at the cost of count intensity. Another disadvantage of double focusing mass spectrometers is the technically limited scanning speed of an electromagnet. Scanning from one analyte mass to the other is slow, the total analysis time becomes longer, and it is difficult to determine dynamic signals with different masses.

Equation 2.33:

$$r = m \cdot v^2 / (q \cdot E) \tag{2.33}$$

As described above, quadrupole mass spectrometers discriminate by generating stable flight paths toward the detector with electrical AC and DC fields. Changing AC frequency and voltage and DC potential can be fast. Filtering according to the m/z ratio is fast but limited in resolution to tenths of an atomic mass unit.

Spectral interferences in mass spectroscopy

It has been pointed out in earlier sections (see AAS, OES) that spectral overlap is a main parameter to accuracy and to the limits of quantitation in spectroscopy. While spectral interferences in AAS can be controlled quite easily they are frequent and complex in plasma optical emission spectroscopy. ICP-MS is in between. The number of possible overlaps is much less frequent compared to OES but some of the classical interferences are very difficult to handle.

Different elements with quasi-identical m/z are one of the challenges. An often-named example is $Fe^{53.940}$ and $Cr^{53.939}$, two different elements with the nominal mass 54. It requires a resolution of about 100,000 to resolve these two peaks. The elements would therefore be measured as the sum. Depending on the concentration of analyte element and matrix element a quantitative determination would be impossible. This type of interference is called isobaric overlap. Isotope tables help to quickly reveal the possibility of interferences. Other isotopes of the interfering element help to determine the true concentration of the elements, however often at the cost of a loss in sensitivity and detection limits depending on the abundance of the respective isotopes. The relative abundances of the Fe isotopes in nature are approximately 5.8% for Fe^{54}, 91.7% for Fe^{56}, 2.2% for Fe^{57} and 0.3% for Fe^{58}. The respective naturally occurring isotopes of Cr are Cr^{50} (4.4%), Cr^{52} (83.8%), Cr^{53} 9.5%, and Cr^{54} (2.4%). In real analytical life Cr would be detected on mass 52 and Fe on mass 54 unless these masses would be burdened by other spectral problems. In most cases, isobaric interferences are easy to predict and control.

More serious, as less predictable, are interferences by molecular ions. They may derive from solvent (water, acids, organics), from gases, mainly nitrogen, oxygen, argon in case of ICP-MS, hydrogen from separation processes, etc. These may form all types of molecules if present in large excess over the analyte ions. Examples from elemental analysis are Ar^+ ions at masses 36, 38, 40, ArO^+ at mass 56. The latter hampers the quantitation of the most abundant Fe line and the most intense ArN^+ peak at mass 54 interferes with the second strongest iron mass. Molecular spectral interferences depend strongly on the ionization mechanism, the sample load into the spectrometer, and the complexity of chemicals in the sample. The sharpest weapon is resolution. However, chemical reactions (matrix or analyte modification) as described in the sections up on AAS and ICP-OES are suitable ways for minimization as well. These will be touched in the following sections.

Most ions entering the mass spectrometer are singly charged. Elements with a low second ionization energy may lose a second electron and appear as doubly charged ion. In this case the doubly charged ion appears at half its original mass as interferent. It is obvious that this effect will mainly appear in elemental mass spectrometry with strong ionization sources. Modification of the operating conditions helps to tackle this kind of interference.

2.2.4 Mass spectrometry of the elements

Elemental analysis requires free atoms or, in the case of mass spectroscopy, free cations. An inductively coupled plasma is an ideal source for ions of most elements. ICP-MS therefore became by far the most widely used atomization and excitation source for mass spectroscopy. In this chapter we will focus mostly on ICP as an excitation source for elemental mass spectroscopy. It should be briefly mentioned though, that other excitation sources are commercially available for special applications.

Glow discharge MS is used mainly for solid metallurgical samples [118]. The ion source is based on an electrical discharge between two electrodes under reduced pressure. The commercially available instruments allow a fast solid sample determination of most elements in metallurgical samples down to the µg/kg range. Another special application for GDMS is detection of samples from gas and liquid separations.

The features of ICP sources for elemental spectroscopy have been described in the section on optical spectroscopy (Section 2.1.8). The plasma sources used in mass spectroscopy are like the former ones. OES features preferentially standing torches where plasma gas and fragments of matrix and analyte are removed by a fume hood. In contrast the flow of sample and sample gas into the mass spectrometer is mandatory. Essential for maximal sensitivity at minimal plasma gas flow into the spectrometer is an efficient interface between the hot plasma at ambient pressure and the cold chamber at high vacuum conditions. This challenge has been tackled with a two-step device from the very beginning. Two water cooled cones with small orifices of about 1 mm diameter are mounted at a distance of a few centimeters from the plasma. The space between the two cones, the sampler and the skimmer cone, is pumped with a rotary pump to a partial vacuum of less than 1 Torr (133 Pa). The pressure behind the sampler cone is that of the high vacuum chamber of the spectrometer at about 10^{-5} Torr (0.001 Pa). The arrangement is sketched in Figure 2.86.

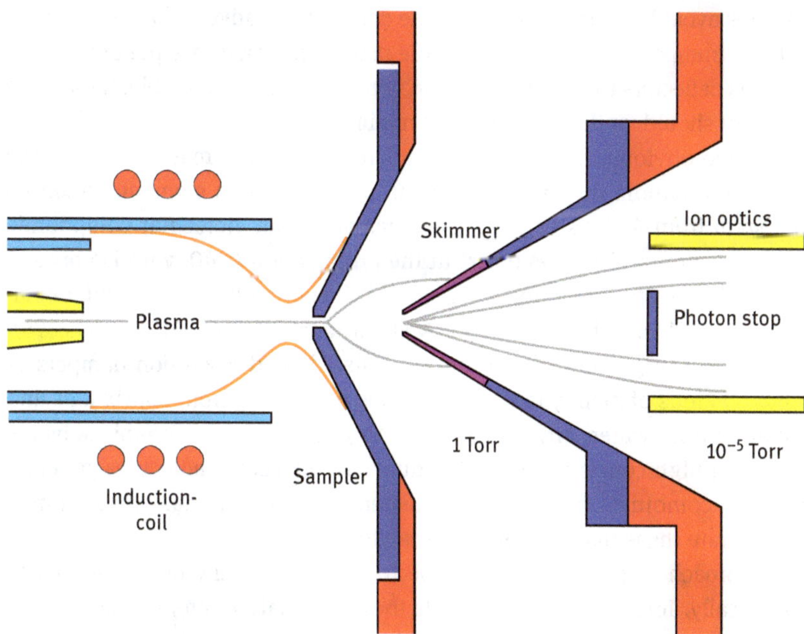

Figure 2.86: Plasma torch, sampler cone and skimmer cone of an ICP-MS spectrometer. Sketch from webmaster@icp-ms.de, modified.

The sampler cone made from nickel or platinum deviates the plasma gas to both sides of the orifice while the sample channel is sucked into the orifice. The sample gas with a fraction of the generated ions is expanding rapidly into the expansion chamber. Collisions within the sampling gas are damping the expansion. Between a zone of shock waves of the collision, a zone of silence forms where the ions are dispersed, and the original plasma conditions remain partially undisturbed. The apex of the skimmer cone protrudes into this zone of silence and further extracts the ions and a fraction of the sampling gas into the high vacuum chamber. The interface is crucial for the proper function of the spectrometer and defines FOM such as sensitivity, repeatability, and matrix load. It suffers from harsh conditions and must be maintained and replaced regularly. The cones must be chemically resistant, mechanically stable and must provide good thermal conductivity. The material of choice would be platinum. Due to cost, however the less expensive choice is nickel. Faster wear of the latter material must be balanced against the significantly higher cost of platinum cones. The dominant applications of the instrument should help to judge the corrosive load on the cones and thus help to decide whether cost or service life is prioritized.

Behind the skimmer cone the pressure is in the high vacuum range. The analyte ions, matrix element ions and molecules, uncharged particles, plasma gas and photons are entering the mass analyzer. Lenses are guiding the cations toward the mass selector. Photon stops avoid background noise on the detector. Negatively charged particles are repelled and pumped off before entering the mass analyzer. These pre-optical systems use various deflectors to keep the mass analyzers as clean as possible. Loss of the cations of interest should, at the same time, be minimized.

It has become obvious that the most efficient and simple mass analyzer, the quadrupole cannot handle the most prominent interferences in elemental analysis which often stem from Ar-ions of the plasma gas and from molecular compounds with Ar. Examples are the Isotopes of Ar^+ at the prominent m/z 40, with isotopes at m/z 36, 38, Ar_2^+ (mass 80) the doubly charged Ar^{2+} (mainly m/z 20), molecular compounds such as ArN^+ mainly at mass 54, ArO^+ at mass 56, and $ArCl^+$ (mass 75). The latter molecules originate from air and solvents. Limited resolution hampers as well handling all kinds of mainly molecular cations originating from matrix ions and atoms originating from water, air, organic and inorganic solvents. Organic solvents or samples with a high concentration of residual carbonaceous molecules generate carbon in lots of combinations with other elements. Ultra-high mass resolution is required to separate these masses from the analyte elements.

Another approach is, just as in optical spectroscopy, to modify matrix or analyte elements chemically. Interaction of ions inside the mass analyzer are, and should be, rare. Therefore, chambers with deliberately forced interaction are added which make modification inside the mass analyzer possible. They are named collision – or reaction cell which only partially describes the primary function. The collision cells are multipole setups (in the easiest case a quadrupole) where gases can be added. These slightly increase the pressure in the vacuum chamber. This results in enhanced collisions with both the atomic and molecular ions. Molecules have a larger cross section for collisions than atoms and loose more kinetic energy upon impact. The electromagnetic field of the cell will therefore prevent these molecules from passing on to the mass selector while the atomic ions will overcome the additional barrier. However, the collisions will lead as well to a reduced sensitivity. If reactive gases are used, the analyte ions or the matrix ions (atomic or molecular) can be modified by chemical reactions such that matrix is removed by the mass filter or the energy barrier, or the ions of interest are modified such that they are detected on a different m/z ratio. Operation of collision cells is challenging and requires experience by the user. The multi-element character of ICP-

MS will be partially lost when more complex gases are introduced. The approach, however, helps to remove some of the most severe interferences in ICP-MS.

The selectivity of quadrupole ICP-MS can further be enhanced by adding a second quadrupole mass filter in front of the collision cell. This can be operated as a focusing ion guide only, or as a second mass filter which separates one mass at a time entering the collision cell and the second mass filter (see Figure 2.87). This allows very selective mass selection and removal of spectral interferences by the proper collision gas.

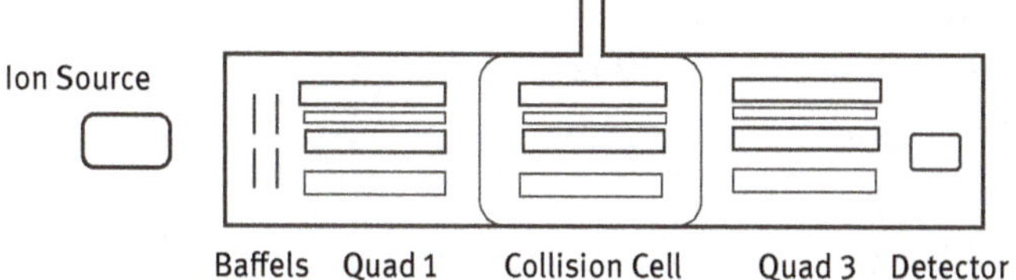

Figure 2.87: Tandem quadrupole ICP-MS set up with pre-mass filter, collision cell and main mass filter.

It becomes obvious that elemental mass spectrometry with ICP-MS requires sophisticated technical and chemical solutions to provide the highest quality of analytical data expected from the gold standard of elemental analysis. A source of potential problems is the interface where unwanted formation of molecules leads to interferences which are difficult to control. The unparalleled sensitivity and detection limits, on the other hand, make it an essential technique for all big laboratories for multi-element analysis in the range of µg/kg and sub-µg/kg determinations. The technique, however, should not be seen as universal replacement tool for optical techniques such as ICP-OES and AAS. In many cases the optical techniques are sufficient in limits of quantitation, comparable in reproducibility, easier to handle and maintain and, hence, more cost efficient.

2.2.5 Mass spectrometry of molecules

Molecules are often determined directly out of a solution or out of a solid which can be volatilized into the vacuum of the ionization chamber. Equally important are applications which include a chromatographic separation followed by a direct introduction into the coupled mass spectrometer. In both cases, a small vapor pressure of the substance is sufficient to transport enough sample into the mass spectrometer. If required, the sample can additionally be heated to increase its vapor pressure. Generation and mass spectrometric separation of ions require low pressure in the range of 10^{-4}–10^{-6} Pa in the ionization chamber and in the mass spectrometer. Molecules should not hit gaseous particles on their way through the spectrometer. The average mean free path must therefore be significantly higher than 1 m. Mass spectrometers for molecular applications require generally lower pressures than, e.g., an ICP-MS.

It is obvious that mixtures of substances can be analyzed only in rare cases as the vapor pressure of organic substances is quite different. The compounds with the higher volatility would mask the involatile partner. Although distillation inside the mass spectrometer is possible, the risk to evaluate the wrong spectrum would be high. Solvent must be completely dried off the sample to avoid dominance of the solvent spectrum and a significant reduction in sensitivity for the sample to be analyzed. The statements in the paragraph above do not hold for the determination of isotopes as these are uncritical with respect to distillation phenomena. Isotope analysis became a main field of mass spectrometry (see below).

Coupling of MS with separation techniques

Mass spectrometry is a widely used, common detector for chromatography. Separation techniques are often coupled with ESI-MS and ICP–MS such that the target analytes can be separated according to their chemical form or oxidation state. The separation techniques mostly coupled to ICP–MS are gas chromatography (GC) and high-performance liquid chromatography (HPLC). Modern methods which can handle very small samples, such as capillary electrophoresis (CE) and field-flow fractionation (FFF) are taking over more and more of the applications.

In ICP-MS, separation is important for element species analysis. The atom remains the target analyte. Its chemical form, not directly accessible to the sample introduction and ionization method, is deduced from chromatographic characteristics of the species. In order to obtain both sections of information of a sample, the output of the chromatographic device must be split into an elemental string via ICP and ESI for the molecular excitation. [119].

Coupling of separation techniques with mass spectrometry requires technically that the sample introduction and the flow from the chromatographic system match. The sample should be transported quantitatively from the chromatographic system to the ionizer of the MS. If this is not possible a representative and constant fraction of the sample should be delivered. Technically the outlet of the chromatographic system is directly coupled to the MS unit. Supplement of a solvent flow or split of the effluent flow may be required to match the sample flow from HPLC to the MS requirement. GC coupling requires often heated transfer lines to avoid condensation of the gaseous sample.

The resolution of the entire analytical process depends on the resolution of the chromatographic process and on the resolution of the mass spectrometer. It is obvious that a two-dimensional resolution – mass and appearance time of the species – will be much more powerful than a process based on one separation parameter only. Specifically in organic mass spectrometry GC-MS and HPLC-MS have become the gold-standard, in particular for biochemical applications. Coupling of techniques, on the other hand, imposes additional parameters to the process which must be controlled.

The mass spectroscopic process, usually based on a constant flow of sample to the mass filter becomes dynamic. The response of the spectrometer are time-dependent peaks which must be handled differently from short and rapidly changing peaks. Obviously, a substantial deterioration of the chromatographic resolution can be avoided only if the transfer is fast, and if the mass section of interest can be scanned and evaluated rapidly. Technically, transfer lines, possible dead volumes of connections, etc. should be avoided to the largest extent possible. Minimizing volumes is advantageous for efficient transfer but may impose strong demands on the speed of the mass selector. A perfect timing between separation and detection requires triggering between the instruments. This is a prerequisite for detection and data interpretation. A coupling between chromatography, ICP-MS and ESI-MS is sketched in Figure 2.88.

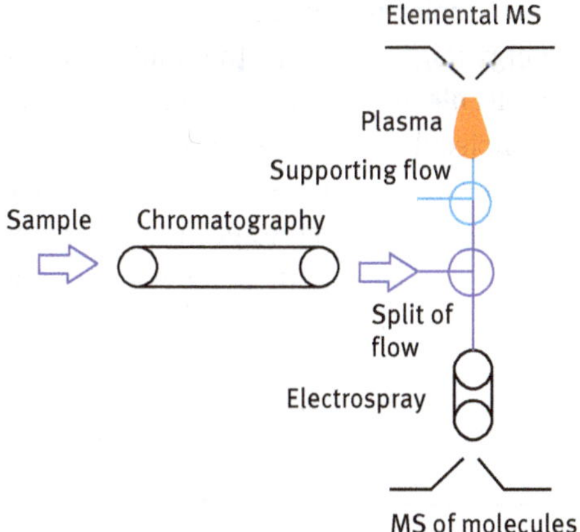

Figure 2.88: Separation techniques combined with elemental and molecular mass spectrometry.

Molecules or fragments of molecules?

In elemental mass spectrometry the aim, and the ideal situation for the spectrum, is the complete destruction of the sample into atoms, followed by ionization. The complete opposite is the case in molecular mass spectrometry. However, this is the main obstacle of this analytical technique. Reactions which take place prior to ionization are defined as "thermal." Reactions may take place as well in the ionization chamber due to bombardment with electrons. The higher the required temperature for volatilization of molecules the more probable is the fragmentation of these molecules. Higher molecular substances are therefore generally more susceptible to fragmentation. Technically, sophisticated methods for gentle vaporization and ionization are extremely important for molecular mass spectrometry (see Section 2.1.6.2).

Catalytic effects may contribute to a faster fragmentation. High inertia material throughout the introduction system of a mass spectrometer helps to suppress thermal fragmentation. Knowledge about the stability of organic compounds helps to predict the probable fragments in the mass spectrum. Examples are hydrogenation/de-hydrogenation, off-splitting of CO_2-groups of carboxyls and carbonyls, isomerization, or retro-Diels-Alder reactions (according to eq. (2.34)). Thermal fragmentation is often a relatively slow process. The mass spectrum becomes time-dependent.

Equation 2.34: Retro-Diel-Alder fragmentation

Diels-Alder

Retro-Diels-Alder

H

H

(2.34)

Fragmentation in the ionization process depends strongly on the electron energy used for ionization. A typical electron energy is 70 eV. This is a much higher energy than required for the ionization of many organic molecules, mainly those in the bio analytical field. Even small changes in this energy might influence the stability of the spectrum due to a changing ratio of original molecule and fragments. Qualitative and quantitative interpretation of mass spectra in molecular mass spectroscopy thus becomes an issue of the correct interpretation of the mass fragments. Balancing masses, positive charges, and electrons of all compounds appearing in the spectrum is a cumbersome and time-consuming process. Table 2.10 may be a guide to a few important and frequent fragments of organic molecules.

Table 2.10: Example of typical fragments in organic mass spectroscopy.

Mass ratio m/z	Fragments
1	H
15	CH_3
17	OH
18	H_2O
19	H_3O, F
20	HF
26	C_2H_2, CN
27	C_2H_3, HCN
28	C_2H_4, CO, H_2CN

Table 2.10 (continued)

Mass ratio *m*/*z*	Fragments
29	C_2H_5, CHO
30	CH_2O, CH_2NH_2
31	CH_3O
32	CH_4O, S
33	CH_2F, HS
34	H_2S
35	Cl
36	HCl
37	Cl
38	HCl
39	C_3H_3
41	C_3H_5, C_2H_3N
42	C_3H_6, C_2H_2O, C_2H_4N
43	C_3H_7, CH_3CO
44	C_2H_4O, $CONH_2$
45	C_2H_5O
46	NO_2
55	C_4H_7
56 57	C_4H_8 C_4H_9
59	$C_2H_3O_2$
60	$C_2H_4O_2$
64	SO_2
79	Br
80	HBr
81	Br

Table 2.10 (continued)

Mass ratio *m/z*	Fragments
82	HBr
91	C_7H_7
127	I
128	HI

The fragments can be charged (ionic) or neutral. Some of the elements or molecules in the list, e.g., Cl, HCl, Br, and HBr form two isotopes at different masses.

2.2.6 Isotopic analysis

Most elements exist in more than one composition of the atomic nucleus. This leads to a different atomic mass and to a different mass to charge ratio. Different isotopes of one element can be easily detected with mass spectrometry. The isotopes are chemically virtually identical. This reduces the number of possible interferences by matrix if the intensities of the different masses of an isotope are compared. The relative abundance of the different isotopes is constant, as a first approximation, and is known. Isotope analysis is therefore an extremely powerful analytical tool accessible to mass spectrometry only. There is an optical pattern of elements with two or more isotopes which can be studied in extremely high resolving spectrometers. In practice, however, isotope analysis is not viable for optical spectroscopy.

The determination of more than one isotope is of interest if:

- Measurement of the mass of the most abundant isotope is disturbed by other masses from matrix
- If addition of an isotope is used for high-performance calibration purposes (isotope dilution analysis)
- If the ratio between isotopes provides analytical information not accessible to other analytical methods

The relative abundance (M_i) of a nuclide of the element is calculated as the number n_i of atoms of the nuclide i divided by the total number n of atoms of the element (the sum of all isotopes of this element). Relative abundances cannot be measured directly, however, the ratio of isotopes can be determined. The ratio of the number of atoms of nuclide i and j, etc. is calculated according to eq. (2.35).

Equation 2.35: Ratio of isotopes.

$$R = n_i/n_j = (n \cdot M_i)/(n \cdot M_j) \qquad (2.35)$$

One of the nuclides of an element is the one with the highest natural abundance which obviously provides highest intensity in the mass spectrum. If ratios are determined, the isotope with the lowest abundance and hence with the lowest intensity will define standard deviation and other figures of merit of the ratio measurement. We have stated already that the number of possible interferences in isotope ratio measurements is significantly reduced as compared to the determination of different elements. The best possible match of measurement conditions would be obtained if the determination of the masses of interest are true simultaneous. The argument is – just like in optical spectroscopy – prevention of temporal differences in the sample transport and ionization processes. Scanning from mass to mass requires time which may contribute to the standard deviation of the measurement.

Isotope dilution analysis is accepted as the most accurate quantitation of elements presently available. It is based on the enrichment of a sample with a specific isotope such that the natural isotope ratio is altered. Determination of the single masses of the two isotopes may be hampered by one or several interferences typical for mass spectrometry. The ratio of two isotopes is much more robust. Additionally, the blended sample will behave virtually identically to the original sample. A spike with a known concentration with enriched ratio of one of the isotopes is added to the unknown sample. The spike concentration of the element under investigation is usually much higher than in the sample such that spiking does not change the sample volume. If it does, the dilution must be calculated. Isotope enriched spikes are expensive and often not sufficiently accurate. In this case, the isotope ratio of a spiked sample must be determined prior to spiking of the sample. A simple example may explain the process. We assume the easiest case of an element with only two isotopes. The abundances consequently add up to 100%. The measured intensities of the isotopes (I and J) in analyte (a), spike (s), and blend (b) are measured as intensities I_a, J_a, I_s, J_s, I_b, J_b. Measured intensities are calibrated with reference solutions and converted into concentrations or masses. Preparation of standards, including isotope enriched standards, requires calculation in masses and concentrations. As you will see below, the isotope ratios $R_a = I_a/J_a$, $R_s = I_s/J_s$, $R_b = I_b/J_b$ are used instead of intensities for concentration measurement. The following simple equations are used (see above; the blend is the sum of analyte and spike).

Equations 2.35.1 to 2.35.3:

1. $R_a = I_a/J_a$ (2.35.1)

2. $R_s = I_s/J_s$ (2.35.2)

3. $R_b = (I_a + I_s)/(J_a + J_s)$ (2.35.3)

J_a and J_s are replaced by the respective ratios from the equations above.

Equation 2.36:

$$R_b = (I_a + I_s) / (I_a/R_a + I_s/R_s) \tag{2.36}$$

Multiplication of eq. (2.36) by $(I_a/R_a + I_s/R_s)$ results in eq. (2.37).

Equation 2.37:

$$I_a + I_s = R_b \cdot (I_a/R_a + I_s/R_s) \tag{2.37}$$

Equation (2.37) is rearranged

Equation 2.38:

$$I_a + I_s = (R_b/R_a) \cdot I_a + (R_b/R_s) \cdot I_s \,; \tag{2.38}$$

I_a and I_s are separated:

Equation 2.39:

$$(R_b/R_a - 1) \cdot I_a = (1 - R_b/R_s) \cdot I_s \tag{2.39}$$

Rearrangement of eq. (2.39) leads to eq. (2.40).

Equation 2.40:

$$(R_b - R_a) / R_a \cdot I_a = (R_s - R_b)/R_s \cdot I_s \,; \tag{2.40}$$

solve eq. (2.38) for the unknown I_a:

Equation 2.41:

$$I_a = (R_s - R_b)/(R_b - R_a) \cdot R_a/R_s \cdot I_s \tag{2.41}$$

In the final eq. (2.41), the unknown (I_a) represents a mass or concentration which is related to the known mass or concentration I_s of the spike. The measurables are isotope ratios of analyte, spike, and blend.

An example:

The isotope ratio of an analyte (a) element with a natural distribution of 0.2 (isotope I) and 0.8 (isotope J) is determined: $I_a/J_a = R_a = 0.2/0.8 = 0.25$.

A spike with known relative isotope concentrations of $I_s = 0.6$, and $J_s = 0.4$ yields a ratio R_s of 1.5.

About 10 µL of a spike, say, 1 mg/L is added to 10 mL of the sample. The dilution factor can be disregarded. The concentration c_s of the spike in the blend is 1 µg/L.

We assume that the isotope ratio R_b is measured as 0.7.

The concentration of the element in the original sample is calculated with eq. (2.41):

$$I_a = 1\,\mu g/L\,(1.5 - 0.7)/(0.7 - 0.25)(0.25/0.7) = 0.63\,\mu g/L$$

As discussed above isotopes higher in abundance will provide higher analytical quality, provided that their mass is not hampered by an interference. The ratio will provide the figure of merit corresponding to the less abundant isotope. All chemical effects happening during preparation (digestion, dilution, etc.), during sample transport and ionization will be calibrated by the isotope spike. An element artificially enriched in an isotope is therefore the best possible internal standard. The only remaining critical effect may be spectral overlap and a time gap in measuring the isotopes. The highest accuracy is expected from a high mass resolution multi collector (simultaneous) mass spectrometer. This instrument equipment is unfortunately the most expensive one. Combined with the high cost of enriched isotope spikes, an isotope dilution determination will be high-prize analytics, and will therefore be limited to fundamental analytical questions, mainly in research.

Apart from isotope enrichment which changes isotopic ratios drastically, the analysis of small variations in the isotope ratio may provide information which is unattainable with other methods. Variations in the isotopic composition of materials may find their origin in natural or industrial processes, such as the decay of naturally occurring and long-lived radionuclides, natural fractionation effects, the influence of cosmic radiation and anthropogenic activities. Significant analysis requires high accuracy and high precision determination of these ratios. Technical features such as simultaneity of mass detection and spectral resolution will become essential. The detection limits possible and the standard deviations attainable follow the rules described above. A lower uncertainty can be obtained if signal intensities are high. Measurement time is therefore a parameter to decrease uncertainty. This, however, holds true only if the isotopes under consideration are determined simultaneously. Next to these fundamental aspects, also the type of instrumentation – and mainly the detection system available – has an important influence on the precision. If all instrumental parameters of high-performing instruments are carefully optimized, relative standard deviations of 0.01% and even below are obtainable. Despite the chemical congruence of isotopes of the same element, there is still a mass difference of at least one mass unit. This may lead to a systematic effect in detection called mass discrimination. The phenomenon is by no means limited to isotopic mass differences but is a general source for a systematic effect. Its source are small differences in ion kinetic energy. As a result, any energy-dependent process that takes place in the instrumentation, e.g., sampling of ions or ion transfer will lead to a slightly different response for different masses. If not corrected for it may generate a small systematic bias between the true and the measured isotope ratio. Mass discrimination may be up to 1% difference in intensity per mass unit in the middle range of, say, mass 100. It is obviously more pronounced at smaller masses (for the determination of light elements or small molecules) due to the higher relative differences of the two masses. Due to the extremely stringent demand in the determination of isotope ratios, the effect must be corrected for. The so-called external calibration makes use of an isotopic reference material with a known isotopic composition. The isotope ratio of this standard is measured before and after the sample. The reference measurements are interpolated linearly, and the ratio of the analyte is corrected with this value (eq. (2.39)). The concentration (measured intensities) of analyte and standard should be not too different.

Equation 2.42: correction of the isotopic ratio R_a using the measured value R_m bracketed between measurements with a known reference solution R_k, experimentally determined as R_{k1}, and R_{k2}.

$$R_a = R_m \cdot R_k / (R_{k1} + R_{k2}) / 2 \tag{2.42}$$

The mass discrimination can be corrected with a stable internal standard with known isotopic ratio as well. A mass doublet close to the analyte isotopes is used. The masses used for correction may be of the same or of a different element. The measurement sequence is like that of the external standard described above. An example may be the stable $^{88}Sr/^{86}Sr$ ratio to correct for mass discrimination of high precision $^{87}Sr/^{86}Sr$ measurements. ^{87}Sr shows natural variations due the radioactive decay of ^{87}Rb into ^{87}Sr.

Isotope ratio detection usually postulate that different isotopes of an element show an identical chemical behavior. However, there may be small differences in chemical and physical behavior which are based on the mass differences. The phenomenon is called isotope fractionation. As described above, the effect is stronger for light elements such as H, C, N, Li, B, but it is not nil for heavier elements. The study of natural variations in isotopic composition provides important additional information in different research areas. It became indispensable for several research areas, among others geochemistry, archaeometry, biomedical and forensic studies.

Carbon, the central element of life, appears in nature as ^{12}C and ^{13}C. ^{13}C contributes with only 1.1% in the distribution of stable isotopes. The abundances of ^{12}C and ^{13}C are therefore 0.989 and 0.011. The ^{13}C isotope will certainly be found in molecules as well. With increasing number of carbon atoms in a big molecule, the probability to find one ^{13}C carbon or even two ^{13}C carbon atoms in the cluster becomes bigger. If the probability is 0.011 ^{13}C relative to ^{12}C for 1 carbon atom in the molecule, it is about 10 times higher for a cluster composed of 10 carbon atoms. The molecule peak will be mass 120 and mass 121 with 10% (0.10) probability. Even the peak with 2 ^{13}C atoms in the molecule will be found at mass 122, however with a probability of only 0.50%. It becomes obvious that the interpretation of big molecules becomes more and more complex due to stable isotopes of the respective elements. Besides the stable isotopes ^{12}C and ^{13}C the instable isotope ^{14}C is present with a constant abundance of about 10^{-10}%. It is generated from ^{14}N by neutron trapping and proton release. Cosmic radiation is the leading energy source for this process. The abundance of ^{14}C is quasi constant whilst carbon is in free exchange with the environment. As soon as the exchange is stopped, ^{14}C decays with a half-life of 5,730 years. This is the basis for radiocarbon dating, one of the most important age estimation methods in science.

2.3 Quantitative detection of atoms: spectroscopy based on X-rays

2.3.1 General considerations and fundamentals of X-ray fluorescence spectroscopy

X-ray fluorescence spectrometry provides qualitative and quantitative elemental information. It is used predominantly for solids but may be used in liquids as well. Main fields of application are metallurgy, construction materials, geochemistry, and oil analysis. It is based on collecting radiation emitted by electrons which return from an elevated level into the ground state. Different to the situation discussed in optical spectroscopy so far, the electrons of inner shells are removed by high-energy radiation in the range of 1 nm wavelength or below. The existing holes are filled with electrons from outer shells. The released fluorescent radiation is short wavelength/high energy. The emitted energy is specific for the element involved in the fluorescence process. The number of collected photons of a certain energy level is used as quantitative measure for the analyte in the sample. Although the technique belongs to optical spectroscopy, the extremely short wavelength range necessitates different and modified processes compared to the techniques using light close to the visible range. The fundamentals have been described in Section 2.1.1 and these should help to follow technical and analytical explanations.

X-rays are waves in the short wavelength range from 0.01 nm to less than 10 nm. Using the known equation which relates energy with wavelength ($E = h{\cdot}c / \lambda$) we talk about 2×10^{-16} J at a wavelength of 1 nm. As 1 J corresponds to 6.2×10^{18} eV, the energy of incident radiation is in the range of 1,200 eV, which is 100 times higher than electron transitions in optical spectroscopy. Matter like atoms and molecules interacting with energy-rich photons in the wavelength range of 1 nm and below may react in different ways:

1. The wave may permeate without interaction. This will become dominant if the energy is high, the material is not dense, and the layer of interaction is thin.
2. X-rays can be scattered as described above. Scattering can occur without change in energy (Rayleigh scattering) or with release of an electron in an outer shell. The latter effect is called Compton scattering. It will result in a lower energy of the original X-ray wave plus an electron in the system.
3. The process providing the analytical information is absorption and fluorescence. Radiation is absorbed by electrons with high binding energy. As discussed above, these are electrons from inner shells. They are removed from the atom which is ionized by the process.

An electron from one of the outer shells fills the vacancy in case 3. The released energy (lower energy than the incident radiation) is evaluated. The emitted wave depends on the type of atom and on the type of electron transition. In X-ray fluorescence the nomenclature of transitions follows the Bohr-Sommerfeld model of electron orbitals. Obviously, electrons from outer shells may be closer or further away from the nucleus (L to K transition or M to K transition, etc.). The number of possible transitions is greatly increasing from the lighter to the heavier elements. Transitions from the L to the K shell are named K–L whereas K-M are transitions from the M shell to the K shell (according to IUPA nomenclature). The electrons in higher shells occupy differently shaped space with different angular momentum. In addition, the electron itself has a spin. The combination of angular momentum and spin is defined as the total angular momentum "J." It can take the values 1/2, 3/2, 5/2, . . . The energies of electron transitions from higher shells, e.g., from the L to the K shell are different if the J-term is different. According to IUPAC they are named with different indices (LI, LII, LIII, etc.) The energy difference is much smaller than that of the main shell transition and they often cannot be resolved by the spectrometer. In addition, not all possible transitions are observed. Selection rules ask for an orbital angular momentum of the transition (a change in the "J" number). Cu, as an example, has 29 protons and electrons. The latter ones are distributed in the K shell (2), the L shell (8), the M shell (18), and the N shell (1). An electron, removed from the K shell can be filled with 2 transitions from the L, 2 transitions from the M shell, and 1 transition from the N shell.

In general, the following rules apply:

- The energy drops from K to L to M lines. K and L lines are usually used for determinations.
- The binding energy increases with increasing atomic number. The energy is proportional to the square of the atomic number ($E \sim Z^2$). The energy of a K-L transition of Ti (22) as an example, is 4 times smaller than that of Ru (44).
- K-L transitions of an element with low atomic number may have the same energy as M-L transitions of an element with high atomic number
- Interaction of X-ray radiation with matter results in an absorption process for the X-rays by the effects described above. The attenuation is depending on the initial intensity of the X-rays, the type of element under investigation, the density of matter and on the penetration depth into the material under investigation. The process is like absorption described in AAS or UV spectroscopy. X-rays can permeate much denser matter, however.
- The fluorescence process is competing with the other photon interaction processes described above. Additionally, the probability for knocking out a K-shell electron is higher than for L or M electrons. This probability factor varies by up to two orders of magnitude from light to heavy elements explaining the much more intense fluorescent radiation of heavy elements.

For quantitation of the absorption the mass attenuation coefficient (MAC) has been defined. Tables of MAC values comprise the element, the transition (e.g., K-L), and initial values of X-ray energy. Similar to optical absorption the formula relates Intensities with the MAC and the absorption length (eq. (2.43)).

Equation 2.43: X-ray absorption as function of MAC

$$I = I_0 \cdot e^{-\mu l \rho_m} \tag{2.43}$$

I and I_0 are the energies of X-ray radiation before and after the transition. μ is the attenuation coefficient normalized with the density of the material, l is the length of the absorption path and ρ_m is the mass density of the element.

Absorption of X-rays becomes generally stronger with decreasing X-ray energy. MAC becomes generally higher (stronger absorption) with increasing number of protons. This process, however, is discontinuous. At certain photon energies, the MAC value reaches a (local) maximum, called absorption edge. Impact of photons at energies below the absorption edge can no longer stimulate a certain transition. Consequently, the MAC value drops as only less energy consuming transition (e.g., removal of electrons from the L shell instead of the K shell) are possible.

2.3.2 Technical setup of X-ray fluorescence spectrometers

Radiation providing analytical information is mainly stimulated by primary radiation from the source. X-rays are produced in a source which operates basically the way elaborated at the end of section 2.1.3: electrons are thermally generated in a filament similar to that in a light bulb. The bulb, or rather the tube, is evacuated and a high voltage in the range of up to 100 kV accelerates the electrons toward an anode. The voltage defines the energy of the electrons, the current in the filament the number of electrons generated per time unit. The electrons impinge on the anode which is usually a pure metal corpus. Deceleration generates a broad spectrum of bremsstrahlung and specific lines emitted by fluorescence of the anode material (one of the processes described above). The generated X-rays are extracted from the tube via a window which is as transparent as possible for X-rays, usually beryllium. With exception of the window, the tube must be carefully shielded from radiation leakage (usually with a lead foil) and must be cooled to remove the heat produced by the process. The high-energy radiation present in the process obviously requires highest safety standards for the equipment.

The schematics of an X-ray tube are sketched in Figure 2.89.

The photon output of the source is a broad band spectrum of waves in the X-ray range with strong fluorescence emissions of the anode material on top (see Figure 2.90). The type of anode material is an important selection criterion for the tube relative to the specified analytical application.

The radiation for stimulation of analytically useful fluorescence may originate as well as secondary effect from the sample itself. High-energy photons, e.g., from

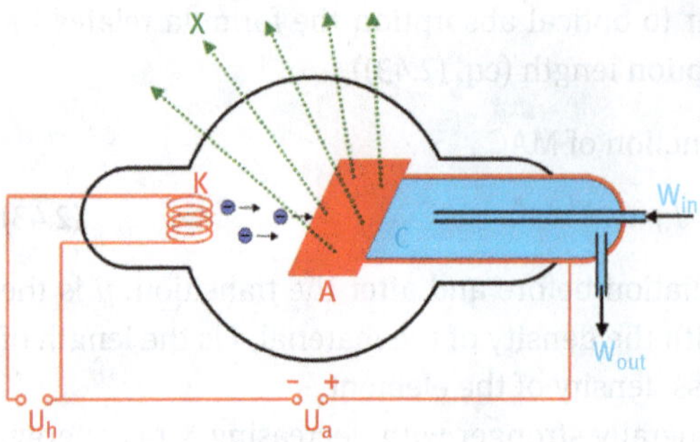

Figure 2.89: X-ray tube used for analytical purposes. [Heine, Ruth. (2008). Digital In-Line X-Ray Holographic Microscopy with Synchrotron Radiation; Researchgate]; K: heated wire for generation of electrons acting as cathode. U_a: high potential for acceleration of electrons toward the anode. X: generated X-rays. The anode is water cooled.

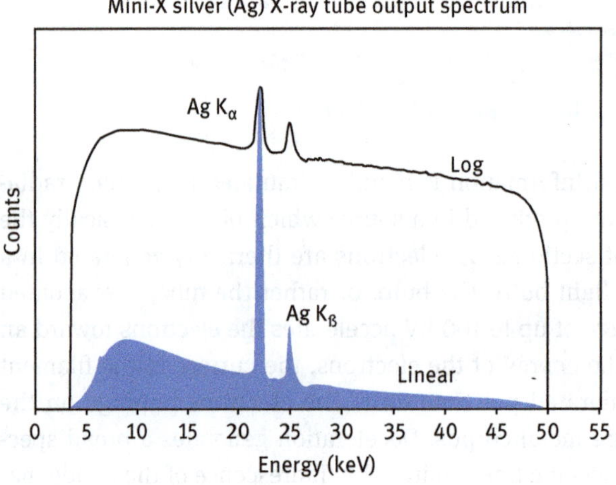

Figure 2.90: Sketch of an X-ray spectrum from a source with Ag-anode. Figure courtesy of Amptek Inc.,Bedford,Ma,USA.

a K-L transition may kick out L-electrons and lead to L-M fluorescence which can be analytically evaluated as well. K-L transitions of a heavier element, e.g., Fe 26, can induce K-L fluorescence of an element with lower atomic number, e.g., Cr 24. Together with fluorescence of the primary radiation the secondary and tertiary fluorescence acts as enhancement of the observed analytical signal.

As mentioned, XRF is predominantly applied for solid sampling analysis. The penetration depth into or through the material is limited due to the absorption effects discussed above. The heavier the element, the smaller the penetration depth. The higher the energy of the radiation, the bigger the depth. Fluorescent radiation coming from the sample can escape only through a certain depth which is called infinite depth. Note that photons of the incident radiation, at higher energy, can penetrate deeper in the material than fluorescent light which cannot escape beyond the infinite depth.

Optics

Just like in optical spectroscopy, photons must be guided from the source to the sample, from the sample to the selector and from the selector to the detector. Handling of high-energy photons is more difficult than photons in the UV and IR range. The former ones will penetrate technical components and may generate unwanted response. The technical components of XRF spectrometers are therefore in part different from the optical systems described so far. The analytical response to the excitation of the sample requires discrimination of the energy of photons and the number of photons per time unit at a given energy level. If detectors are available which can put both parameters into effect at a time, the system setup becomes much simpler than if energy or intensity separation is required. In XRF the discrimination of energy and intensity is possible at the cost of specificity. If the result is sufficient for the analytical requirement, the instruments are much simpler, smaller, and cheaper than instruments with wavelength selection. The major difference in XRF is between energy dispersive X-ray fluorescence (EDXRF) and wavelength dispersive X-ray fluorescence (WDXRF). We will start with a description of latter one.

WDXRF operates similar to a sequential optical spectrometer. Radiation is generated in a source (described above) and focused on a sample. The response of the sample is collected, separated, and focused on a suitable detector. As the interaction with the environment, air, water vapor, etc. is much more pronounced in the short wavelength range, the path of the photons must be as short as possible. Additionally, it must be protected from all kinds of matrix which leads to unwanted interaction, including the ambient atmosphere. The components of a WDXRF system are:

Function of component	Specification
Source	X-ray tube (Figure 2.81); various embodiments
Optical path	Airtight chamber; vacuum or gas filling (He, N_2)
Sample holder	Devices for various sample types: solids, powders, liquids
Filter, beam stop, collimator	Beam shaping and wavelength restriction before and after sample
Dispersion	Crystal with appropriate interatomic distance
Wavelength selection	Goniometer; angle of reflection for wavelength selection
Detector	Collector for radiation; different detectors dependent on WL

The samples analyzed with XRF systems are often solid bodies. They can be placed into massive metal holders with suitable aperture and can be placed very close to the X-ray source without the risk to contaminate the window of the source. Keeping the sample in place is more difficult if powders or liquids are analyzed. In this case the sample cup may need a seal to prevent sample from escaping into the vacuum or gas filled chamber. Filters and masks restrict unwanted background radiation or specific wavelengths. Usually, metal foils are used for this purpose. The beam is further shaped with a collimator. This is like the mechanical slits in optical spectroscopy. Resolution is compromised against photon intensity for best specificity and signal to noise.

Wavelength selection requires a very short distance between the reflecting elements. While in optical spectroscopy the spacing between grooves on a ruled grating are in the range of 0.5 µm, the requirement for spectral resolution of X-rays is 3 orders of magnitude smaller. This can be accomplished by crystals with a lattice spacing in the range of 0.5 nm. Various crystal types are in use, e.g., LiF, Ge. They offer the required resolution. With the help of a crystal changer the optimal crystal lattice for the type of sample can be selected. Just like in the case of the grating, the angle of reflection defines the selected wavelength range. The Bragg equation (eq. (2.44)) holds for X-rays.

Equation 2.44:

$$n \cdot \lambda = 2 \cdot d \cdot \sin\theta \tag{2.44}$$

where n is the order of the diffracted beam, λ the diffracted wavelength, d is the lattice distance, and θ the angle between the incident X-ray beam and the surface of the crystal. The selection of different crystals with different lattice spacing allows an optimized selection of the incident angle for optimal signal to noise for a specific wavelength. For the correct setting of wavelength, the crystal must be positioned on a precise goniometer. This, again, is comparable to the wavelength drive accuracy in optical spectroscopy.

The detectors for X-ray fluorescence follow principles of operation not discussed in optical spectroscopy so far.

Flow detectors operate with a permanent slow flow of a suitable mixture of gases (e.g., 10% methane in argon). The flow is in the range of 10 mL/min passing a cylindrical cell with a suitable window collecting the X-ray photons. The photons ionize gas atoms and thus generate positively charged ions and electrons. In the case of an Ar/methane filling, Ar atoms are ionized, and methane quenches the process of additional unwanted ionization. The cell consists of an electric wire in the center and a negatively charged housing. The voltage between the electrodes is in the range of 2 kV. The electrons are strongly accelerated toward the wire and generate an avalanche of additional ions. The electrons hitting the wire generate a charge pulse. This pulse is proportional to the energy of the X-ray generating electrons, if the quenching and avalanching processes are mutually optimized by gain, cell dimensions, and flow. However, the detector (as every detector in optical spectroscopy), is wavelength limited. It is mainly used in the lower energy range up to 8 keV. Flow detectors have the big advantage that their response is not ageing due to the continuous flow.

Sealed detectors without this constant flow can be adjusted to respond proportional to higher energy levels up to 15 keV.

In a scintillation detector the primary reaction of X-ray photons takes place in a scintillation crystal. A solid crystal, e.g., NaI, is doped with an activator, e.g., Tl. The energy-rich photons generate free electrons, holes or pairs of electrons and holes. These charges move through the crystal and hit an activator atom. This leads to light emission near the visible range. The energy is proportional to the flow of secondary photons, the intensity is measured by the number of scintillations per time unit. The detector housing opens to the incoming photons with a beryllium window. The photons hit the crystal; the energy is converted to longer wavelength light which is amplified in a photomultiplier type of detector. This detector is used in the high-energy range between 8 and 32 keV.

As discussed previously, the main energy selection in wavelength dispersive XRF is the combination of crystal lattice distance and incident angle. The photons reaching the detector are not strictly monochromatic but span a range of energies, like the spectroscopic window in optical spectroscopy above 150 nm. The response is a pulse height distribution with a main peak and possible secondary peaks at slightly lower or higher energy. Different types of high- and low-energy "stray light" may occur from fluorescence of initial photons or excitation of secondary fluorescence, e.g., in the scintillating crystal. These photons have different energy compared with the analyte fluorescence, which is expressed by a different detector response. As the number of photons per second in the X-ray range is much smaller than for light, fast electronics is able to resolve the energy deposited by each individual photon (the number of photons per second gives the intensity). This raw data can be displayed in the form of an energy/time diagram (see Figure 2.91). Suitable mathematical algorithms can convert the detector response into an intensity/energy (wavelength) chart. This is similar to the information obtained from a CCD line in optical emission spectroscopy recording the vicinity of the peak. Suitable filtering of incident radiation and fluorescent radiation further reduces the occurrence of non-specific pulse intensity.

With respect to mechanical and thermal stability requirements, wavelength dispersive XRF is like high-resolution OES. The systems are usually thermostatted. Water cooling is required for the X-ray tubes. The gas in the flow through detector must be ultrapure.

The detectors used in XRF have a finite energy response, provided the incident photon avalanches can be separated in time. This makes it possible to simplify XRF to optical systems without wavelength selector. These instruments are named energy dispersive XRF (EDXRF). The instrumental setup becomes quite simple (see Figure

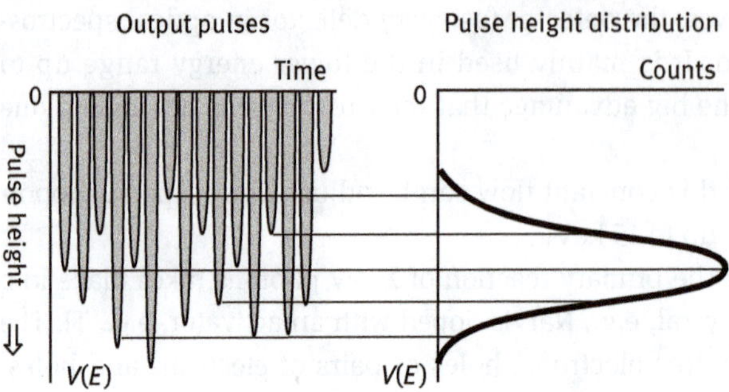

Figure 2.91: Detector impulses over time. Histogram and intensity versus energy plot. Figure courtesy of Malvern-Pananalytical, a Spectris company.

2.92). The beam of a suitable X-ray source is directed straight on the specimen of interest. The fluorescent X-rays are collected on a detector and processed by a suitable analyzer.

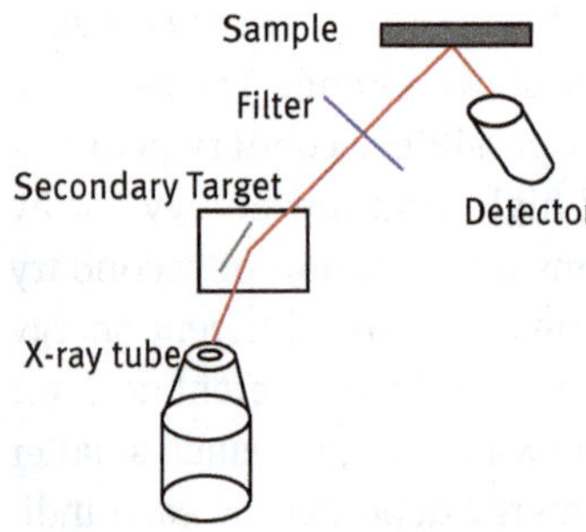

Figure 2.92: EDXF setup: the simplest device works without any installations between tube and specimen, filters or a secondary target will improve resolution and signal to noise ratio.

The analyte-specific lines in EDXRF are on top of a prominent broad background continuum. Resolution and signal to noise are limited (see Figure 2.93).

The background continuum can be substantially reduced by making use of suitable filters. The principle of this effect is comparable to WDXRF but the improvement in signal to noise is higher in EDXRF. If the X-rays from the tube are used to generate secondary fluorescent X-rays from a secondary target, this fluorescence has a much sharper energy distribution than the primary radiation from the X-ray source. Additionally, it is polarized. This further reduces the background. However, much higher primary energy from the tube is required in the latter setup for obtaining the desired analyte fluorescence. The latter design is mainly used for clearly defined samples with a limited number of elements to be detected.

EDXRF is often used when smaller laboratory instruments or hand-held devices are pursued. This requires smaller components, predominantly smaller tubes, which are simpler in handling. While the excitation potential of the tube is comparable to

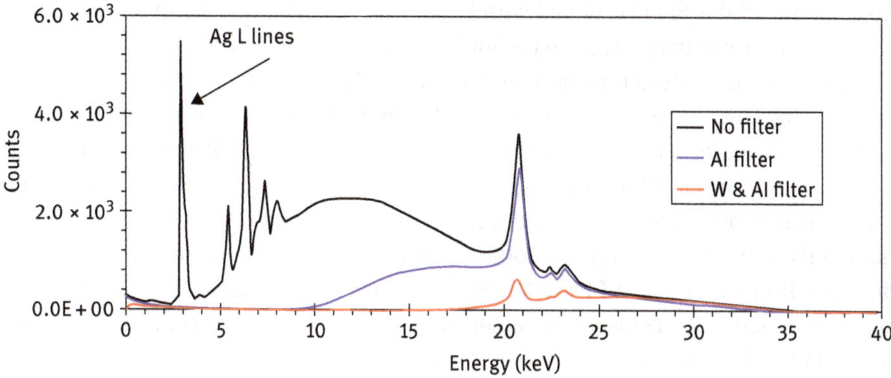

Figure 2.93: Analyte-specific lines and continuum spectrum of a simple EDXRF device. Figure courtesy of Amptek Inc.,Bedford,Ma,USA, modified.

that described already for WDXRF, the tube current and output power is much smaller. The tubes are usually air cooled further reducing instrument mass and power. Fine-tuning of excitation energy and power according to the requirements of the detector is important for the performance of the system. In hand-held instruments suitable radio-isotope sources are an attractive alternative source for X-ray primary radiation. Sources of, e.g., 55 Fe or 137 Cs and other isotopes are in use. Although these sources make smallest design of the instruments possible, the strict regulations for isotope handling limit the use and distribution of these types of instruments.

Unlike WDXRF instruments, EDXRF instruments usually work in air. This holds true for hand-held instruments in particular, whereas lab instruments often offer the option for vacuum sample chambers.

Detectors in EDXRF

Like in WDXRF sealed gas detectors may be used in EDXRF as well. However, their resolution is poor.

In semiconductor detectors, XRF photons are producing electron hole pairs. Unlike in gas detectors, where ion pairs are formed in in a comparatively diluted medium with rather high ionization energies, the high density of the semiconductor material assures a high probability of interaction with the photons. Due to the small band-gap, a lot of free charge carriers are generated. The charges are separated and detected. The ultimate aim is single photon counting, i.e., a temporal selection of the incoming charges. Higher energy photons create a bigger number of charges and thus a higher peak. The energy separation of this type of detectors is in the range of hundred V at a line of 5,000 V, comparable to an optical resolution R of about 50 (the resolution has been normalized to the Mn K_a fluorescence peak at 5,900 eV). The detector material is Si or Ge doped with Li, for example. To keep the shot noise small, the detectors are usually cooled with liquid nitrogen to very low temperatures.

Semiconductor detectors are classified according to their specific function. Silicon PIN detectors are widely used in EDXRF. They offer a resolution of about 30 and very good signal to noise under moderately cooled conditions (Peltier cooling). This detector type consists of an undoped or minimally doped layer between the p and n layers. This is only intrinsic conductive. It is therefore named I-layer, and lends the name PIN to the detector type. Another popular design is the so-called silicon drift detector. It operates like a PIN detector and consists of circular P-N layers with the collector anode in the center. It is Peltier cooled and permits good signal to noise at a resolution R of about 50.

Just as described in WDXRF the highly energetic radiation of X-rays may induce fluorescent radiation in the detector material itself, leading to Si Kα radiation from within the detector. The generated peaks are named escape peaks and may result in an interference. The detector material may as well operate as a diffracting crystal for X-rays and thus result in additional peaks.

As described already, the electrical pulses of the detector represent different energies and different photon flux due to a higher or lower amount (concentration) of the analyte elements in the sample. For discrimination an electronic pulse analyzer is required which finally provides energy resolution. The output is a spectrum displaying energy versus counts per second (see Figure 2.93).

From the discussions above it becomes obvious that EDXRF is simpler and lower cost than WDXRF at the cost of performance with respect to detection limit and specificity. The entire information on the sample is obtained much faster as the method is intrinsic multi-element capable while WDXRF requires wavelength selection from line to line (sequential multi-element analysis). Specification of the exact analytical question is therefore an essential part of method selection.

Total reflection XRF

The technique became popular as powerful variant of XRF. The incident beam is guided onto the surface of the analyte sample at an extremely flat angle. The sample itself is mounted as thin film on a reflective base, such as polished quartz. The fluorescent radiation is read directly above the sample under a 90° angle to the surface in EDXRF mode (Figure 2.94). The sample may be a thin film, powdered solid or dried liquids. As the film is very thin and the carrier of the sample is exactly known, the number of possible interferences becomes very small. This makes detection below the range of tenth of picograms possible with a wide dynamic range of the working curve. The angle can be modified between total reflection and grazing incidence reflection which allows to modify penetration into deeper depths of the material. Just as in other techniques used for direct analysis of solids, the accuracy

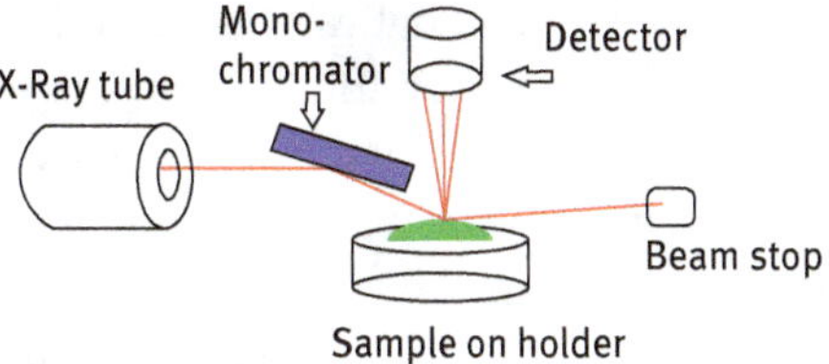

Figure 2.94: Sketch of the TXRF setup.

of quantitative determinations is based on suitable reference materials with known concentrations of analyte for calibration.

2.3.3 Quantitative analysis with XRF

Like in other analytical methods, several interdependent steps must be followed to obtain the desired figures of merit (FOM) using XRF techniques. These are:

- Sample preparation
- Optimization of signal to noise ratio
- Interference correction including the use of algorithms
- Calibration and calibration algorithms

Samples and sample preparation

Though all types of samples, i.e. particles in gases, liquids and dried liquids, film-type samples, etc. can be analyzed with XRF, the predominant types of applications are solids. These may be metals, rocks, pressed powders, etc. While liquids are usually homogeneous and thus representative for the bulk of sample, representativeness is a critical issue for solids. While element distribution in metals is often homogeneous, this is usually not the case in rocks and powders.

Liquids are placed between thin special polymer cups with a polymer film holding the liquid in place. The films may be interacting with the X-rays and their thickness and composition will influence the active X-ray intensity. Obviously, the polymer material must not react chemically with the liquid sample. Due to vaporization effects, the sample chamber with the liquids must not be subjected to vacuum. The analysis takes place in air or helium atmosphere. The liquid layer should always have the same thickness in the cup. Samples can be mixed for standard addition calibration purposes. Liquids can be dried on suitable film material and can then be analyzed in vacuum.

Solid samples can be directly analyzed or scanned with hand-held EDXRF instruments. The result will be an approximate element value used mainly for identification, quality control issues or scanning for further laboratory analysis. For the latter samples must be taken which are representative for the bulk but can be placed into the sample holder. The samples may be grindable (e.g., rock) or grinding may not be required (e.g., metals). If grinding is required, it should not contaminate the sample with the elements of interest and should produce a fine powder which can be homogenized. Additionally, the grain size has an influence on the measured intensities of the analyte elements [120]. The particle size where further reduction does not change the intensity response is depending on matrix and element. Powders can be handled like liquids (positioning between polymer film material, determination in air or helium atmosphere). The density of the powder and the atmosphere, however, are limiting the achievable signal to noise ratio. Like in the case of IR determinations, samples can be pressed to pellets. These can be placed into the sample cup without applying film closure. Usually, binders from softer material such as starch, boric acid or graphite are required to glue the solid powders into compact disks which do not disintegrate in vacuum. The binder should consist of light material not to generate intense X-ray response. Under these conditions the optimal XRF conditions, vacuum, no additional layers, are achievable. Pressing of disks requires experience with respect to the amount of glue and the pressure required to reach a plateau in the X-ray output. The disk should be thick enough to provide an infinitely thick sample with constant output. If the sample is too strong for proper grinding it can be chemically treated by fusion like the process of rock digestion. In general, the required sample mass is usually in the range of 5 to 10 g.

In the case of metals, disks must be cut from larger pieces. The final sample must have an even surface which equals that of the standard in use. In metallurgical applications the samples are often cast in the appropriate amounts for analytical purpose. The processes of casting and cooling must be optimized for the analytical use.

Optimization of signal to noise ratio

XRF is not too different from other emission methods with respect to signal to noise ratios. The lines used for evaluation have a certain intrinsic sensitivity. Provided sampling and measurement conditions have been optimized, the selected line is often much stronger than the background signal. The noise on the line is the criterion for standard deviation for concentrations significantly higher than the detection limit. The background noise is the criterion for the detection limit. Both effects depend on counting statistic or measurement time. Obviously, the time for measurement in WDXRF sums up for each individual element, which can be individually optimized with respect to measurement time. EDXRF as simultaneous multi-element technique is much faster but the measurement time must be optimized for the most critical situation, the lowest required d.l., the best r.s.d., the most critical line with respect to signal to noise. These considerations hold true if matrix effects including spectral interferences are not the limiting magnitude for the detection capability in the sample for measurement.

Dealing with matrix effects

Like in optical elemental spectroscopy the biggest source of error is spectral overlap. The rules for correction are therefore similar. Background is an issue in general, if traces or low analyte concentrations are determined in an excess of matrix which potentially overlaps analyte signals. The energy discriminating capability of detectors allows only moderate resolution. In WDXRF the region around the analyte line of interest can be scanned, stepping the angle of the resolving crystal. The result is a spectrum including the analyte peak of interest (see Figure 2.95). From the spectrum, one or several background correction points can be selected which provide a good impression of the true baseline underneath the peak of interest. Software and suitable algorithms support the user. A reference material with composition as close as possible to the unknown sample helps to set the background points correctly. The best choice would be a reference matching the matrix but without analyte. This, however, is seldom obtainable.

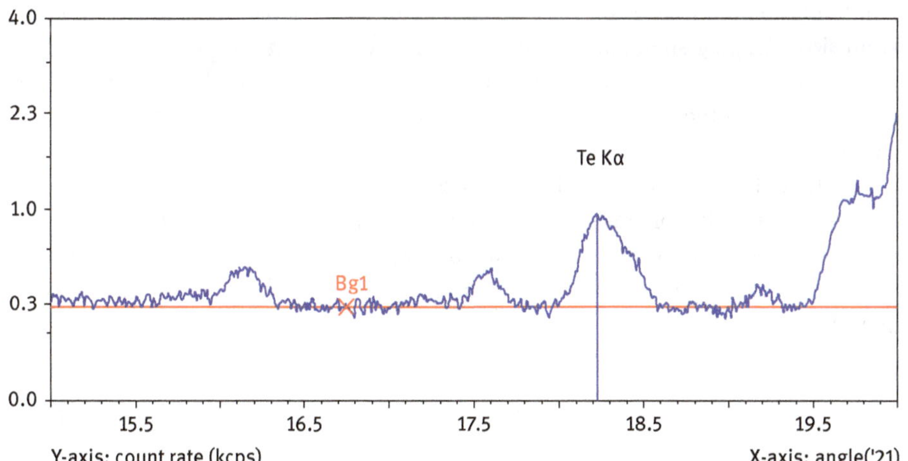

Figure 2.95: Stepped spectrum of energy (abscissa) versus intensity (ordinate) in WDXRF. The Te K_a line is the analyte peak. Several peaks occur in the spectrum, but the baseline intensity (red correction point) can be set accurately. Figure taken from [122].

In EDXRF the accuracy of background correction depends on the resolution of the detector. Like in WDXRF, the peak of interest is defined in energy counts and the average of the intensities between a specified energy minimum (left of the peak) and maximum (right side of the peak) is subtracted from the analyte peak intensity. The process can be extended to the total spectrum obtained from an EDXRF measurement. Peaks and background are stored along the energy axis with the resolution possible by the detector. Peaks and background are then modeled by a mathematical deconvolution method until the model shows best agreement with the spectrum.

Direct spectral overlaps of analyte and matrix peaks can no longer be corrected by modelling the spectrum or setting background correction points. Knowledge about the sample is required in such cases and tables (often inherent in XRF instruments' software packages) are required to recognize these cases. Other available lines must be selected in these situations, probably at the cost of a loss in signal to noise ratio. The same holds true if lines from the primary source interfere with the analyte element. In this case, however, the material of the excitation source is usually well characterized and possible interferences are easier to predict.

Random instrumental errors, such as instrument and wavelength stability, output stability of the source or dark noise and amplification stability of the detector are comparable to the elemental methods of optical and mass spectrometry. This is not the case for systematic error due to the sample itself. It has been outlined that X-ray interaction with matter follows various paths. Compared to dissolved matter, the structure and chemical composition of solids is much more demanding for the analytical measurement process. As the method is non-destructive, contrary to solid sampling in graphite furnace AAS (see Section 2.1.6) for example, the surface structure, grain size, etc. play an important role for the analytical response. The result is a high scatter on the correlation line between measurement intensity and true concentration of the analyte in the sample. Correlation factors R^2 of 0.7 are often typical. Sophisticated correction algorithms are required to make XRF a quantitative analytical method, with FOM comparable to destructive optical and mass spectroscopy. These correction factors take individual care of effects specific to XRF.

The general procedure is to add factors (K_i, M_i) to the proportionality between concentration of the analyte elements c_i and intensity of the respective lines for quantitation I_i (eq. (2.45)).

Equation 2.45:

$$C_i = I_i \cdot K_i \cdot M_i \tag{2.45}$$

Correction factors are applied for the specific instrument response to certain elements M_i. They must be used to correct for the absorption of X-rays by matrix (lead, as an example absorbs X-rays much stronger than iron or tin). In this case the correction factor M is replaced by the sum of all mass attenuation coefficients (see above) of the matrix.

Enhancement by secondary and tertiary excitation is a positive correction > 1 which partially can balance the absorption but is always correlated with it. Other factors are the chemical composition of the matrix and particle size effects.

A lot of painstaking research work was necessary to feed the software packages of XRF instruments with the required data for good correction. Still, lots of knowledge about the unknown sample and its composition is necessary to feed the software with the parameters that lead to an accurate and precise analytical result.

Calibration

Calibration in XRF requires a close match between the composition of standard and sample. For unknown samples this is generally difficult to achieve. However, XRF is often used in application fields where the principal composition of the sample is known and constant within tight limits. This is the case in metallurgy and, to a certain extent, for geological samples as well. In these cases, calibration works like described before using linear or non-linear curves for correlation of concentration/mass and measured intensity. Typical for analysis of solids are external or laboratory internal reference materials. The latter ones can be produced according to matrix and working range and are usually long-term stable and usable. The calibration curves usually span one or two orders of magnitude. Often bracketing of the expected concentration range by standards is applied for highest measurement precision. Correlation coefficients of 0.9 and above, which are typical for quantitative elemental analysis, can only be reached when the correction algorithms mentioned above are applied.

Often the exact concentrations of elements in a sample are not required but an estimate defining the type of alloy or rock, etc. is aimed at. As described for other techniques (AAS, ICP-MS), semi-quantitative analysis or even standardless estimates (GFAAS) will provide this information. In XRF these calibration procedures are based on a kind of universal reference, a sample which includes a number of elements in a concentration of few (below 10) percent. The elements of interest are entered into the software and the algorithms in the software calculate and set the best parameters for this suite of elements. With the help of the correction methods discussed earlier, an estimate of the concentration in the unknown samples is calculated. Semi-quantitative methods are usually only used for minor and not for trace element determination.

2.3.4 Typical applications

As application examples we will discus [121] (taken from Elemental Analysis [37]), which show weaknesses and strengths of XRF: an application where environmental issues and question of process control are examined, and an application where product purity with highest accuracy is the mission. EDXRF and WDXRF are presented.

In the first case As and Ag in the range of ppm to below 1% and Zn in the low percent range are determined in a bulk matrix of PbO and PbS with high sulfur concentrations. The matrix is dominant with respect to absorption and spectral intensity. The sample is broken up with appropriate mills and pressed into pellets with a 1/10 addition of wax. Reference materials are prepared from similar rocks. The concentrations of the elements of interest in the reference samples are determined with spectroscopic analytical methods (ICP-OES, AAS).

The samples are measured under vacuum conditions. Although EDXRF would allow simultaneous detection of the three elements, the complexity of the matrix requires an optimization of the settings for the primary X-ray intensity. As and Zn are optimized for lower X-ray tube settings (40 kV, 0.11 mA), whereas Ag requires the maximum output at 50 kV and 1 mA. Due to a direct energy overlap between the Pb Lα line and the most sensitive As Kα line, a less sensitive secondary line, the As Kβ1 line is used for evaluation. This is still surrounded by Pb lines. A long integration time of 600 s! is required to obtain a reasonable signal to noise ratio for the expected per mill concentration range. Zn, in the low percent range can be detected on the most sensitive K_a line without spectral overlap and good signal to noise ratio. The same applies for Ag. The spectrum for the determination of As and Zn is sketched in Figure 2.96.

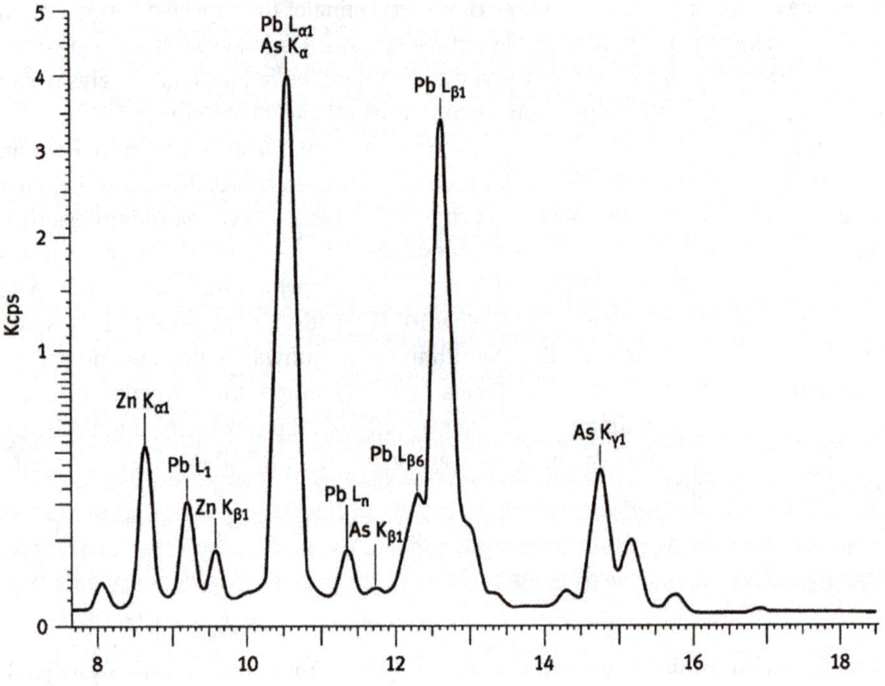

Figure 2.96: EDXRF spectrum for the determination of As and Zn in lead concentrate. The energy range of 8 to 20 keV is imaged. Courtesy of Alexander Seyfarth, SGS North America.

EDXRF in this case offers the advantage of direct solid sample analysis following a routinely used sample preparation method. Requirement is a good reference material. The disadvantage is a long analysis time of almost 20 min per sample/reference.

The second application originates from noble metal analysis [122]. Elements at high concentrations are determined with highest demand on precision for major and minor elements in the sample. In the described case the bulk element Ag is determined with its alloying partner, Cu. An addition of P is also determined. The latter one is used for optimization of the casting process. The precision requirement for Ag is 0.03% which cannot be met with optical spectroscopy (ICP-OES). For WDXRF this requirement is tough as well. It requires highly stable temperature conditions, warming up of instrument and X-ray tube and keeping the measurement conditions between reference and sample as similar as possible. The samples are from cast metal which is polished following an exactly defined procedure. The reference samples are obtained following the same process. The concentration of the bulk element Ag was determined with classical chemical precipitation (Gay-Lusac method); P in the reference sample is determined with ICP-OES. The counts for the main element are very high. A filter is used for this element to reduce the count rate including the background of a nearby line of the rhodium X-ray tube. A sensitive scintillation detector is used together with a less sensitive Xe flow detector. The Ag Kα peak is evaluated without background correction as it is by far the most intense peak in the spectrum. Cu is determined with the same detector combination and wavelength selecting crystal. P is determined with a flow detector and a different wavelength selector. The integration times are in the range of about a minute, one replicate measurement of three elements in one sample/standard requires around 3 min.

2.4 Quantitative analysis based on electrochemical methods

2.4.1 General considerations

Analytical methods based on phenomena of electrical current, charges, or potential differences have been in use for more than hundred years. They are the first ones that used physical indicators for quantitative chemical information. Walther Nernst, Robert Behrend, Wilhem Böttger combined classical chemical methods of titration with the response of an electrical measurand. Based on these findings voltammetry and polarography became available in the first half of the last century. Various sophisticated technical options were developed based mainly on these principles and became commercially available. For some time, electroanalytical methods made the detection of the lowest concentration of elements in solution possible, outperforming spectroscopy. GFAAS and ICP-MS were finally able to reach these ultra-trace levels of detection. In addition to methods for elements, cations and anions, specific methods for molecular compounds are available as well.

Basics for detection are predominantly phenomena at boundary layers between two or three electrodes and the solution for measurement in an electrochemical cell. To a lesser extent the electrical conductivity of the solution itself is determined and used for quantitation of the analyte. Measurand is the difference of the electrode potential, the current flowing over time at constant voltage or varying voltage. The electrodes are either permanently installed in the cell or continuously refreshed in the form of mercury drops.

Electroanalytical methods make use of charged ions in solution. This may simply be the conductivity of solutions due to additional charges. It may be based on redox reactions, complexation of ions, precipitation, etc. The prerequisite is that the analyte is available in a known chemical form. If it originates from a digestion process, the decomposition must be complete. Thus, the requirement for presentation of the sample to the system are high. Matrix which – under these conditions is usually in ionic form as well – will often interfere with the analyte. The specificity will be the lower, the less specific the interaction of ions with the electrodes is (e.g., in the case of conductometry). Ruggedness, specificity, and figures of merit in general depend strongly on type and sophistication of the electrodes in use.

Basic laws of electrochemistry are Faraday's law and the Nernst equation. Faraday's law relates the amount of an ion deposited in an electrolytic process to the electrical charges necessary (eq. (2.46)).

Equation 2.46:

$$Q = n \cdot z \cdot F \tag{2.46}$$

The charges (Q) necessary for deposition equal n (moles) of a substance with electrical charge z multiplied by the Faraday constant F. The mass m deposited is the molar mass M multiplied with n.

Nernst's equation relates the potential between electrodes in an electrolyte system to the redox potential of ions present in the system (eq. (2.47)).

Equation 2.47:

$$E = E_0 + R \cdot T / (z_e \cdot F) \cdot \ln (a_{ox} / a_{red}) \tag{2.47}$$

where E is the measured potential between electrodes, E_0 is the standard potential between the electrodes of the system; R is the universal gas constant, T the temperature in Kelvin, and z_e the number of electrons transferred by the process. a_{ox} and a_{red} are the ion activities of the redox system under consideration.

The measurement cell

The simplest and cheapest setup contains two inert electrodes of a certain area at a defined distance. The solution for measurement covers the electrodes completely. The dimensions of the cell define the cell constant. An alternating current is applied between the electrodes to determine the resistivity or the conductance (Siemens, $S = \Omega^{-1}$) of the solutions. The dimensions of the cell and electrodes are included into the calculation via the cell constant. The temperature during measurement must be kept constant as conductivity changes strongly with the temperature. The instruments usually contain an automated temperature compensation such that the conductance becomes a magnitude specific only for the solution of measurement referenced to a temperature of 25 °C.

If two electrochemically different electrodes are used in a cell filled with an electrolyte, a potential builds up between the cells. For analytical purpose, an electrode with well-known potential is used as reference. The potential of this electrode should be constant and independent of the solution for measurement. It consists of a stable redox system inside the electrode and a small and well controlled flow of ions out, providing the constant potential to the solution for measurement. A typical so-called electrode of the second kind is the Ag-AgCl electrode sketched in Figure 2.97.

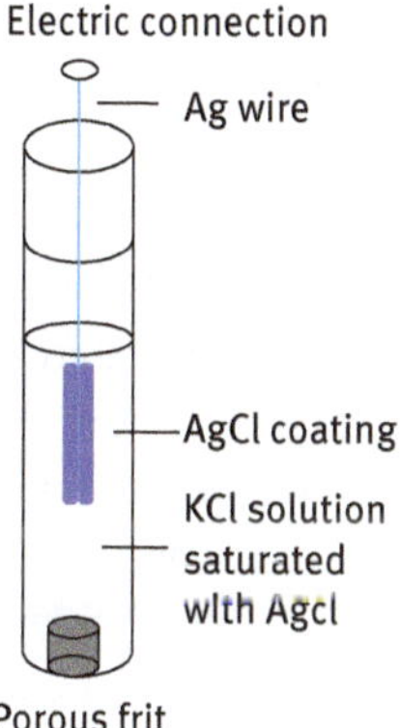

Figure 2.97: Ag/AgCl electrode: a silver stud is covered with a layer of AgCl. The stud is immersed in highly concentrated KCl solution. The system is enclosed into glass but has a connection to the outside solution by a porous material, e.g., sintered MgO. The magnesia stud allows a minimal ion flow to the measurement solution of the electrochemical cell. The potential is defined by the high excess of Ag/AgCl electrolyte inside the electrode.

The second electrode, the indicator electrode, is specific for the type of ion to be measured: redox, ion-sensitive, acid-base, etc. It may be made of platinum or other metals or it may be a glass electrode, again filled with a redox system in limited contact to the solution for measurement. A special type of electrode is the dropping mercury electrode. The working electrode consists of a capillary filled with mercury reservoir and constantly fed by mercury from a reservoir. The capillary and thus the mercury drop are charged relative to a reference electrode of the second kind. Often a third electrode of an inert metal is used to deflect flowing currents. The drop forms at the end of the capillary and is renewed continuously. The electrode surface is kinetically inhibited from the reduction of H_3O^+ ions usually present in the solution for measurement. This allows – to a certain extent – to reduce elements with chemical nobility lower than that of hydrogen.

Stimulation and response

The electrochemical cell, which contains the electrodes, and the electrolyte (composed of analyte, solvent, other matrix, and possible buffers) settles down in an electrical state. This state can be modified by changing the electrical conditions (addition of an AC or DC field) or changing the chemical conditions, e.g., by titration. The electrical response to these changes is measured and evaluated to obtain the qualitative and quantitative analytical information. The cell may run statically or dynamically (without or with a flow of charges). The electrical conditions in dynamic mode can be voltage controlled or current controlled. These options generate technically different instruments and electroanalytical methods with different names.

2.4.2 Classification of electroanalytical methods

Conductometry

A normalized cell (mathematically characterized by an apparatus constant) is used to determine the conductivity of a solution for measurement. Electrical stimulation is an alternating current with selectable frequency up to the kHz range. The conductivity of the solution for measurement is determined with a precise electronic circuit. The AC circuit with higher frequency is applied to prevent additional resistance generated at the boundary layer between electrode and electrolyte (resistance due to polarization). The conductivity in S/cm (Siemens per cm) depends on the activity of all ions in the cell for measurement. The correlation is nearly linear for low concentrations. Doubly charged ions contribute twice to the conductivity. As pointed out already, the temperature influences the conductivity substantially. Direct conductometric measurements are very non-specific and are used often for control of purity of solutions (ultrapure water) or quality control of solutions with roughly defined salt content.

Conductometry becomes substantially more specific when a chemically specific titration process is used. The conductivity of the solution is changing with the chemical composition of the solution for measurement. The most well-known process is an acid/base titration where the hydronium ions of an acid are neutralized by the hydroxy ions of a base. The conductivity of the solution for measurement reduces to a minimum value at the point of complete neutralization and increases again beyond this point. The minimum of conductivity defines the end point of titration and is used to quantitate acidity. The same reduction of charges is used for titrations where the conducting ions are precipitated as an insoluble salt with the help of a suitable solution. Cl^-, Br^-, J^-, CN^-, SCN^- are determined with an $AgNO_3$ solution precipitating the insoluble silver salts. Conversely, Ag^+ can be titrated with an ammonium thiocyanate solution of known concentration (argentometry). Other typical applications are the determination of SO_4^{2-} with a $BaCl_2$ solution or the determination of thiol (-SH^-) with a $HgCl_2$ solution of known concentration. In the latter cases, the conductivity is decreasing as long as free ions are reduced by precipitation. Beyond the point of equivalency, the conductivity is increasing again. A typical response of the electrical circuit to the titration of Cl^- ions with a $AgNO_3$ solution is depicted in Figure 2.98.

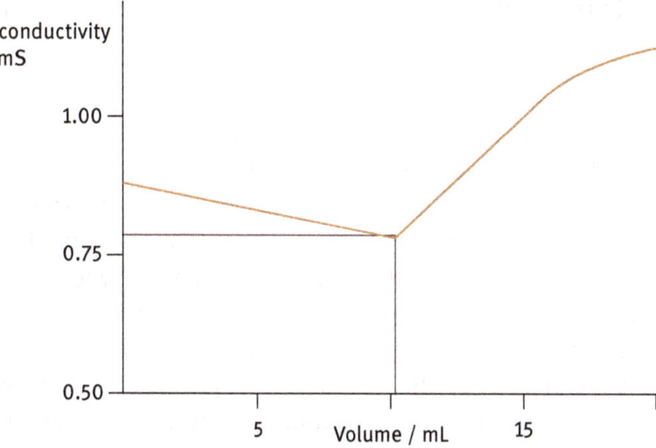

Figure 2.98: Conductivity (ordinate) of a solution containing approximately 0.4 g/L Cl^- with a $AgNO_3$ solution of 0.05 mol/L.

The conductivity of the chloride containing solution is decreasing linearly with small slope upon addition of the silver nitrate solution. Beyond the point of equivalency, the conductivity increases more rapidly. The curve is linear for a short concentration range before it levels off. The point of equivalency is calculated by the intersection of the two linear portions of the conductivity curves. The method of linear regression is used for calculation of the curves.

Potentiometry

In contrast to the method described above, potentiometry is based on the measurement of an electrical potential between two electrodes connected by the electrolyte which contains the analyte. One electrode has a known and constant potential (see the section about electrodes above) the other electrode (indicator) changes its potential depending on the composition of the electrolyte. Electrical currents between the electrodes are avoided as far as possible. The more specific the indicator electrode is toward the analyte element in solution, the higher is the specificity of the determination and the analytical quality (FOM). Like in conductometry, the mere potential in an electrochemical cell can provide quantitative information about an ion or a sum of ions in solution. The result is obtained by calibration with a known concentration of the analyte of interest.

When a titrant is added to the solution of measurement, the potential changes slowly, until close to the point of equivalency a rapid change is observed. Usually, the first derivative of the curve is used for its determination. Thus, the concentration of the analyte can be determined with high analytical quality. Potentiometry is used for acid/base determinations, for redox titrations, and determinations based on complexation and precipitation. Potentiometry allows titrations in strongly diluted solutions.

Voltammetry

The denomination contains "volt" and "ampere" and indicates that determinations are run on the base of voltage and current. While current is determined, the voltage between two or three electrodes is controlled (altered) with time. The electrical field between electrodes forces a movement of ions (electrolysis). This is partially inhibited at the electrodes and in the boundary layer between electrodes and electrolyte solution. To increase the polarization the surface of the indicator electrode is usually very small, in the range of a square mm. At specific voltage levels individual redox reactions are strongly accelerated accompanied by a step upward in current through the cell. The respective ions are moving from the boundary layers into the bulk solution. With steady or stepped increase of the voltage several steps of current, each one specific for a defined reaction, are plotted and provide analytical information in a fast sequential process. The process results in a precipitation of elements on the counter electrode which can be used as a concentration step as well. In this case, precipitation is forced prior to the measurement process. After preconcentration the voltage is inversed, and the voltammetry is run with electrodes coated with a metal layer. This way sensitivity and detection limit enhancement of a factor of 1,000 or higher may be reached. Detection in the ultra-trace range becomes possible. In case of a mercury electrode (see above) the preconcentration is usually amalgamation. The latter allows ultra-trace determinations, e.g., of the environmentally important element cadmium. Voltammetry is often used as three-electrode setup. This avoids changes in the potential while current is flowing. The third electrode acts as counter electrode balancing the current produced by the redox reactions. Its surface is much larger than that of the working electrode.

A typical diagram of the redox system Fe^{2+}/Fe^{3+} is plotted in Figure 2.99. In this case two plots of voltage (abscissa) versus current (ordinate) are recorded at different speeds of voltage ramping.

Coulometry

The basic idea of the method is the determination of the entire charges of an analyte in solution. The oxidized and the reduced substance may be deposited on the respective electrodes or may remain in solution. The end point of the reaction is determined by indicators, or precise control of voltage and current. Using Faraday's constant (see above) the analyte mass can be calculated, provided the conversion is complete and no other electrochemical conversions are taking place in parallel. The electrodes are separated by a diaphragm to avoid that reaction products are blocking the working electrode.

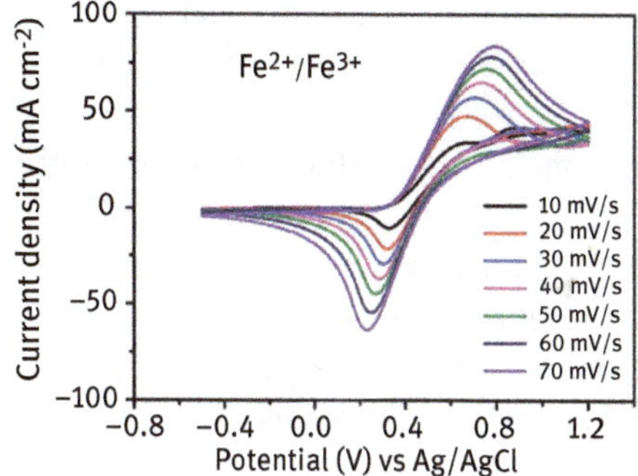

Figure 2.99: Voltammogram of the redox system Fe^{2+}/Fe^{3+}. Source: https://creativecommons.org/.

The voltage between the electrodes can be kept constant (potentiostatic coulometry). The advantage is good specificity for the ions of the respective redox potential. The analyte concentration is calculated by the sum of the charges during the process (eq. (2.48)).

Equation 2.48:

$$Q = \int I \cdot dt \tag{2.48}$$

With decreasing charges in solution, the current flow drops. The determination is usually stopped when the current flow drops to 0.1% of the initial value. This process is usually very time consuming and this limits the analytical applicability.

If the current flow during electrolysis is kept constant, the speed of determination becomes much faster. As the voltage between the electrodes changes during the process, the specificity may be smaller, depending on the matrix. The indication of the end point is more critical as well.

Coulometry is usually applied for specific determinations where all parameters are clearly defined by an SOP or international standard. An example is the determination of water content with the Karl Fischer method. In this case the comparatively complex chemical procedure is simplified by the electrochemical method [123].

The Karl Fischer method is based on the reaction of Iodine with SO_2, which requires the presence of water according to eq. (2.49).

Equation 2.49:

$$2\,H_2O + SO_2 + I_2 \rightarrow SO_4{}^{2-} + 2\,I^- + 4\,H^+ \tag{2.49}$$

The titrant (I^-) is added to the solution for measurement (with traces of water as the analyte) in water-free methanolic solution. In this environment SO_2 reacts to an acidic ester (CH_3OSO_2) which must be neutralized with a base (e.g., imidazole) to enable the titration with iodine. The process remains still the same: methyl sulfite is oxidized to methyl sulfate and iodine is reduced to iodide. The water in the system is consumed. As soon as no more water is available, colored iodine indicates the point of equivalency.

The coulometric method is based on the same principle. However, iodine is produced by a generator electrode, equipped usually with a diaphragm, where a defined iodide solution is injected. In the coulometric determination iodine is generated and subsequently reduced again by the chemical reaction. The end point is detected by voltammetry imbedded in the same apparatus. Absolute masses of water are determined starting from about 10 µg of water and spanning three orders of magnitude of dynamic range.

A second typical coulometric application is the determination of the sum of adsorbable and organically bound halogens in water.

2.4.3 Apparatus and software

Instrumentation for electrochemical applications is comparatively simple. Depending on the number of functions available they can be hand-held for applications in the field or laboratory instruments with small footprint. They do not require special installations of electricity and/or special gas supplies. An example is featured in Figure 2.100.

The sophistication of individual instruments are application-specific electrodes, electronic circuits, and software for special applications often following international standards and providing quality control parameters and instrument validation.

Electrochemical methods are usually applied for routine testing of individual parameters in laboratories, e.g., for water quality or quality control of products.

Voltammetry is often applied for ultra-trace determinations in specific matrices such as water, seawater, and ultra-pure chemicals. The number of elements is usually limited but detection limits in the range of 100 ng/L are possible. An example of a Voltammetry system is featured in Figure 2.101.

Sum parameters like conductivity and oxidative potential can be obtained by easy and fast determinations which are very well standardized. Quantitative anion determinations, e.g., Cl^-, SO_4^{2-}, S^{2-}, CN^- are typical applications using titration with precipitation and registration of the voltage curve. All ions which form insoluble salts with the titrant are potentially interfering. In the case of $AgNO_3$ as titrant these are Br^-, J^-, CN^-, etc. The more insoluble the precipitate with the respective ion, the earlier it will be precipitated. However, the presence of the interfering ions is much bigger in waste waters and process water, than in surface or drinking water. Despite possible interferences, the methods are well developed and described.

Quantitative ultra-trace analyses of cations require complex sample preparation, buffer and modifier reagents and meticulous blank control of all reagents. DIN 38406-

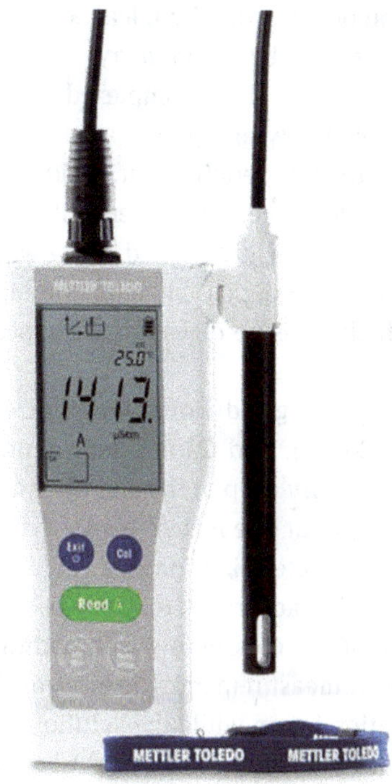

Figure 2.100: Electroanalytical instrument for conductometry and determination of redox potential. Courtesy of Mettler Toledo GmbH, Greifenberg, Switzerland.

Figure 2.101: Voltammetry instrument. Courtesy of Metrohm AG, Herisau, Switzerland.

16 [124], as a typical example, describes determination of zinc, cadmium, lead, copper, thallium, nickel, and cobalt in water, wastewater, and sludge by voltammetry.

It has been emphasized that the analyte must be ionic and not complexed by organics. The solution for measurement must be extremely low in organic carbon content. Drinking water and mineral waters usually fulfil this requirement. Oxidative digestion is required for natural waters, waste, and process water. The water is filtered. Filtrate and solid residues are treated separately. The water is acidified 1/100 with strong nitric or hydrochloric acid, or with hydrogen peroxide, possibly supported by UV light. The residue is digested by microwave or high-pressure digestion at temperatures of 200 °C or higher.

The analyte elements are determined separately using individually optimized methods. The reason is improved specificity. Zn, Cd, Pb, and Cu are determined with a hanging drop mercury electrode (HDME). A mercury drop at the end of a capillary in this case acts as the working electrode throughout one measurement cycle. The drop is only removed and renewed before the next determination starts. Standards of the analyte elements are prepared for calibration according to the expected working range. A buffered KCl solution consisting of the conducting salt, sodium hydroxide and acetic acid, is added to the solution for measurement. The electrolyte is buffered to approximately PH 4.6. The determination is run while the solution for measurement is mixed by stirring. Four current peaks are measured and determined at clearly discriminable voltages: Zn at −0.98 V, Cd at −0.58 V, Pb at −0.38 V and Cu at −0.12 V. The concentrations are determined using the method of standard additions. One individual replicate of four elements requires about 2 min. If we assume sample and two additions each with three replicates, the total analysis time per sample is 18 min per determination.

Tl must be measured separately. Its characteristic potential is at −0.45 V in between Cd and Pb. The latter ones would interfere with the determination. The interference is blocked with an EDTA solution. This must be added together with the electrolyte and the buffer to the solution of measurement.

Ni and Co are often present in very low concentrations. They should be preconcentrated before measurement. Additionally, Zn is usually present in much higher concentrations in these waters and would interfere with the determination. Ni and Co are therefore determined using their dimethylglyoxime complex. The solution for measurement contains the buffered electrolyte and the complexing agent. The elements are determined at the characteristic potentials −0.97 V (Ni) and −1.13 V (Co).

As usual in standardized methods the relative maximum concentrations for interference free analysis of the analyte elements as well as possible other concomitants are stated in the method.

Literature

[32] Mulliken RS: Spectroscopy, molecular orbitals, and chemical bonding. Nobel Lecture, December 12, 1966, pdf research gate.net

[33] Welz B, Becker-Ross H, Florek S, Heitmann U. High-Resolution Continuum Source AAS, Wiley VCH, Weinheim, Germany.

[34] Lahiri A. Basic Optics, 2016, Elsevier Inc. Amsterdam, The Netherlands.

[35] Geboren M, Wolf E. Prinzipien der Optik. Cambridge University Press. ISBN 0-521-63921-2.

[36] Richardson gratings, Technical Note 11.

[37] Schlemmer G, Balcaen L, Todoli JL, Hinds M. Elemental Analysis. De Gruyter, ISBN 978-11-050108-7, 2019.

[38] Sangeeta B, Sunil N, Narendra K, Kothari DC, Ganguli AA, Joshi JB. Flow and Temperature Patterns in an Inductively CoupledPlasma Reactor: Experimental Measurements and CFDSimulations. AIChE-Journal. 2014, 60, 10, 10.1002/aic.14547.

[39] Pfeifer T, Janzen R, Steingrobe T, Sperling M, Franze B, Engelhard C, Buscher W. Development of a novel low-flow ion source/sampling cone geometry for inductively coupled plasma mass spectrometry and application in hyphenated techniques. Spectrochim. Acta Part B, 2012, 76, 48–55.

[40] Agatemor C, Beauchemin D. Towards the reduction of matrix effects in inductively coupled plasma mass spectrometry without compromising detection limits: The use of argon–nitrogenmixed gas plasma. Spectrochim. Acta Part B, 2011, 66, 1–11.

[41] Bilgic AM. et al, A new low-power microwave plasma source using microstrip technology for atomic emission spectrometry. A M Bilgic et al 2000 Plasma Sources Sci. Technol. 2000, 9, 1.

[42] Andrew J, AJ S, Yan C, Jovan J, Velibor P, Ashok M, SJ R, GM H. A New inductively coupled plasma for atomic spectrometry: The microwave-sustained, inductively coupled, atmospheric-pressure plasma (MICAP). J. Anal. At. Spectrom. 2015, 10.1039/c5ja00418g.

[43] ASTM D8322: Elemental Analysis of Crude Oil and Residual Fuels using MP-AES, Agilent Technologies, Inc. May 4, 202, 5994-3230EN

[44] Massmann H. The comparison of atomic absorption and atomic fluorescence in the graphite cuvette. Spectrochim. Acta. 1968, 23B, 215.

[45] L'vov B. Recent advances in the theory of atomisation in graphite furnace atomic absorption spectrometry: The oxygen-carbon alternative. Plenary Lecture. Analyst. 1987, 112, 355.

[46] Schlemmer G, Palladium WB, Nitrates M, More Universal A. Modifier for graphite furnace atomic absorption spectrometry. Spectrochim. Acta. 1986, 41B, 1157.

[47] L'vov B. Atomization from a platform in graphite furnace atomic absorption spectrometry. Spectrochim. Acta. 1978, 33, 153.

[48] Welz B, Wiedeking E. Determination of trace elements in serum and urine with flameless atomization. Z. Anal. Chem.. 1970, 252, 111.

[49] Welz B, Melcher M. Investigations on atomisation mechanisms of volatile hydride-forming elements in a heated quartz cell. Part 1. Gas-phase and surface effects; decomposition and atomisation of arsine. Analyst. 1983, 108, 1283.

[50] Shuttler I, Feuerstein M, Schlemmer G. Long-term stability of a mixed palladium–iridium trapping reagent for in situ hydride trapping within a graphite electrothermal atomizer. J. Anal. At. Spectrom. 1992, 7, 1299.

[51] Hattendorf B, Guenther D. Laser Ablation Inductively Coupled Plasma Mass Spectrometry (LA-ICPMS)) in Handbook of Spectroscopy, second edition ISBN: 9783527321506, 2014.

[52] Wagner B. et al., Topochemical investigation of ancient manuscripts. J. Anal. Chem.. 2001, 369, 674–679.

[53] Fenn JB 2002 Nobel Lecture, Electrospray Wings for Molecular Elephant, www.nobelprize.org/uploads/2018/06/fenn-lecture.pdf
[54] Hillenkamp F, Karas M. Matrix-assisted laser desorption/ionisation, an experience. Int J Mass Spectrom, 2000, 200, 71–77.
[55] Kramida A, Yu R, Reader J, and NIST ASD Team (2018). NIST Atomic Spectra Database (ver. 5.5.6), [Online]. Available: https://physics.nist.gov/asd [2018, June 14]. National Institute of Standards and Technology, Gaithersburg, MD.
[56] Raposo J, Costa L, Barbeira P. Simultaneous determination of Na, K and Ca in biodiesel by flame atomic emission spectrometry. J. Braz. Chem. Soc. 2015, 26.
[57] Guo T, Baasner J, Gradl M, Kistner A. Determination of mercury in saliva with a flow-injection system. Anal. Chim. Acta. 1996, 320, 171.
[58] Slavin W, Carnrick G, Koirtyohann R. Background Correction in Atomic Absorption Spectroscopy (AAS). C R C Crit. Rev. Anal. Chem. 2008, 19, 2, 95.
[59] Smith SG, Hieftje D. A new background-correction method for atomic absorption spectrometry. Appl. Spectrosc. 1983, 37, 419.
[60] Wibetoe G, Langmyhr FJ. Absorption spectrometry caused by zeeman splitting of molecules. Anal. Chim. Acta. 1987, 198, 81.
[61] Wibetoe G, Langmyhr FJ. Spectral interferences and background overcompensation in inverse zeeman-corrected atomic absorption spectrometry: Part 2. The effects of cobalt, manganese and nickel on 30 elements and 53 elements lines. Anal. Chim. Acta. 1985, 176, 33.
[62] De Loos- Vollebregt MTC. Background Correction Methods in Atomic Absorption Spectrometry. Encycl. Anal. Chem. 2013, https://doi.org/10.1002/9780470027318.a5104.pub2.
[63] Browner RF, Boorn AW. Sample Introduction Techniques for Atomic Spectroscopy. Anal. Chem. 1984, 56, 7, 875A–888A.
[64] Wenzel N, Trautmann B, Große-Wilde H, Schlemmer G, Welz B, Marowsky G. Cars temperature studies of the gas phase in a massmann-type graphite tube furnace. Opt. Commun. 1988, 68, 75.
[65] Kurfürst U ed, Solid Sample Analysis, Direct and Slurry Sampling using GF-AAS and ETV-ICP 1998, Springer-Verlag, Berlin Heidelberg.
[66] Miller-Ihli N. Slurry sampling for graphite furnace atomic absorption spectrometry. J. Anal. Chem. 1990, 337, 271.
[67] Fang Z, Ruzicka J, Hansen E. An efficient flow-injection system with on-line ion-exchange preconcentration for the determination of trace amounts of heavy metals by atomic absorption spectrometry. Anal. Chim. Acta. 1984, 164, 23.
[68] Welz B, Sperling M. Atomic Absorption Spectrometry, ISBN: 978-3-527-28571-6 Wiley-VCH, Weinheim, 1998.
[69] Frech W. Non-spectral interference effects in platform-equipped graphite atomisers. Spectrochim. Acta. 1997, 52B, 1333.
[70] Hein H, Klaus S, Meyer A, Schwedt G. Richt- und Grenzwerte im deutschen und europäischen Umweltrecht;Luft – Wasser – Boden – Abfall – Chemikalien, Springer VDI-Verlag, 2007.
[71] D'Ulivio A, Dedina J. The relation of double peaks, observed in quartz hydride atomizers, to the fate of free analyte atoms in the determination of arsenic and selenium by atomic absorption spectrometry. Spectrochim. Acta Part B At Spectrosc. 2002, 57, 12, 2069–2079, DOI:10.1016/S0584-8547(02)00165-9.

[72] Sperling M, Yin X, Welz B. Determination of ultra-trace concentrations of elements by means of on-line solid sorbent extraction graphite furnace atomic absorption spectrometry. J. Anal. Chem. 1992, 343, 754.
[73] Schlemmer G, Radziuk B. Analytical Graphite Furnace Atomic Absorption Spectrometry, ISBN 3-7643-5770-3 Birkhäuser Verlag, 1999.
[74] Winefordner JD. Principles, methodologies, and applications of atomic fluorescence spectrometry. J. Chem. Educ. 1978, 55, 72.
[75] Greenfield S. Atomic fluorescence spectrometry; progress and future prospects. TrAC Trends Anal. Chem. 1995, 14, 435.
[76] Butala SJ, Scanlan LP, Chaudhuri SN. A detailed study of thermal decomposition, amalgamation/atomic absorption spectrophotometry methodology for the quantitative analysis of mercury in fish and hair. J. Lebensmittelschutz. 2006, 69, 11, 2720–2728, 10.4315/0362-028x-69.11.2720.
[77] Water quality – Determination of mercury – Method using atomic fluorescence spectrometry. NEN-EN-ISO 17852: 2008
[78] Chemical analysis – Decision limit, detection limit and determination limit under repeatability conditions – Terms, methods, evaluation. DIN 32645 2008; 11
[79] Grotti M, Magia E, Leardib R. Selection of internal standards in inductively coupled plasma atomic emission spectrometry by principal component analysis. J. Anal. At. Spectrom, 2003, 18, 274–281.
[80] Nölte I. ICP Emissionsspektrometrie für Praktiker. Wiley-VCZH Verlag GmbH, ISBN 3-527-30351-0, 2002.
[81] Mermet J. Ionic to atomic line intensity ratio and residence time in inductively coupled plasma-atomic emission spectrometry. Spectrochim. Acta B, 1981, 44, 1109–1116.
[82] Botto RI. Long term stability of spectral interference calibrationsfor inductively coupled plasma atomic emission spectrometry. Anal. Chem., 1982, 54, 1854–1859.
[83] van Veen EH, de Loos-vollebregt MTC. Application of mathematical procedures to background correction and multivariate analysis in inductively coupled plasma-optical emission spectrometry. Spectrochim. Acta Part B. 1998, 53, 5, 639–669.
[84] Kalman RE. A new approach to linear filtering and prediction problems. J. Basic Eng. 1960, 35–45.
[85] Barnett WB, Fassel VA, Knisely RN. Theoretical principles of internal standardization in analytical emission. Spectrochim. Acta, 1982, 37B, 1037.
[86] US EPA. (2018) Method 6010D (SW-846 Update VI): "Inductively Coupled Plasma – Optical Emission Spectrometry", Revision 6. https://www.epa.gov/sites/production/files/2015-12/documents/6010d.pdf
[87] Cui J, Traynor T; (2021) Thermo Scientific application note AN74146-EN 0621S
[88] Analytik Jena AG (2021), Applikationsschrift PlasmaQuant 9100 Elite, Bestimmung von Spurenelementen und Mineralien in Speiseölen und -fetten mit HR ICP-OES.
[89] ISO 21033 (2016). Animal and vegetable fats and oils – determination of trace elements by inductively coupled plasma optical emission spectroscopy (ICP-OES).
[90] European Pharmacopoeia (Ph. Eur.) 10th Edition (2022)
[91] United States Pharmacopeia, USP <857> (42 nd edition) (2019)
[92] Norm DIN EN ISO 18412:2007-02; Beuth-Verlag Berlin
[93] Noelle A, Hartmann GK, Fahr A, Lary D, Lee YP, Limao-Vieira P, Locht R, Martin-Torres FJ, Orlando JJ, Salama F, Vandaele AC, Wayne RP, Holland DMP Hrsg. UV/Vis+ Spektrendatenbank, 2015. 10. Aufl..
[94] UV atlas of organic compounds", published in collaboration with the Photoelectric Spectrometry Group, London, and the Institute for Spectrochemie und Angewandte

Spectroskopie, Dortmund, volumes 1-5, London, Butterworths and Weinheim, Chemie, 1966-1971

[95] ISO 6878 (2004) Water quality – Determination of phosphorus – Ammonium molybdate spectrometric method; reviewed and confirmed (2019)

[96] Nagul EA, McKelvie ID, Worsfold KSD. The molybdenum blue reaction for the determination of orthophosphate revisited: Opening the black box. Anal. Chim. Acta. 2015, 26, 890, 60–82.

[97] Council Directive 98/83/EC of November 3, 1998. On the quality of water intended for human consumption.

[98] ISO 6777 (1984) Water quality – Determination of nitrite – Molecular absorption spectrometric method

[99] Morés S, Monteiro GC, da Silva Santos F, Carasek E, Welz B. Determination of fluorine in tea using high-resolution molecular absorption spectrometry with electrothermal vaporization of the calcium mono-fluoride CaF. Talanta, 2011, 85, 2681–2685.

[100] ASTM E308 – 08. Standard Practice for Computing the Colors of Objects by Using the CIE System

[101] Grundlagen Instrumentation und Techniken der UV VIS Spektroskopie; Analytik Jena AG, Jena

[102] Günzler H, Heise HM. IR-Spektroskopie – Eine Einführung, VCH Verlagsgesellschaft mbH, Weinheim, 3rd edition.

[103] Barbooti M, Al-Jibori S, AlJanabi A, AH A, Sami N, Aziz B, Basak S (2017). Synthesis, characterization and thermal studies of mixed ligand mercury(II) complexes of N-hydroxymethylsaccharin (Sac-CHOH) and phosphine or heterocyclic amine co-ligands. Researchgate.net

[104] Fanelli S, Zimmermann A, Gandolpho Totóli E, Nunes Salgado HR. FTIR spectrophotometry as a green tool for quantitative analysis of drugs: Practical application to amoxicillin. J. Chem. 2018, 11-12, 1–8. DOI:10.1155/2018/3920810.

[105] Naftaly M. et al., THz (IEEE), Cambridge, England, 140.

[106] Golay MJE. Theoretical consideration in heat and infra-red detection, with particular reference to the pneumatic detector. Rev. Sci. Instrum. 1947, 18, 347, 10.1063/1.1740948.

[107] Raman CV. A new radiation. Indian J. Phys, 1928, 2, 387–398.

[108] Fleischmann M, Pj H, AJ M. Raman spectra of pyridine adsorbed at a silver electrode. Chem. Phys. Lett. 1974, 26, 2, 163–166.

[109] Bell SEJ. et al., Towards reliable and quantitative surface-enhanced raman scattering (SERS): From key parameters to good analytical practice. Angew. Chem. Int. Ed. 2020, 59, 5454–5462. 10.1002/anie.201908154.

[110] Zhang W, Yeo BS, Schmid T, Zenobi R. Single molecule tip-enhanced Raman spectroscopy with silver tips. J. Phys. Chem. C. ASAP Article. 10.1021/jp064740r.

[111] Stahlkopf F (2018) Qualitative und quantitative Untersuchungen von Designerdrogen und berauschenden Mitteln mittels Raman Spektroskopie; dissertation Kiel University

[112] Stella P, Kortner M, Ammann C, Foken T, Meixner FX, Trebs I. Measurements of stickoxide and ozon fluxes by eddy covariance at a Wiese. Biogeosci. 2013, 10, 5997–6017. 10.5194/bg-10-5997-2013.

[113] patents.google.com/patent/WO2009015640A1/de

[114] Eberl R, Parthey B, Wilke J. Fluorimetrische Schnellbestimmung von Bitterstoffen in Bier und Würze. Brauwelt, 2006, 27, 788–790.

[115] Abramowitz M, IA S. Pocketbook of Mathematical Functions, H. Deutsch, 1984.

[116] Hu Q, Noll RJ, Li H, Makarov A et al (2005) The Orbitrap: a new mass spectrometer. J Mass Spectrom 40:430–443.

[117] Briois C. et al., Orbitrap mass analyser for in situ characterisation of planetary environments: Performance evaluation of a laboratory prototype. Planet Space Sci. 2016, 113, 33–45.

[118] Derrick QC Jr., Kenneth MR, (2017) Glow discharge mass spectrometry – An overview. Encyclopedia of Spectroscopy and Spectrometry (Third Edition).

[119] Xie D, Mattusch J, Wennrich R. Separation of organoarsenicals by means of zwitterionic hydrophilic interaction chromatography (ZIC®-HILIC) and parallel ICP-MS/ESI-MS detection. Eng. Life Sci, 2008, 8, 6.

[120] Takahasi G. Sample preparation for X-ray fluorescence analysis; pressed and loose powder methods. Rigaku J. 2015, 31, 10, 26–30.

[121] Seyfarth A. X-ray spectroscopy. Jw R, Skelly FEM, GM F. Eds, Undergraduate Instrumental Analysis, 7th ed. CRC Press, Baton Roca, FL, USA, 2014.

[122] Hinds M. Elemental Analysis, An Introduction to Modern Spectrometric Techniques, Walter de Gruyter GmbH, ISBN 978-3-11-050107-0, 2019.

[123] Standard Test Method for Water in Organic Liquids by Coulometric Karl Fischer Titration; ASTM E1064-12, (2016).

[124] DIN 38406-16 (1990) German standard methods for the examination of water, waste water and sludge; cations (group E); determination of zinc, cadmium, lead, copper, thallium, nickel, cobalt by voltammetry (E 16)

3 Analytical information using chemical and biochemical methods

Without proper method no scientific knowledge! Aristoteles suggested to dissect a complex scientific question such that the individual elements of the problem become manageable. Aristoteles was one of the first known people to suggest how to base scientific progress on methods. About 2,300 years later Enrico Fermi challenged his students with the so-called Fermi-questions. Queries from daily life where the ability of dissection of a problem and estimation of magnitude were required to obtain a reasonable answer. A simple example: what fraction of time in human life is 1 ppm? Dissection of a complex mix into manageable components is one of the core competencies of analytical chemistry and the main topic of this chapter.

3.1 Separation techniques

3.1.1 General remarks

The goal of development of the methods described in Chapter 2 is qualitative and quantitative information with a minimum of chemical sample handling. The presumed silver bullet is the interaction of an energy field with the sample of interest and the detection of response of interaction between analyte species and energy field. The latter one is evaluated by computational methods. A perfect example is the laser-induced plasma spectroscopy (LIPS; see Chapter 2) where laser light is used to release and excite matter to respond with characteristic optical radiation. This system can even be exported to other luminaries for exploration. We have seen, however, that the applied methods for excitation, interpretation of the answer received, and mathematical correction methods are often not specific enough to obtain the required accuracy and detection levels. Examples are the complex Raman spectra of proteins, spectral interferences in optical spectroscopy, or the interaction of ions when subjected to electrochemical processes. In all these cases, sample preparation with subtle chemical processes offers solutions for applicability, improvement of specificity, and improvement of analytical figures of merit. Analytical information is therefore usually based on a method where chemical competence and optimized systems based on physical methods are combined in a clever way. Although analytical scientists in research and application should overlook the entire analytical requirement, there is still a distinct classification or assignment of “spectroscopists” and “separation scientists.” In this chapter, we will discuss the methods which are mainly based on chemical and physicochemical processes for segregation of wanted species from unwanted species. The detection methods required to quantitate these separated species have mainly been thematized in Chapters 1 and 2.

https://doi.org/10.1515/9783110689662-003

3.1.2 Sample preparation

Sample preparation depends on the nature of the sample, the analytes to be determined and their concentration and on the desired FOM for the analysis. In the scope of this book, we will not be able to discuss the wide field of sample preparation thoroughly. Comprehensive literature on the topic is available [125].

In this paragraph we will provide a few very general considerations on the topic only. Most analytical methods require liquid samples which are fed to the place of measurement manually or with automatic sampling devices. Clearly defined specifications as to the goal of the required information are needed for the total analytical method in general and for sample preparation in particular. Often sample preparation is the most important part of the total analysis. In the easiest case the sample may be just taken as it is. It may have to be diluted or preconcentrated by evaporation of solvent. Supporting liquids, such as modifiers or buffers, may have to be added. Reference solutions (see, e.g., isotope dilution, Chapter 2) may be added to obtain quality control. The aim of the process may be a reproducible extraction of one or several elements or element species out of a solid which is still partially present and may be discarded after the digestion procedure. It may be a complete dissolution of the sample with a mix of inorganic and organic compounds or a completely mineralized sample with only element ions or coordinated ions in solution. In any case, the composition of the final solution for measurement will have an influence on the measurement process and hence on the analytical information obtained. It must be optimized together with the analytical method for determination. An exactly defined sample preparation may take a lot of burden from interference control within the measurement device, however, may be at the cost of a lengthy sample preparation process. Thus, sample preparation plays an important role in economic planning of the entire analytical method.

Standards describing the analytical methods for environmental, medical, consumer protection, life science, and product quality applications usually define exactly the type of sample pretreatment. Even if more modern methods are yet available, the standards should be followed.

An example may explain the statements: Assumed an agricultural soil is polluted with chromium. The analytical question might be total content of chromium in the soil. The method of choice would therefore be a complete dissolution of the soil without any remaining residues. This would be accomplished by a strongly oxidizing mix of acids active at high temperatures, possibly at elevated pressure. If residue is accepted, a time limited boiling under reflux with an acid mix defined in concentration will give an answer which may be different from the complete dissolution. If only the content of chromium is of interest which can be extracted into the ground water and plants an extraction of the soil with a mild acid or buffer may provide the answer of interest. If only the species with high toxicity to the biological organism (Cr^{6+}) is of interest, yet a different mix of buffer solutions, e.g., acetate, sodium carbonate, and phosphate may be the extractant of choice [126].

In general, sample preparation starts from a gas, a liquid, or a solid. Gases are often not treated at all and are analyzed directly. If they contain a high load of matrix, e.g., organic compounds and sulfur, they may be condensed for precipitation or crystallization. They may be bubbled through liquids where distributed compounds are adsorbed. The sample becomes a liquid by this process. They may be filtered to extract small particles from a defined volume and be treated consecutively as a solid sample. Like gases, liquids may contain high matrix or be very dilute in matrix (e.g., blood versus liquor, tap water versus sea water). The question whether sample pretreatment is required, depends again strongly on the analytical question and the analytical method used consecutively. Blood analysis using mass spectroscopy, for example, requires a completely different pretreatment than that required in electrothermal AAS. Liquids may be analyzed directly or oxidized with acids at elevated temperatures to reduce or remove the amount of organically bound matter. Excess of matrix may be removed by precipitation and filtration. Particulate matter in the original liquid may be filtered and treated separately as a solid. Pretreatment of the solid starts with sampling of an amount which is representative for the type of sample investigated and the analytical question (is it bulk analysis? is it profiling?). Often the sample must be broken up and milled before treatment with chemical agents for extraction of the substances of interest or partial or complete dissolution of the sample. Further treatment by dilution or the removal of unwanted chemical compounds from the sample may follow in the process making sample pretreatment to a multi-step procedure.

The process of dissolution depends very strongly on solubility of a compound (a halide, an oxide, a sulfate, a phosphate, an organic polymer) in a solvent. Compounds of mainly organic structure are often dissolved in non-polar solvents or strong acids which are not active via the H_3O^+ ion (e.g., concentrated H_2SO_4). Some polymers, such as fluorinated hydrocarbons (Teflon) are almost inert against non-polar and polar solvents. They must be decomposed by application of high temperatures.

Inorganic compounds, salts, rocks, and ceramics are often attacked by acids, such as HCl, HNO_3, and H_2SO_4. The general rule of thumb of chemical reactivity (doubling of speed of reaction by an increase by 10 degrees in temperature) holds true for digestion processes as well. Digestions are therefore run at least at elevated temperatures; more often, under high pressure, which result in high temperatures, often limited by the mechanical stability of the pressure container and/or the thermal stability of the container material used. The mix of acids at elevated temperature often results in generation of compounds with high chemical activity. Aqua regia, as an example, reacts via the Cl radical and NOCl, formed by HNO_3 and HCl. It reacts as well by complexation of gold or platinum with Cl radicals. If solids are formed of extremely involatile compounds, ceramics, stable oxides, and other stable and insoluble salts, treatment by active liquids may not be able to completely break up the solid. In these cases, melting with basic or acidic salts such as Na_2CO_3 or $KHSO_4$ transforms the insoluble compound into a soluble solid which can be treated by water or acid in a consecutive digestion step. These melts must be run in specific vessels, such as platinum crucibles or other high-temperature stable vessels at temperatures in the range of up to 1,000 °C.

The general trend in digestion is to keep the number of chemically active compounds small. If possible, only one acid should be used. The reason is obvious: if not removed by another sample pretreatment step later, the chemically active agent is present as highly concentrated matrix in the analytical measurement process. It has been discussed earlier (see Chapter 2, ICP-MS and/or graphite furnace AAS) that chlorine, originating from highly concentrated HCl or $HClO_4$ causes potential spectral (ICP-MS) and non-spectral (ETAAS) interferences. Every additional agent added to the sample is a potential source of contamination of the original sample. HNO_3 is a widely used very active compound which can be cleaned effectively, is relatively easy to handle and does not show pronounced effects for many of the analytical methods. At high temperatures above 200 °C it develops a very high oxidation potential. The possibly unwanted generation of NO_x in the process of digestion can be reduced by the addition of H_2O_2 which can be cleaned easily as well, and which is not a complex matrix for analytical methods. The burden of chemical activity is put on the apparatus which requires highly stable and inert vessels, carefully controlled heating, high pressure tolerance, and sophisticated technical safety control features. Apparatus for sample pretreatment became as important as the analytical instrument in the last decades. The systems include different ways of heating (in particular microwave energy), other sources providing chemical activity (e.g., UV radiation), and extensive logging and documentation of the entire digestion process including all relevant physical parameters.

Transformation of the analyte into a chemical form easy-to-manage by the selected analytical method is of utmost importance to the total analytical process, and to its figures of merit. Chemical reactions play the essential role for the success of the entire method. However, they are strongly supported by sophisticated apparatus which make the process easier, safer, more controllable, and often reduce the number of active chemical substances. All agents added to an analytical process are "matrix" which may complicate the detection process. Even more important is the control of contamination of the sample by the reagent or losses of analyte in the process of transformation. The most straightforward approach is therefore the silver bullet.

3.1.3 Separation

Separation of different species of analyte, separation of matrix from analyte or from other matrix compounds, and enrichment of analyte are core strategic tasks of analytical chemistry. Making use of physical effects and chemical reactions in kinetically or thermodynamically controlled processes is collectively seen as the sphere of "separation sciences." The use of separation sciences requires mixtures of substances. Usually, all samples that are handled in analytical chemistry are mixtures. The separation processes convert these mixtures partially or completely into their constituent parts. The processes may be intended for analytical purposes only. In this case the separated substances are discarded after analysis. The process may be preparative to supply the separated and often enriched substance to an analytical method or to an industrial process. Separation science is often named "chromatography" – "writing with colors" – originating from first separation of plant pigments in the early days of the twentieth century. While qualitative detection of different substances may not require sophisticated detection as a subsequent step, quantitative detection requires physical means of measurement. After the separation or enrichment process detection techniques are used which have been discussed already in Chapter 2: optical spectrometry, i.e., absorption, emission, fluorescence, mass spectrometry, or other detection methods based on electrochemistry, temperature, acoustic waves, etc. Generally, gases or liquids can be subjected to separation. Ions, small organic molecules up to proteins of high mass, can be subjected to the various separation processes. No surprise that the use and the commercial market for chromatographic instrumentation is by far the biggest one in the field of analytical chemistry.

Separation can be achieved by:
- filtering and/or precipitation (partition of solid and liquid)
- separation of the gaseous space above a liquid, including boiling (isolation of a gaseous phase from liquid or solid phases)
- chemical phase separation (chemical generation of a gas, see chemical vapor generation Chapter 2)
- solvent extraction (distribution into different solvents, e.g., polar/nonpolar)
- adsorption/desorption to a solid phase (temporary or permanent fixing of a compound to a solid phase)
- adsorption/desorption to a liquid
- complexation (prevention of a response of the complex to the analytical method)

Besides these general principles, a lot of modifications and special techniques are in use for analytical purposes.
- Preconcentration of an analyte species is often achieved by similar processes as described above. Solvent or solid phase extractions are most popular examples.
- Amalgamation is a widely used technique to perform extreme ultra-trace detection of mercury.
- Adsorption of gaseous metal compounds to metal coated carbon surfaces allows sub picogram detection of species.
- Exponential preconcentration methods use reaction mechanisms (predominantly in the field of biomolecules) which uses a complex chemical molecule (an enzyme) to double the amount of the analyte, the DNA under investigation, in an exactly defined thermal process. The products of the doubling are directly used for a second and third doubling in the next thermal sequence. Ten cycles

which take about 2 h of time will amplify the original amount of DNA by 2^{10}, amounting to 1,024, three orders of magnitude. This PCR method (polymerase chain reaction) was implemented into modern bioanalytical chemistry in 1983 by Kary Mullins and became one of the most important amplification processes for analytical and preparative purposes.

Bioanalytical methods such as TALEN or CRISPR/CAS meanwhile allow to cut DNA sequences at selected and exactly defined places and "edit" the gene sequence by cutting out parts or introducing new parts into the sequence. This modification process is extremely important in biochemical design. It is, however, beyond the topic of quantitative analytical methods.

In the following sections we will discuss the basic principles, instrumental realization, and typical application of methods primarily based on separation science.

3.2 Basic concepts in chromatographic processes

Chromatography is the separation of matter by repeated distribution of species between a stationary phase and a mobile phase. The mobile phase usually defines the type of chromatography: gas chromatography, liquid chromatography, supercritical fluid chromatography, field flow fractionation. Moreover, the state of matter of the stationary phase as well as the physicochemical mechanisms of separation serves to further distinguish the different separation techniques.

In the following a few fundamental parameters will be described which are used to describe the analytical performance of species separation.

The concentration of the species of interest is distributed between two phases: the stationary and the mobile phases. The distribution between the phases demands chemical equilibrium (eq. (3.1)).

Equation 3.1:

$$K_c = c_{st}/c_m \tag{3.1}$$

K_c is the distribution constant of the analyte species of interest; c_{st} is the concentration of the analyte species in the stationary phase; and c_m in the mobile phase.

1. How is the species of interest distributed between the phases?
   The distribution depends on several parameters; one of the most important ones is the bonding force of the phases for the species of interest. The concentration in the mobile phase is calculated per unit volume of the phase. The concentration in the stationary phase may be related to volume, mass, or surface. This will have an influence on the distribution coefficient. Equation (3.1) changes to eq. (3.2).

Equation 3.2:

$$K_c = (m_{st}/V_{st})/(m_m/V_m) \tag{3.2}$$

In this case the mass of analyte in the stationary and in the mobile phase m_{st}, m_m are related to the volumes of the phases V_{st}, V_m.

2. How bulky is the mobile phase compared to the stationary phase?
3. The volumes of the mobile phase relative to the stationary phase are defined as the phase ratio β. Equation (3.3) describes the simple relation.

   Equation 3.3:

$$\beta = V_m/V_{st} \tag{3.3}$$

It is obvious that the distribution is correlated with the phase ratio. The phase ratio is always bigger than unity. Typical values are in the range of 10 for high-pressure liquid chromatography and 100 for gas chromatography (see below).

4. How much of the species will fit on the stationary phase?
   This will depend primarily on the analyte concentration in the mobile phase and the volume of the mobile phase. Analytically and technically important as well is the number of active centers per mass or volume of the entire stationary phase. The latter is defined as capacity of the stationary phase. It may be indicated in milliequivalent per mass or volume of the stationary phase. The higher the capacity, the higher can be the analyte concentration in the mobile phases before the separation column (or medium) is overcharged. The capacity can be factorized (eq. (3.4)).

   Equation 3.4:

$$k' = c_s V_s / c_m V_m \tag{3.4}$$

   where k' is called “factor of capacity.”
   k' is obviously associated with β and K (see above). The capacity of the stationary phase will influence the residence time of the analyte on the stationary phase. The higher k' and the longer the analyte will reside on the solid phase, the longer the retention time.
5. How will the species be separated from each other? Chromatographic resolution, similar to optical resolution discussed earlier (see Chapter 2), becomes finally visible in the chromatogram. Can the peaks of the analyte species be clearly separated without substantial overlap? The detector response in chromatography is orders of magnitude faster than the separation process. Thus, the latter one defines the resolution R' of the process. Movement of the eluent through the stationary phase requires a time t_0 under the selected technical conditions (type of column, pressure, temperature, etc.). The species of interest will leave the column later, after retention times t_{r1}, t_{r2}, etc. These gross retention times include the time for the eluent and can be corrected by t_0, yielding the net retention times. The peaks are not

infinitely narrow but have a width ω which depends on many parameters and is different for each analyte. Assuming a normal (Gaussian) peak defines the distribution of the peak, the value of expectation is found at peak maximum, the standard deviation σ defines the inflection point to the left and to the right side of the maximum ($+\sigma$, $-\sigma$). If we construct tangent lines to the points of inflection, they will intersect the abscissa (the timeline of the chromatogram) at value of expectation with a standard deviation of $\pm 2\sigma$. In this case 95.49% of the values will be within the area of the curve. This defines width ω of the elution peaks. The resolution of two species is defined as double the difference of retention times (at peak maximum) divided by the sum of the peak widths (eq. (3.5)). $R > 1.5$ is assumed as sufficient resolution for two species. Resolution is sketched in Figure 3.1.

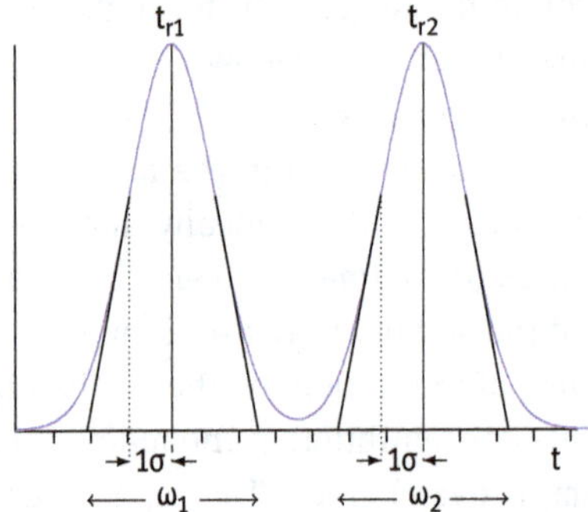

Figure 3.1: Resolution in chromatography.

Equation 3.5:

$$R = 2(t_{r2} - t_{r1})/(\omega_1 + \omega_2) \tag{3.5}$$

6. One more parameter used in separation processes is the efficiency. Efficiency depends on the type of carrier material. Typically realized as a resin, it may be inorganic, mainly based on silica gel or it can be polymer based. The particle size plays an important role as well as the porosity of the material. These parameters define how the mobile phase is moving through the column and how the contact with the functional groups is achieved. The concept of separator stages helps to explain efficiency. Distillation changes the ratio between two components. Per stage the ratio is changed by a certain ratio. Multiple distillation starts with the ratio of components after the first step to proceed to the second step. This is defined as a theoretic separator stage. The concept can be transferred to separation columns. The total length is separated into theoretic separator stages where a thermodynamic equilibrium between the stationary and the mobile phase is obtained in infinitely short time. The efficiency of the column is defined as the number of theoretic separator stages N (eq. (3.6)).

Equation 3.6: Number of theoretical separator stages (N) of a column with length l. H is the length of one theoretical separator stage:

$$N = l/H \tag{3.6}$$

The number of theoretical stages for a species can be calculated with the help of the gross retention time t_r. For the calculation, the half width of the elution peak of the species must be taken into consideration. The longer the gross retention time and the narrower the elution peak, the higher becomes the number of theoretical stages. This simple model assumes ideal flow conditions and Gaussian-shaped elution peaks. Non-ideal conditions, typical for chromatographic processes, have therefore been developed to refine the simple model of theoretical stages. The model developed by van Deemter et al. [127] takes imperfections of flow and mass transport of the mobile phase into account. These are mainly:

- channel formation leading to different flow speed through the solid phase
- retarded mass transport between mobile and stationary phase
- retarded and non-spontaneous attainment of thermodynamic equilibrium
- analyte diffusion phenomena between mobile and stationary phase

The consequences of variations of the flow speed through the column (see above), the "imperfections" are described by mathematical terms. Three terms are used in van Deemter's equation which can be traced back to physical and technical conditions in the chromatographic process. The height of the theoretical stage increases additively by each of these parameters, the number of theoretical stages decreases, correspondingly.

One parameter describes the difference in distance of the mobile phase through the stationary phase due to channels and diffusion. This can be traced back to the particle diameter of the stationary phase and a factor which describes the quality of the packing in the column. The smaller the particle diameter, the smaller becomes the parameter of imperfection. The same is true for the packing quality. The latter factor can be modelled by mathematical simulation of liquids flowing through a porous solid material [128].

The parameter describing diffusional processes depends mainly on the mobile phase. Diffusion is usually small in liquids and can often be neglected. However, for gaseous mobile phases the porosity of the solid phase is of importance. The higher the porosity, the larger becomes the theoretical height of the stages in the column.

The factor describing retarded mass transfer depends strongly on porosity again. Species on the surface of the solid phase are transported faster than those which are trapped into pores. Lower flow rates through the solid as well as higher temperatures both increase diffusional processes and thus minimize the effects.

Parameter optimization in chromatographic systems aim to keep the "parameters of imperfection" small and approach the ideal situation of the theoretical stages described above. All imperfections finally result in broadening of elution peaks and thus in a reduction of the resolution of separation.

The elution process substantially influences the FOM of the analytical process. The optimal eluent often consists of a subtle mixture of two or more components which assure the required chemical condition for elution. During method development the mixture is subject to optimization for best analytical quality. The process may be cumbersome as the eluent reservoir may have to be changed frequently until the final mix is achieved. If the eluent is not changing in composition during the elution process, it is named "isocratic." Only one high-pressure pump is required for the procedure. The method is straightforward but not flexible. The matrix may be strongly changing, and matrix influence would require a correction of the eluent mix. The mixture of the composition may change during composition, e.g., from a more polar to a less polar component, etc. Two or more pumps are required for this "gradient elution process" which pump with different flow speed ratios. This will change the chemical conditions of elution from the first to the last analyte peak. The elution process can be optimized with respect to chromatographic separation and time of analysis. The flexibility of the method increases substantially as optimal conditions can be pre-programmed into the instrument software and run automatedly. In particular, method development becomes much more flexible. The instrumentation becomes more expensive though as the high-pressure pumps are a substantial cost of the entire instrument. In addition, there is one more active parameter to be controlled. Changing mixtures may lead to temperature or viscosity changes by the mixing process which must be known and taken into consideration.

3.3 Ion chromatography

3.3.1 General remarks

Separation of ions increases the specificity of electrochemical methods (see Section 2.1.15). Ion chromatography, however, is not limited to electrochemical detection only. The central component is a solid stationary phase which adsorbs ions dissolved in a liquid phase passing through the column under elevated or high pressure, followed by elution with the solution for measurement. This central unit is usually backed up by a column for pre-cleaning of the sample and a column for conditioning of the solution for measurement for the electrochemical measurement process. The methods of ion chromatography are distinguished by the mechanisms of specificity. The ions may be exchanged (classical ion chromatography). Adsorption of strong ions can be minimized by formation of a kind of membrane enfolding specific solid phase material (ion exclusion chromatography). Modification of the bonding efficiency on the solid stationary phase may be added to the mobile phase. In this case, the term ion pair chromatography is used.

The experimental setup of a system for ion chromatography consists of a column with the sorbent selected for the type of application. Often a pre-column of a different kind of material is used to remove high loads of unwanted matrix from the sample. A suppressor column modifies the eluent such that it is suited for the detector, e.g., conductive measurement. The column system is fed by a pump generating pressures in the range of 10^5 kPa and more. An injection valve serves to inject a volume of typically 10–50 μL into the system (see Figure 3.2).

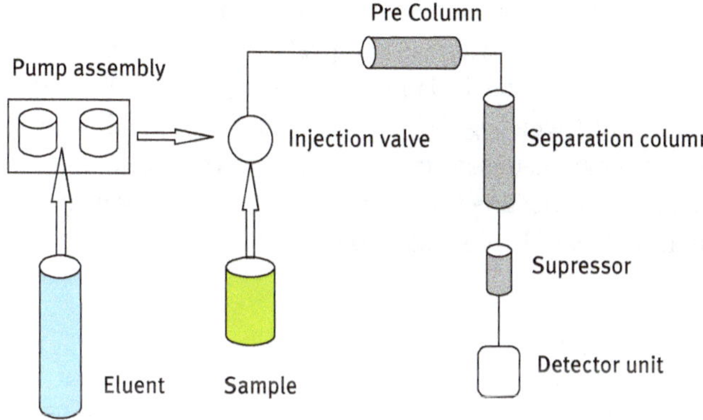

Figure 3.2: Components of a system for ion chromatography.

The injected analyte compounds are separated on their way through the column and reach the detector at times which are characteristic for the system and the analyte species (see Figure 3.3).

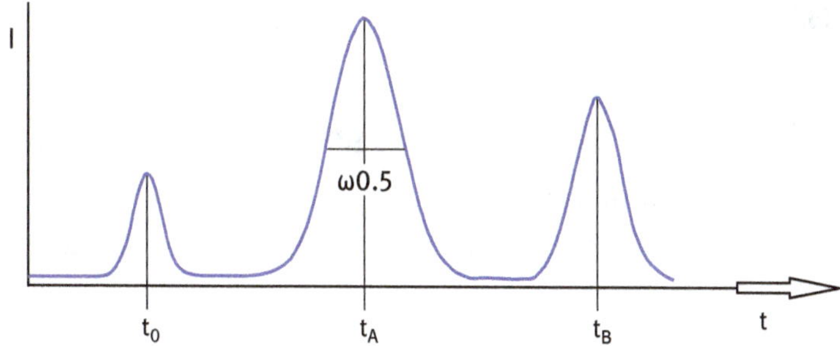

Figure 3.3: Sketch of an ion chromatogram of substances A and B.

The intensity (ordinate) is plotted against the time from sample injection to the signal decay of the substances. The minimum time of the eluent passing the entire system (dead time, t_0) can be calculated by the flow of the eluent through the system. It becomes visible, however, by a small peak (the injection peak) which is a good approximation of t_0. This peak is due to a change in conductivity between solvent of injected sample and eluent. The first analyte appears with a delay of t_A compared to the injection peak. This is the net retention time of analyte A, the total time of A in the system diminished by t_0. Analyte B appears with a delay to A. Diffusion processes in the column broaden the peaks which should be Gaussian shaped under ideal conditions. Broadening and shape of the peaks is mainly depending on the sorbent and, to a lesser extent on the elution conditions. Broadening has an influence on specificity and chromatographic resolution. In this sense it is similar to optical resolution in spectroscopy. The width of a peak is therefore a quality parameter. It can be defined as width at half peak height ($\omega_{0.5}$) or as width of the peak between the inflection points if Gaussian shapes predominate. The deviation from Gaussian shape is characterized by the factor of asymmetry T which is defined as the ratio of peak tail to peak front at 10% peak height. These parameters describe the qualitative process of separation. Quantitation of the analyte is obtained by the measurement of peak height or signal integration by comparison with reference solutions.

3.3.2 Chemistry of ion exchange

Ion exchange is a concerted action between functionality of the solid phase, and ions of the eluent reacting with the surface. The latter ones are exchanged against the analyte ions of interest. If anions are separated and determined, the solid phase must provide cationic functionality. A typical functional group coupled to the synthetic resin of the column is NR_3^+, a trialkyl-ammonium-group. The column is flushed with a solution containing HCO_3^- which reacts with the ammonium groups of the resin. Analyte anion A^- is in a chemical equilibrium with HCO_3^- in coupling to the surface (eq. (3.7)).

Equation 3.7:

$$SP - NR_3HCO_3 + A^- = SP - NR_3A + HCO_3^- \qquad (3.7)$$

SP is the solid phase which is functionalized by NR_3^+.

The separation is achieved by different affinity (activity) of the anion analytes toward NR_3^+ relative to the exchange ion HCO_3^-. For analyte A^- the distribution constant is defined by eq. (3.8):

Equation. 3.8: Distribution coefficient of analyte A^- concentration and exchange ion HCO_3^- concentration in the mobile phase (m) and in the stationary phase (s):

$$K = (A^-)_s(HCO_3^-)_m/(HCO_3^-)_s(A^-)_m \qquad (3.8)$$

The coefficient for a given analyte ion system can be experimentally measured by addition of an activated resin into a known volume of analyte and exchange ion of known concentration. The higher K for the analyte ions, the longer is the elution time in the chromatographic process.

For the determination of selected cations, a resin with functionalized anions is used. A popular example is the SO_3^- functionalization with benzenesulfonic acid. Elution of the cation A^+ is usually performed with the H^+ ion from dilute mineral acids.

As the analyte may contain ionic matrix which may quickly degrade the performance of the exchange column, often a small pre-column or guarding column is added to it. This can be economically replaced when high matrix in the sample result in fast degradation of the resin.

Even more important for the analytical detection is the suppressor column. In case of conductometric detection, which is the most widely used detector technique in chromatography, the eluent may possess high conductivity, and may hamper sensitive detection of additional conductivity. This is comparable to high background emission or absorption in spectroscopy. The function of the suppressor column is to remove or substantially reduce this background. The suppressor column will have an influence on the analyte ion as well. This should be reinforcing, however, and not lessening the analyte. Coming back to the example of anion exchange above, the eluent is HCO_3^-, to a lesser extent CO_3^{2-}, the analyte A^-. The counterion is Na^+. The suppressor column, in the easiest form, exchanges Na^+ against H^+. Na^+ and HCO_3^-, the eluent from the separation column, is replaced by H_2CO_3 which is almost non-conductive. The analyte ion A^- and H^+ from the suppressor column remain in the eluent and generate the response to the change in conductivity when A^- is eluted. The suppressor column would soon be depleted and would require frequent replacement. Therefore, systems are in commercial use, which use small electrochemical cells. In these, water is hydrolyzed in channels separated by diaphragms from the "reaction channel" where the eluent is treated. H^+ passes through the membrane protecting the anodic channel, Na^+ passes through the membrane near to the cathode. The supply channel produces the required exchange ions constantly, while the eluent remains constant in chemical composition. A sketch of a suppressor column is displayed in Figure 3.4. The electrochemically supported suppressor column can provide additional analytical information by counting the transferred charges (see the section "Detectors").

3.3.3 Ion exclusion chromatography (HPICE)

This technique is based on the specific characteristic of a porous and charged solid.

- Ions resident in the pores of a matrix or membrane with the same sign of charge are significantly lower in concentration compared with the surrounding mobile phase. This effect is called Donnan exclusion, the respective membrane equilibrium Donnan equilibrium [129].
- Ions of different size elute with different speed from pores of suitable dimensions.
- Other adsorption phenomena support or enable separation.

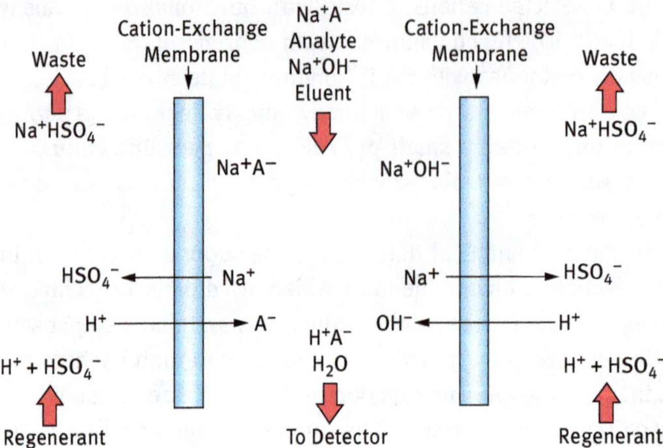

Figure 3.4: Permanently regenerating suppressor unit. Dionex CRS 500. Dionex cCorporation, Sunnyvale, Ca, USA.

While strong, completely dissociated acids are predominantly in the eluent, undissociated molecules or weak acids (the undissociated fraction) may penetrate the pores. The strong acids are passing the column quickly as a sum peak while weak acids like acetic acid, alcohols, aldehydes, amino acids, and carbohydrates are released from the pores significantly later. The separation mechanism is non-ionic. The separation from the pores is following steric aspects. The resin in use is completely functionalized with sulfonic acid groups. These groups tie water via bridges around its surface and act as a kind of membrane. The positively charged membrane prevents dissociated cations from attachment to the surface. Elution of the columns is performed with water, or slightly acidified solutions or long chained carboxylic acid.

The combination of specifically designed columns and eluents allows ultratrace determinations for analytes which are difficult to determine with other techniques, e.g., boric acid in the sub µg/L range.

3.3.4 Mobile phase ion chromatography (MPIC)

The name of this alternative to ion chromatography already indicates that the mobile phase is the essential part of chromatographic separation. The substrate actively supporting separation is added to the eluent.

Standard IC uses hydrophilic stationary phases which absorb ionic species and require polar solvents for elution. The so-called reversed-phase chromatography uses alkyl chains covalently bonded to the stationary phase to create a hydrophobic stationary phase. The resin is quasi neutral with high tolerance against acids and lies. It shows strong affinity for hydrophobic or less polar compounds. The term "reversed-phase chromatography" indicates that the polarity of the stationary phase is opposite to standard IC.

MPIC is used to separate weak acids and alkaline solutions with hydrophobic characteristic. These may be, for example, long chained sulfonic acids, surfactants, or long chained substituted amines. The eluent becomes a rather complex mixture of solvents with hydrophobic and hydrophilic parts, solubilizer, and the reagent specific for the anionic or cationic separation (a kind of chemical modifier to the system).

An alternative name for this separation technique, ion-pair chromatography, indicates the possible mechanism of analyte separation. The analyte ions form pairs with the "modifier" in solution, binding to the solid phase as a pair [130, 131] or binding to the stationary phase which is already functionalized by the modifier [132, 133].

In addition to these direct binding mechanisms between analyte and modifier, the formation of boundary layers with adsorptive functions at the solid phase may play a role for the separation process as well. It has been observed that in particular in the case of MPIC the temperature of the column plays an important role for the retention time of the analyte species. The retention time is decreasing with increasing temperature. This allows, on the one hand, to optimize the conditions for the chromatographic process but requires, on the other hand, a careful control of temperature during analysis.

3.3.5 Instrumental setup and technical parameters of ion chromatography

Separation techniques based on a solid stationary phase are using almost exclusively high-pressure mobile phase propulsion. Optimized separation efficiency requires tightly packed resin material with small particles and large surface. High pressure is a requirement for chromatographic resolution and adequate speed of the analytical cycle. Pumps in particular, but also valves, back pressure protection, connectors, etc. are core components for analytical performance. Instrument cost, quality of all mechanical parts, and long-term stability of the entire system are closely connected. The propulsion of the eluent must be smooth with low pulsation. Dual pump systems are commonly used which are designed for minimal pulsation and uninterrupted flow. The solution for measurement, often resident in an autosampler vessel at ambient pressure must be injected into the flow through the column. This is often realized by a rotary injection valve which is expensive and critical for the instrument performance. The influence of the pump system on the analytical performance may range over:

- Stability of retention time and chromatographic resolution
- Noisy and fluctuating baseline
- Spiking
- Drift

The column is the centerpiece of the instrument. Its characteristic determines the scope of application. Combination with the sample preparation, the eluent, and the settings for flow speed (pressure) and temperature makes up the analytical application. The pre-column, if required, and the suppressor column are parts of the column system.

Detectors in IC

Conductometric detection is by far the most widely used technique in IC. The technique was described in Section 2.1.15. The eluent is usually strongly conductive itself and thus attributes significantly to the background response. Depending on the conductivity of the analyte the elution peak may even become negative. Conductometric detection in IC is therefore almost always combined with the use of a suppressor unit (see above). Only in specific applications with eluents of low conductivity, the suppressor column can be set aside (single column ion chromatography). IC without suppressor column usually does not offer the lowest detection limits. The development of suppressor columns was therefore a permanent field of research in separation sciences [134].

The eluent of the conductivity detector coupled to an electrochemically supported suppressor column can be further evaluated by adding a downstream electrochemical cell. Two membranes, for anion exchange and cation exchange separate the eluent from the regeneration liquid flowing to the suppressor column. The latter channels are in contact with the anode and cathode of the charge detector. Ions eluted from suppressor column and conductivity detector will move through the membranes and will be discharged there. The discharge can be integrated and provides surplus analytical information in addition to conductivity.

Detection by amperometry (Section 2.1.15)

If the analyte ions are not strongly dissociated conductometry is not sensitive enough for ultra-traces. In this case detection by amperometry provides detection limits in the low µg/L range.

Detection by coupling with optical methods (UV, Section 2.1.10 and fluorescence, Section 2.1.13)

Coupling of separation techniques with spectroscopy will add analytical value to the determination if spectroscopy alone is not able to handle the complex matrix in the sample. Separation techniques will improve the specificity of the determination. In the case of UV detection most ions from IC are absorbing at wavelengths in the far UV range and the respective absorption coefficients are low. In samples where chloride concentration is high (seawater, wastewater, biological samples, food, and feedstuff) nitrite, nitrate, iodide, and bromide are typical application examples [135, 136].

The effluent of the column can be modified for high UV activity. A well-known reagent is “PAR,” an azo compound and reagent for chelate complexation. Several metal ions can be complexed and detected with IC-UV. The addition of the reagent requires an additional pump for injection of the reagent to the effluent of the IC column.

Ion separation coupled to IC becomes particularly important if the interesting species must be separated selectively from the other species of the element. Cr^{3+}/Cr^{6+} is an example where Cr in low oxidation state is essential for the human body whereas Cr^{6+} is carcinogenic and toxic at low concentrations. Cr^{6+} traces can be selectively detected after complexation with DPC (diphenyl carbazide) at 520 nm using the UV detector.

The same considerations, namely activation of the spectroscopic effect, hold true for detection with fluorescence. Most ions do not fluoresce. Fluorescence in IC became popular for trace determinations of amines, amino acid, and peptides long ago. These are activated by phtaldialdehyde and mercaptoethanol fluorescing at 455 nm. The fluorescence is activated at 340 nm [137]. Important products such as Memantine, a medical product against Alzheimer's disease, or Glyphosate, an herbicide, can be analyzed with this method.

The effluent of IC columns can be coupled with other spectroscopic techniques such as ICP-MS, ICP-OES, or AAS. In all these cases the detector is no longer a part of the ion chromatograph but an individual instrument which is using the sample separated by chromatography. A lot of research literature describes successful applications. Few of them, however, found their way into routine laboratory procedures.

3.3.6 The analytical process of an IC analysis

The samples analyzed in liquid chromatography must be free of particles. The requirement is tight: filters in use should have a porosity in the range of 0.50 µm. Depending on the particle amount in the sample, the filters can be an integral part of the flow system. An increase in backpressure will indicate the requirement for a filter change. If required, suitable filters must be used to remove long chained hydrocarbons, surfactants, and fatty acids from the sample.

Compared with direct spectroscopic methods, the samples processed through chromatographic systems need to be more strictly controlled with respect to their chemical activity. The chemical activity of strong acids or brines from sample digestion must be minimized. This can be realized with ion exchange cartridges. These are available to compensate for typical types of samples. It must be kept in mind that this is an additional step in the sequence of necessary actions potentially prone to contamination, adsorption and other phenomena which may influence the FOM. It was mentioned above already that often small pre-columns, filled with the same material as the exchange column are used. They can be easily exchanged, are low cost, and often protect the actual column from "dirt" of the sample. Pre-columns increase the retention time of the analyte by 10–20%.

The status of packing and surface of columns are an additional parameter which influences the analytical performance. Columns must be maintained and, if not in use, stored and kept properly. The figures of merit are used to validate the equipment on a regular basis. The characteristic performance figures have been discussed in the general section.

Outside the solid phase in the column, changes in the pressure of the system must be avoided. These changes are often the result of dissolved gases in the eluent. Pumps are producing negative pressure when sucking liquids through tubes with small diameter. Degassing of the liquids is therefore often mandatory to obtain stable pressure and flow conditions. Systems making use of ultrasonic energy or vacuum are commercially available.

The quality of the eluent is equally important for the chromatographic performance and for potential introduction of contaminants. The criteria for water, acids, and organic solvents are similar to those described in earlier chapters. Additionally, the total organic content (TOC) is critical for the long-term function of the stationary phase. For water, as example, the TOC does not become visible from its conductivity, the control parameter for ultrapure water.

3.3.7 Application example

IC is a technique with a wide application range. It proves successful in particular for those ions, where other analytical methods are weak. IC is almost indispensable for anion analytics, whereas metal cations are usually determined with atomic spectroscopy. Applications are found in almost all fields of analytical problems, from environmental to clinical chemistry and control of industrial processes.

The application example describes the determination of seven anions, namely bromide, chloride, fluoride, nitrate, nitrite, phosphate, and sulfate, in drinking water, surface water, marine water, and wastewater. The method is standardized internationally [138].

Depending on the number of ions and the type of matrix the determination can be fast. Three anions in drinking water take less than 5 min for a determination. The description of the method follows the norm closely.

The scope of the method asks for a limit of quantitation (LOQ) of 0.05 mg/L for Br^- and NO_2^-, and of 0.1 mg/L for the other anions. These levels are obtainable with conductometric detection. Br^-, NO_2^-, and NO_3^- can be detected at lower levels using UV detection. Concentrations below 0.01 mg/L (LOQ) can be obtained by preconcentration. The standard makes reference to other ISO methods discussing the handling of water, general guidelines for calibration etc.

The wide range of matrices, from drinking water to industrial sewage asks for adapted sample pretreatment. The sample needs to be free of particles and is passed through a 0.45 µm filter. If high concentrations of cations are present, a cation exchange unit is used to remove matrix. Higher sulfite concentrations must be removed as well by reverse osmosis or ion exchange. Higher concentrations of aliphatic organic acids may also bias the result. The "clean sample" is slightly acidified to about the same level as the eluent. Usually, 20 µL of sample are injected.

The eluent is a mix of $NaHCO_3$ and Na_2CO_3 at concentrations of 0.001 mol/L and 0.027 mol/L in ultrapure water. As described in the general section, the water may only contain a very small concentration of ions and hydrocarbons. It is degassed prior to pressurization and must be prepared freshly to avoid algae growth. All vessels, pumps, tubes must be suited for ultra-trace analysis. The flow rate through the system is around 1.2 mL/min. The schematic of the system setup is like the one sketched in Figure 3.2. The pre-column may be omitted if clean waters and soft tap water are analyzed only. An UV detector is optional for providing lower detection limits for the anions mentioned above.

A definite quantitative detection of the anions requires an adequate resolution in separation. The resolution shall be better than 1.3. It is defined by the retention time of the analytes 1,2,3, . . . compared to the width of the peaks (eq. (3.9)).

Equation 3.9: Chromatographic resolution as defined by ISO 10304–1

$$R_{2,1} = 2 \cdot (t_{r2} - t_{r1})/(\omega_2 + \omega_1) \tag{3.9}$$

$R_{2,1}$ is the resolution of the two analyte ions, t_{r2} and t_{r1} their retention times and ω_2 and ω_1 the peak widths. For the latter Gaussian shape is assumed (see Figure 3.1).

The sequence of sample measurement is as usual. The system is started until the baseline is stable. Reference solutions, samples, and blanks are run. The reproducibility of signal intensity and retention time is determined. The results are reported.

International standards usually provide information on FOM of interlaboratory testing. The result provide data on agreement with the expected mean value of the analyte (spiked or natural) as well as the reproducibility (the standard deviation of all laboratory data) and the mean repeatability of the laboratory determinations. In the case of anions in water, typical coefficients of variation are in the range of 1% to 5%. From a closer inspection of the values, it becomes obvious that clean samples are determined more accurately and precisely compared with dirty waters. Low analyte levels obviously result in higher coefficient of variation.

The ratio of interfering species to analyte is listed as guideline. Freedom from interferences is usually provided if the ratio is between <50 and <500 depending on the type of species and analyte. Some species are problematic at almost any excess concentration. In this case it is the SO_3^{2-} anion which must be controlled carefully or removed prior to analysis (see above).

Unlike in environmental regulations only few species are strictly regulated by threshold levels and tested in food chemistry on a regular basis. Extensive determinations are required for occasional quality control and product development. In this field of application reference is made to an application paper by one of the leading manufacturers. The example stems from the determination of inorganic and organic anions in beverages [139].

3.4 HPLC

Many aspects of HPCL have been discussed in the section "Ion chromatography" above. HPLC is quasi-generic term for high-pressure elution of a solid stationary phase. In this section the separation of organic compounds will be explored in more detail.

The solid phase

Organic substances are mostly non-polar. The processes of adsorption and elution as well as the underlying physicochemical mechanisms are therefore different from IC. The same is often true for detection. Conductivity is seldom a sensitive parameter. Optimization of instrumentation and method development concerns mainly the stationary phase, the eluent, and the detector. Just like in IC, the aim is to process small sample volumes within a moderate elution time. The stationary phase consists of small particles of about 5 μm diameter packed in thin columns of less than 5 mm inner diameter and roughly 10 cm length. A fundamental difference in nomenclature is made between "normal phase" and "reversed phase." The former is more polar than the eluent, the latter is less polar than the eluent. Organic analytes are mostly separated by reversed-phase columns.

The substrate of the solid phase may act as the adsorbing phase without further functionalization. It often consists of silica gel (colloidal SiO_2) with an even particle size distribution. The material is quasi-insoluble, has a very high melting point (can be thus regenerated at elevated temperature), and offers a big surface of about 600 m^2/g. SiO_2 is often spiked with other particles of ceramic origin for optimized adsorptive functionality or structured packing for optimized flow. SiO_2 is polar, however, offering free –OH groups for hydrogen bridges. For reversed-phase chromatography it needs to be functionalized with organic groups. Functionalization with, e.g., alkyl or phenyl groups allows a tuning of the chromatographic functionality. The interaction of solid phase and analyte is mainly based on Van der Waals forces. The adsorptive force is the more pronounced, the more similar analyte and adsorptive phase are. Functionalized solid phases may consist of a polymer carrier. Functionalization with partially polar groups, such as amines or long chained alcohols, allows a subtle adjustment of polarity. Design of columns for HPLC today allows high flexibility concerning the analytical capability. Separation mechanisms such as polar/nonpolar, hydrogen bonds, and steric configuration may be used in parallel to enable best chromatographic performance.

Competition of solid and liquid phases

The analyte is injected into the flow of the liquid phase. At that time the solid phase is loaded with the eluent which itself has some bonding forces to the surface. The analyte must displace the eluent to occupy the stationary phase. This can take place via a simple competitive process (solid-eluent-analyte), or the analyte may interact and bind temporarily with the eluent next to the solid phase and may be temporarily retarded by these forces. Often the retention cannot be explained by mere interactive forces between analyte and surface, but the characteristic of the eluent and analyte-eluent interactions are crucial for the entire process of separation.

The eluent

The eluent must possess several qualities for a good analytical result. It should be of "suprapure" grade, i.e., clean of analyte ultra-traces. The boiling point should be high enough that negative pressure produced by the piston pumps during uptake does not produce gas bubbles. The viscosity should nevertheless be low. It must be suitable for the detector (no strong absorption for UV detection, suitable electrochemical behavior). It should be partially miscible for gradient elution (see Section 3.2) and must be a good match for the stationary phase and the analyte.

Columns with a polar stationary phase (normal phase chromatography) are eluted with non-polar liquids or eluents with moderate polarity and a large organic moiety. The eluent can solve non-polar analyte solutions or readily mix with non-polar solutions. Elution is a competitive process again (see above) and so, a stronger polar eluent, such as acetonitrile, elutes the analyte faster than the non-polar hexane. The same competitive binding activity holds true for the analyte: the retention time of saturated hydrocarbons is shorter than that of, e.g., carboxylic esters.

By contrast, columns with a non-polar stationary phase (reverse phase chromatography) are eluted with polar eluents. Less polar solvents, such as acetonitrile (C_2H_3N), elute faster than methanol. The analyte must be soluble or miscible with the eluent or suitable solubilizers must be added.

Detection

HPLC is working with small pump rates (e.g., 1 mL/min) and the analyte ions are often eluted with a time difference of only half a minute. The mix of eluent and each individual sample must be detected with at least the resolution of the separation process. Mixing of analyte in the detector cell must be avoided. The detector cell volume needs to be small, and the process of detection must be fast. The detector systems described already in IC are used in HPLC as well; however, the UV/VIS detector is much more standard, the conductivity detector much less standard than in IC. It should be emphasized that the UV/VIS detector is selectively determining the analyte while the conductivity detector is quantitating eluent and analyte as a sum. Other frequently used detector systems are other electroanalytical methods, molecular fluorescence, determination of the refractive index, and mass spectrometry. Suitability of a detector again depends on the analytical task (FOM specification, type of analyte, and matrix). The detector technique will significantly contribute to the total cost of the instrument.

The refractive index detector (Figure 3.5.) has not been considered so far. It is used mainly in chromatography for quantitative analytical purposes. The detection capability (signal/noise, detection limit, working range) is generally limited compared to, e.g., UV detection. It is, however, a cost-attractive alternative for specific applications where the analyte is neither UV active nor conductive enough to be detected.

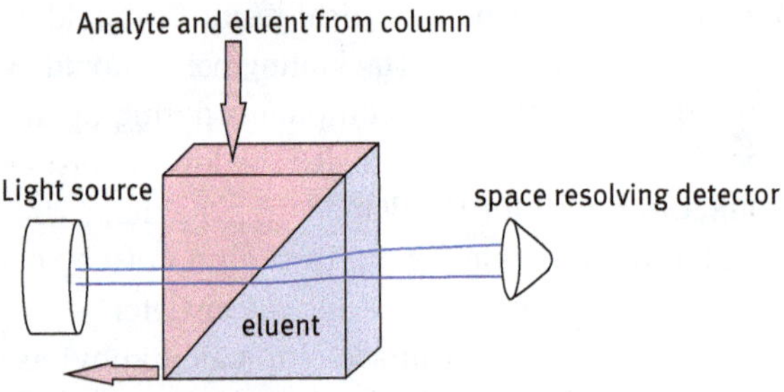

Figure 3.5: Refractive index detector: Light from a source is passing through a cuvette with two compartments. The light beam is detected with a space resolving detector.

The refractive indices are compared to quantify the effect of the analyte. Using a space resolving detector the angle of the incident light beam can be tuned by an adjustable quartz plate. By adjusting this quartz plate, the illumination of the detector is set to maximum. Each analyte, eluted from the cell will change the refractive index in the left part of the cell and will "detune" the light beam from the detector. The change in light intensity or location depends on the species and the concentration of species in the eluent. Both effects must be referenced by an appropriate solutions. As the refractive index strongly depends on temperature, the RI detector unit, in particular the cell, must be temperature-controlled.

3.5 Gas chromatography

Gas chromatography is the separation of gases moving through a thin column. The analyte may be in gaseous form at room temperature, it may be vaporized by elevated temperature, or it may be resident above a reservoir of liquid and sampled from there (head space GC). The distribution of analyte between the stationary phase and the moving phase as well as the technical requirements are substantially different from HPLC/IC. GC is the probably most widely used analytical technique with respect to number of instruments in the market. Its detection capability is in the range of 10^{-9} g or below depending on the detector. One reason is that distribution between mobile and stationary phase of the molecules is significantly faster in GC compared to HPLC, the number of separation stages is higher, the resolution is better in GC. One slight disadvantage is that the compressibility of gases requires control of an additional parameter.

Gases can either be adsorbed directly on a solid phase or exchange can take place with a liquid immobilized on the wall of a capillary or in a packed column. These variations result in a different embodiment of the column.

The mobile phase

The mobile phase consists of pressurized gas where the gaseous sample is injected via a syringe or an automated coil system. Frequently used gases are helium, nitrogen, or hydrogen. Non-reactive inert gases are often preferred, as reactions with the analyte are excluded. The gas is provided by a pressurized bottle, a suitable pressure reduction unit, and a mass flow controller which provides a precisely stabilized flow. Accurate handling of the injected sample is assured by an injection and mixing unit. The mobile phase without and with sample is guided through the column which is encased in a temperature-controlled compartment. The mobile phase is finally guided to the detector. The headspace technique requires a flushing of the gaseous space above the sample to be analyzed with a specific valve system which flushes the head space gas to the column. A sketch of the fundamental setup of GC is featured in Figure 3.6.

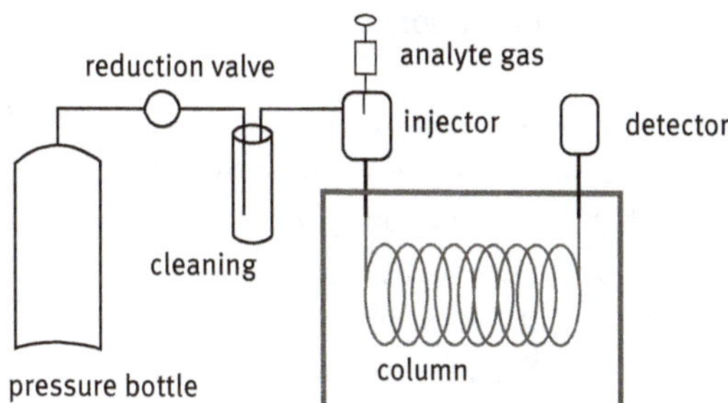

Figure 3.6: Setup of a gas chromatograph.

The analyte will influence the method of sample processing and introduction:

- It may be volatile at room temperature without the tendency of condensation. No elevated temperature during sample introduction and separation is required.

- It may be semivolatile or liquid such that evaporation at higher temperature is required. Sample introduction and separation/elution to the detector need to run at strictly controlled elevated temperatures.
- The sample may be dilute enough to match the detection range and not overload the column. The injected analyte is directly moving to the column.
- The sample is too concentrated or viscous for column and/or for the detector. It needs to be diluted by the eluent prior to charging the detector. The sample flow is split, and a defined portion only is measured volumetrically and mixed with the eluent.

Important basic conditions are the complete evaporation of liquid samples (no droplets to the column!) and a focused analyte cloud moving to the column (substantial dispersion of the analyte generates broader blurred signals).

The stationary phase

First experiments with GC by Erika Cremer et al. date back to the 1940s. The column was packed with charcoal. Gases such as air and CO_2 were separated by a mere adsorption/desorption process. With the development of the technique in the 1960s and 1970s the columns became significantly smaller and technological progress ultimately lead to capillary gas chromatography. Open capillaries generate a lower back pressure compared to classical packed columns and make a significantly higher number of separation stages possible. Early in this process, analyte exchange with an immobilized liquid phase was found to be much more flexible. A liquid or viscous phase is immobilized to the inner wall of the column or to a widely inert solid which covers the inner wall. The chromatographic process is mainly based on distribution between the mobile and the stationary phase not excluding adsorption/desorption processes. The most widely used columns today are from silica:

- Columns from fused silica, coated on the outside with carrier polymer, and functionalized on the inside with a thin coating of the viscous exchange phase. The inside coating is a thin layer with a width of usually less than 1 µm only. This type of column is known as Wall Coated Open Tubular column (WCOT).
- Columns from fused silica, coated on the outside with carrier polymer, and on the inside with an inert solid which carries the viscous exchange phase, called as Support Coated Open Tubular Column (SCOT).
- Columns from fused silica, coated on the outside with carrier polymer, and on the inside with a porous solid stationary phase, referred to as Porous Layer Open Tubular column (PLOT).

The type of column defines the analytical application. Whereas PLOT columns are used to separate very volatile gases, such as CO_2, CH_4, and NO_x, the other column types separate volatile organic compounds in general. Figure 3.7 shows a chromatogram of a bio diesel fuel mix separated with a capillary column. The high resolution of the method is characterized by narrow peaks, and a separation time of about 17 min from the first to the last eluted compound.

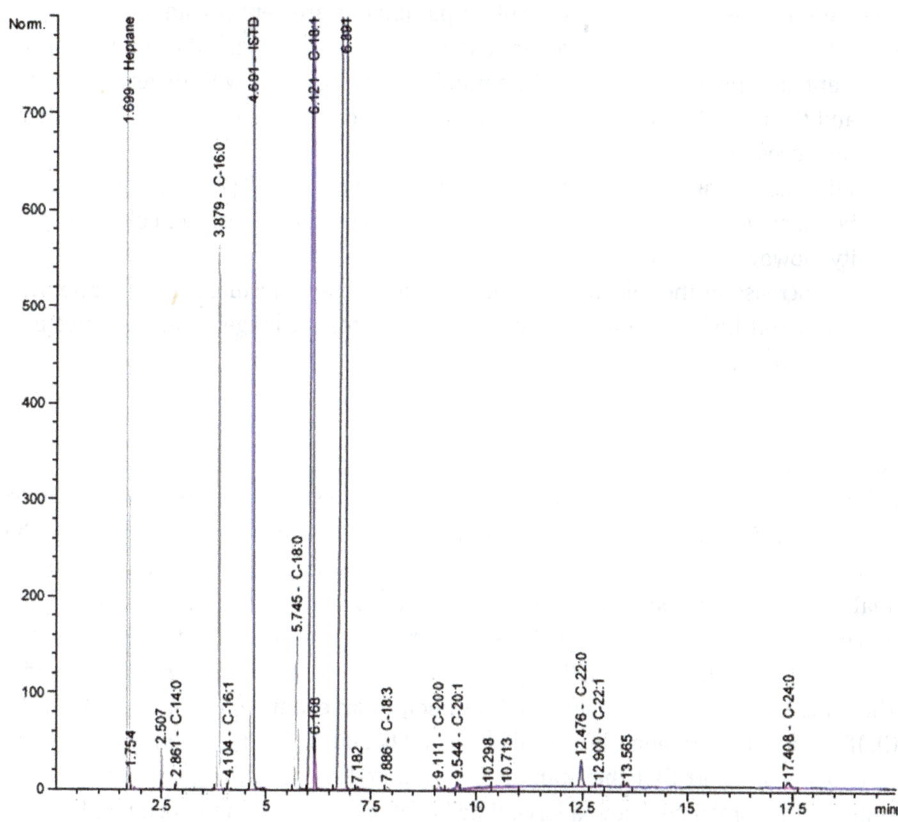

Figure 3.7: Separation of compounds in fuel with a capillary column. Sunflower biodiesel. Picture from Research Gate: Mustafa Z, Surchev S, Milina RS, Sotirov S (2015) Erdöl und Kohle 57(1):40–47.

Selecting the right column and analytical conditions for separation are one of the core competencies in chromatography. In addition to the type of column described above, additional parameters to consider are:

1. Length of the capillary
2. Diameter of the capillary
3. Type and thickness of the inner wall coating
4. Temperature
5. Speed of the mobile phase

An important role in the separation process is played by the ratio between the volume of the mobile phase V_m and the volume of the stationary phase V_s (see Section 3.1.3): β, the phase ratio is in the range of 5 to 200 for packed columns and 20 to 1,000 for capillary columns. The following general statements hold true if only one parameter is changing, and the other parameters are kept constant:

- When length and/or diameter are changed the ratio between the volumes of the stationary phase is changing. The number of stages rises with increasing length and the resolution increases as well. The time for elution and hence the analysis time goes up.
- With increasing diameter of the column, the phase ratio β goes up and the number of stages goes down. The bigger inner surface allows a higher column capacity, however.
- An increase in the thickness of the immobilized layer results in longer retention times and higher resolution, however, at the cost of longer analysis times and wider peaks.

The influence of temperature

It becomes obvious that several parameters must be optimized with respect to a strictly specified analytical task. In addition to the columns, the temperature of the process is a parameter of utmost importance. It has been discussed above that the analyte species must be gaseous and must not condense during the chromatographic process. The columns are usually heated to temperatures above or shortly below the boiling point of the least volatile species. The range of temperatures may be wide. The difference between methanol and decanol, as an example, is 164 °C (65 °C/229 ° C). If in such a separation the column is heated to 200 °C, the retention of the volatile alcohols is so short that they cannot be separated. Decanol, on the other hand is eluted with a broad peak and a large time lag of more than 5 min to octanol. In these cases, the chromatogram is run with a temperature gradient by heating the column from, say, 50 to 240 °C. The majority of separations are run temperature gradient controlled. Temperature gradient separation is somehow equivalent to gradient elution in HPLC.

Detectors

Flame ionization detector (FID)

Hydrocarbons burned at high temperatures (≈2,000 °C) generate radicals and ions. Compared to the carrier gas (He, H_2, etc.) the number of ions becomes orders of magnitude higher when these analyte species are eluted. The ions can be detected in a potential field. An FID consists of a unit controlling and mixing the gas from the column with the burner gases H_2 and air, a burner, and an electrode unit which collects the ions and electrons generated in the flame. It belongs to the oldest and still most widely used detectors in gas chromatography. The system is cost-economic, has a wide dynamic range of up to 7 orders of magnitude, and detects hydrocarbon species in the picogram mass range. It is not suitable for species which do not generate enough radicals and ions. These are mainly gases such as CO_2, NH_3, and some hydrocarbons with COOH- and CN-groups. A special layout of the FID suppresses ionization of the hydrogen flame by using a lean flame. A ceramic bead with imbedded rubidium or cesium, additionally heated by an electric wire, generates a rubidium plasma around the bead which reacts predominantly with PO double bonds and CN triple bonds. This version of the FID is called nitrogen-phosphorus detector (NPD).

As discussed in the section on flame emission spectroscopy, the generated spectra can be used to quantitatively determine specific elements. S and P can be determined with a photomultiplier tube at 394 nm and 526 nm, respectively. This general principle is used specifically for these two elements in GC and is named flame photometric detector (FPD). It is based on FID but requires an additional photometer for detection.

FI detectors do not respond to all types of species from the column, and the analyte is decomposed after detection.

Ionization can be induced by radioactive radiation. Ni-63 is a weak β emitter. It is used in the electron capture detector (ECD) which essentially consists of a nickel metal foil with an electroplated layer of Ni-63. The gas from the column is guided through this nickel foil cylinder which acts as cathode. β-Radiation from the foil ionizes the carrier gas (in this case Ar or N_2) which generates a constant current of ions and electrons between foil and a metal anode located in the center of the cylinder. Eluted analyte molecules can absorb electrons ($R{-}X + e^- = [R{-}X]^-$). The constant current will therefore drop when specific analyte species are eluted. These are mainly molecules with electronegative groups such as organic halides, peroxides, and nitrogen-containing compounds. The detector is most powerful for halogenated pesticides.

The radioactive activity inside the detector is low. The exhaust of the column is no significant radioactive burden. The detector itself, however, is encapsulated and must be handled according to safety regulations. Like the FI detector, the ECD is selective, cannot be used universally, and changes the eluted species. It is very sensitive for halides ($\approx$ 0.1 pg/s) and it offers a working range of about 4 orders of magnitude similar to the range of FID and NPD.

Thermal conductivity detector (TCD)

The gas from the column consists of the carrier (e.g., He) and the flush of species. The carrier gas has a high thermal conductivity (He: 222 mW/mK) which is usually several times higher than that of the analyte species. The difference in conductivity can be measured electrically. Pure carrier gas kept at the temperature of the column is flowing around a resistor. The gas from the column is flowing around a second resistor. The flow in the two cells must be equal. Species with lower heat conductivity will result in higher resistor temperatures and hence in lower currents. This simple and cheap device is remarkably sensitive down to the low ppb range. It offers good linearity over 5 to 6 orders of magnitude. Other than the FID, TCD responds to all kind of species. Determination of the analyte is non-destructive.

Mass spectrometry (MSD)

By far the most powerful – though most expensive – detector is the mass spectrometric detector. GC-MS systems are mainly based on the quadrupole technology. If required, mass selectors can be connected in series (GC-MS-MS). Most aspects of the technology have been discussed in the section on mass spectrometry already (Section 2.1.14.2). In GC-MS systems the coupling of the column (packed/capillary) with the ionization device comprises sophisticated control of gas flows and vacuum conditions to achieve best analytical quality. MSD allows pg-detection and a dynamic range of 6 orders of magnitude. It includes one more very powerful and selective means of resolution. It has become an indispensable tool for analytical and bioanalytical questions.

Other couplings

Gas chromatography can be and has been coupled to many other analytical techniques such as AAS, ICP-OES, ICP-MS, and FTIR. These so-called hyphenated techniques are subject of innumerable scientific applications and publications [140, 141].

Fields of application of GC

The analysis of volatile compounds depends strongly on GC methods, today frequently GC-MS. GC is a standard equipment in most analytical laboratories. Consumer protection, laboratory medicine, drug and doping control, and control of drinking water are only a few examples of routine applications. As one example the determination of flavoring agents and potentially toxic amyl alcohols in wine shall be outlined. This application example is of specific interest as the analytical equipment is supporting the powerful human sensorium "nose." The taste of wine is defined by trace concentrations of aromatic compounds. The total amount of these is not more than 0.5 per mill. Still there are almost 1,000 different compounds which define the entire complexity of smell and taste. Their concentration may vary between mg/L and pg/L. These compounds are, e.g., alcohols, esters, terpenes, organic acids, phenols, and thiols. Some of the compounds are present in the grape; some are formed during pressing, fermentation, and storing. Off-odors are smells which are unwanted such as mercaptan or H_2S. The cork-like taste in wine is another unwanted off-odor generated by a naphthalene. The important criterion is the recognition of the compound by nose and taste.

The first step in wine assessment is therefore organoleptic. The second step is the quantitative chromatogram. An example of the complexity of chromatographic wine analytics has been described by Brandes, Wendelin, and Eder [142]. They determined characteristic aromatic compounds in barrique wines using GC-MS and HPLC. The matrix load on the column and low concentrations of the species of interest do not allow a determination without sample pretreatment/preconcentration. The authors suggest steam distillation for the compounds which are volatile under these conditions. Compounds with low volatility under conditions of steam distillation are preconcentrated by solid phase extraction and determined using HPLC. Here only the GC part will be further considered.

The steam distillation allows to remove and clean the oily aromatic compounds with higher boiling point at roughly the evaporation point of water. These water insoluble species can subsequently be separated by liquid–liquid extraction with dichloromethane, concentrated by vacuum distillation and then absorbed by less than a mL chloroform. The preconcentration factor by this procedure is at least 200. Standards of the substances under investigation, e.g., guaiacol, whiskey lactone, cresol, ethyl phenol, must be handled like the sample. The final concentrate is clean and can be directly injected into the gas chromatograph. One μL of the concentrate is injected at an injector temperature of 255 °C and a transfer line temperature of 250 °C. Separation commences with a suited column at roughly 70 kPa. A suitable separation column is selected, in this case a polyethylenglycol column with an inner diameter of 0.53 mm with 0.25 μm coating. The carrier gas is helium. The separation is run with a temperature program starting with a moderate ramp from room temperature to 110 °C, a slow rate of 2°/min from 110 °C to 226 °C, and a relative fast ramp from 226 °C to the maximum temperature at 255 °C. The total time of separation is 78 min. The mass spectrometer is directly coupled to the column. The characteristic mass fragments of the compounds of interest are read out and quantitated by signal integration. The mass fragments of ethyl phenol, for example, are read out at m/q 77, 107, 122. The detection limits of all species of interest are in the range of less than 10 μg/L; the relative standard deviation at the concentrations typical for the samples is smaller than 10%.

3.6 Other separation methods

3.6.1 Supercritical fluid chromatography (SFC)

Supercritical fluid extraction methods are used mainly in industrial processes, less frequently as an analytical separation method. We will briefly discuss the principles of supercritical fluid extraction with CO_2 as used in quantitative analytical separation science. Possessing an eluent with the density of a liquid but the viscosity of a gas is an attractive alternative for chromatographic processes. CO_2 forms a supercritical fluid under technically manageable conditions. These can be varied within relatively wide limits allowing different conditions for separation. CO_2 is gaseous at room temperature up to pressures of about 7 MPa (70 bar). At higher pressure it becomes liquid. At slightly elevated temperature (304.1 K = 31 °C) and pressures above 7.4 MPa, it forms a so called supercritical fluid (these conditions define the critical point). Supercritical CO_2 may be an alternative to separation problems which cannot be solved satisfactorily with either LC or GC. The unique conditions for chromatography are high diffusion coefficients allowing fast separation due to fast equilibration, high density assuring high solubility, and low viscosity allowing higher flow rates under conditions similar to gas chromatography. The stationary phase and the columns, are mostly the same as in HPLC. Many technical solutions such as valves, injection loops, and pumps are as in HPLC as well. The same holds true for the temperature settings for eluent, injector, and columns. As the mobile phase should move through the column as a dense phase, backpressure regulation is required at the outlet as an additional technical component.

CO_2 is largely non-polar. It is therefore not suitable for solution of polar compounds. In these cases, "modifiers" or "co-solvents," mostly alcohols, are added to the mobile phase. The concentration can be modified in a gradient and lies usually below 5%. In addition to the modifier, variables which control the separation are temperature, pressure, and flow. The conditions must match the conditions of the critical point. Whereas temperature modifications are usually in moderate ranges of around 10 °C, pressure conditions require significant steps of several MPa to effectuate changes in retention times. The flow conditions are optimized like those in HPLC.

3.6.2 Field flow fractionation (FFF)

Flow dynamics is the driving force of separation in FFF. The technique works without stationary phase and is exclusively based on separation due to dynamic processes in a flowing stream. A flow driven by force in a tubular channel generates a flow profile. The speed in the center is faster than close to the wall of the channel (Figure 3.8).

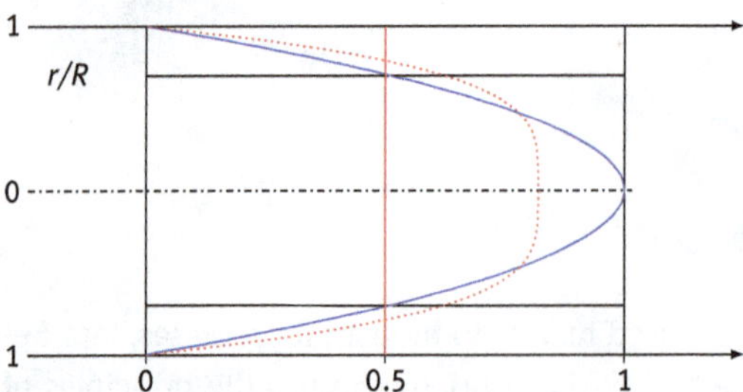

Figure 3.8: Undisturbed flow in a tube; turbulent and laminar flow profile. Velocity as a function of the radial position.

Like molecules or atoms flying through a vacuum in a mass spectrometer within an electrical field separating according to mass-charge ratio, flow dynamics generates a separation in terms of particle size. The separating fields may be simply gravitation, centrifugal force, temperature gradient or a second flow rectangular to the main flow. The naturally occurring molecular movement of the species is acting against the separating force and adjusts to an equilibrium. The separation is based on molar mass, as smaller particles move faster and are found in the center of the flow channel with higher probability than heavier particles. They are eluted from the channel earlier.

The separating fields used in FFF are different. A popular version is the asymmetric flow FF. A second flow is generated orthogonal to the main flow by a separate pump or by a precise split of flows via suitable valves. The modification is called AF4. The sample is injected into a channel where it is transported in a laminar flow with pronounced laminar speed profile (see Figure 3.7). The upper part of the channel allows the generation of a cross flow which facilitates concentration of the sample at the lower part of the channel. The cross flow leaves the channel via a frit with membrane at the lower part of the channel. The membrane allows passage of the small solvent molecules but not of heavy analyte particles. The membrane enables the adjustment of the minimal molecular mass which is removed from the main flow (molecular weight cutoff). The molecular movement of the molecules pushes the molecules with lower mass against this cross flow into higher and faster regions of the main flow. These lower mass flows are eluted and detected earlier (Figure 3.9).

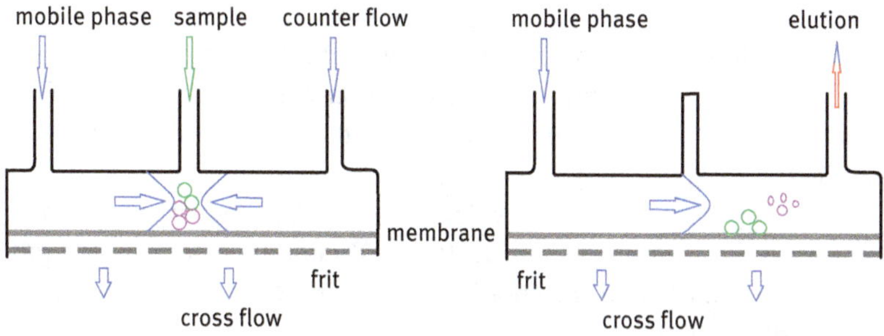

Figure 3.9: AF4 field flow fractionation. The sample is injected and concentrated in the first step by a cross flow. In the elution step, smaller particles are moving faster in regions of higher laminar flow speed. Driving force is the different speed of diffusion of particles with various sizes.

FFF is mainly used for big molecules and particles in the nm range. This includes the separation of virus. The conditions of separation can be selected in wide ranges of the two flows, temperature, and type of membrane. One hardware system thus allows high flexibility for different separation applications without the need to change hardware such as columns. Detection is usually spectroscopic, via UV, molecular fluorescence, angle of diffraction, detection of light scattering, or methods based on multiangle laser light scattering. The latter is used mainly for nanoparticles [143].

3.6.3 Electrophoresis

Separating force in this case is the influence of an electric field on charged particles. Similar to mass spectrometry, molecules with higher charges or lower molecular weight move faster in an electrical field. The electric force F_c is proportional to the electric field strength E and to the number of charges $z \cdot e$ (eq. (3.10)).

Equation 3.10:

$$F_c = E \cdot z \cdot e \tag{3.10}$$

The movement of the molecules is restrained by the friction of the moving molecule in the liquid medium. The size of the molecule and the viscosity of the fluid are the parameters controlling friction (eq. (3.11)).

Equation 3.11: Slowing force in electrophoresis

$$F_R = 6 \cdot \pi \cdot r \cdot h \cdot v \tag{3.11}$$

where r is the size of the molecule, h is the viscosity of the medium, v is the speed of movement. Each molecule will attain a certain speed of movement v where F_c is equal to F_R.

The method is mainly applied for big molecules, often biomolecules. The anionic or cationic character of these molecules depends strongly on the acidity of the solvent. They may bear anionic as well as cationic character and move to anode or cathode in the electric field.

Technically electrophoresis could be realized in a tube between an anode and a cathode. However, as the strength of the electric field must be moderate to avoid denaturation of the molecules and the speed of movement is limited due to warming of the liquid medium, the separation is slow. Additionally, diffusion processes reduce chromatographic resolution. Therefore, several technical embodiments are used to attain analytical economy and quality.

In capillary electrophoresis a capillary is located between two reservoirs filled with a buffer or a polymer solution. A voltage drop of around 20 kV is effective between the reservoirs. The capillary is relatively short (about 1 m) and very thin (about 100 µm). The filling volume is less than 10 µL, the injected sample volume below 0.1 µL. The electric resistivity of the solution is high so that only small currents in the range of 1/10 mA are flowing. Heating effects are thus minimized. Separation is achieved by the different speed of movement (see above) but also by an osmotic effect. The latter is due to negative polarization of the surface of the quartz capillary which exchanges cations with the buffer solution. As a result, anions of the buffer are slowed down close to the wall by electric attraction while anions are repelled and are moving faster in the center of the capillary. The so-called electro osmotic current is superimposed to the movement of the analyte molecules. The analyte cations are moving faster than the buffer as the buffer flow and the electric force are moving in the same direction. Analyte anions of bulky molecules are exposed to a movement in the opposite direction but the electro osmotic current exceeds their slow movement toward the anode. Thus, they are moving toward the cathodic reservoir as well, much slower, however. Neutral molecules are moving with the buffer solution and are not separated if they are not imbedded into transporting molecules. Capillary electrophoresis offers an extremely high number of separation stages per length of the capillary. The method allows the detection of ultra-trace concentrations in the pg/L ranges.

The detectors used are the known candidates from HPLC: electroanalytical methods, UV detection, fluorescence, and mass spectrometry.

3.6.4 Thin layer chromatography (TLC)

In this type of liquid chromatography, the mobile phase (solvents similar or equal to LC) moves through the solid phase by capillary forces. Except for the particle size in the stationary phase the separating principles are similar to LC. The congeniality of the method is its simple technical setup, relatively short separation time, and high chromatographic resolution [144]. Its main limitation is the basically qualitative character of the process. Quantitative determination is possible, but detection is either not very powerful and specific (in the case of densitometry) or matching of the separation part with the detection module is complex (in the case of mass spectrometry).

The stationary phase consists of functional particles in the range of 10 nm or below. The latter are fixed on a glass-, polymer-, or metal-plate. Both the particles and the carrier plate are commercially available. The stationary phase has a defined starting line. The dissolved sample is pipetted or sprayed onto it such that the distribution around the starting line is minimal. The amount of sample required (and also the maximum amount permissible) is in the range of about 10 µg. After sample application the solvent is removed by drying. The analytes are now fixed to the plate.

In the second step (elution/separation) the plate is mounted in a closed narrow-body compartment with solvent saturated atmosphere. The very bottom of the plate resides in the solvent/eluent. The eluent is sucked upward by capillary forces. Analyte separation starts as soon as the eluent reaches the starting line. This process is stopped when the eluent is reaching the top of the plate. The plate is then removed from its container and dried. The chromatogram is printed on the plate like on a photographic plate. As described above, the movement of the solvent, diffusion phenomena, the number of transitions between solid and liquid phase are criteria for the quality of separation. For each separation problem one will find an optimum in flow speed. TLC can be used in normal as well as in reverse mode.

Both qualitative evaluation of the chromatogram and conclusions on analyte quantity often requires activation of the sample. This is usually obtained by reactions which result in fluorescence of the separated species or, in some cases, in absorption by the generated compound. As pointed out, the method is more often used for questions like absence/presence of species or an overview over the number of species in an unknown sample than for exact quantitation of species.

3.6.5 Paper chromatography

This is the traditional form of obtaining a chromatogram and is the long established form of TLC. The stationary phase is not just cellulose, but a thin film of water is covering the fibers. Like in TLC there is a starting line where the sample is applied. The end point is reached when the eluent is reaching the end of the paper strip. The eluent should be miscible or partially miscible with water. Just like in TLC the sample is applied on a minimal area close to the starting line. The eluent is then applied from the bottom, or in a container from the top or, horizontally from the center of the plane. Due to gravitation, the descending paper chromatography from top to bottom allows the fastest separation. Paper chromatography as well as TLC allows separation in two dimensions. The stationary phase is removed and dried after the first separation step and, in a second step, the direction of eluent flow is changed by 90°. This way the sample is separated for a second time.

Just like TLC paper chromatography is a mainly qualitative technique. Quantitative detection via fluorescence or mass spectrometry would be possible but cumbersome [145].

3.7 Preconcentration and amplification

The importance of analyte/matrix ratios with respect to quality of the analytical result has been described in Chapter 1 as well as in the first paragraph of this chapter. Hardly any other technique has attracted so much interest and has been discussed so extensively in all the media as PCR, the polymerase chain reaction. Be it a reliable test for a dangerous diseases (remember SARS-COV-2), the detection of an offender thirty years after commitment of the crime, or knowledge about the gene mix in man's ancestors who went extinct ten thousand of years ago, it is the possibility to amplify genetic make-up out of ultra-traces of DNA which makes previously unthinkable things possible. The PCR method is quite young. About 40 years ago first promising results were reported in the literature [146]. PCR is a qualitative testimony of presence or absence of a certain excerpt of a DNA sequence. It is not a quantitative analytical method and would therefore not fit into the scope of this book. However, the progress of the amplification process is usually followed by typical quantitative analytical procedures. The so-called "real-time detection PCR (RTD-PCR)" will therefore be briefly discussed in this section. As the terms used to describe the quantitative aspect stem from the amplification process itself, we will have to start with a description of the non-quantitative part of real-time PCR, the PCR process. If LIPS (see Section 2.2.6.2) represents an almost pure approach of physical methods used to obtain chemical information, real-time PCR is the opposite. The method is based on sophisticated chemical interaction, backed up by relatively simple hardware, and implemented using developed mathematical algorithms.

As stated above, PCR is foremost a sample preparation method. It is based on a quasi-natural process, the doubling of DNA by an enzyme, the DNA-polymerase. This enzyme is present and active in living organisms. Depending on the host organism the enzyme is slightly different and has a different thermal stability. The enzyme must be isolated as pure substance and used as reagent. The process of amplification uses the target DNA, quasi the analyte. Not the entire DNA can be amplified but certain fractions of it. The starting and end points of the DNA fraction for amplification are defined by nucleotides which are complementary to the target template at the beginning and at the end of the string. These nucleotides are called "primer." Two of them – synthesized and well known – are required for the process. The preparation, which is added to the amplification device, the "thermocycler," consist of the analyte DNA, the primer, nucleotides for amplification and the enzyme.

The steps required in analytically controlled PCR are the following:
1. Collection and cleaning of the target DNA
2. Thermal separation of the strings of the double helix
3. Attachment of the primers to the template, the fraction of the original DNA to be amplified
4. Elongation of the DNA by the enzyme
5. Analytical control of the DNA yield
6. Separation of the copy DNA into two strings like in step 2.

In the first step, the DNA is separated from typical compounds of the biological matrix, such as proteins and small molecules. DNA is a long chain poly nucleotide. The nucleotide consists basically of a base (adenine, guanine, cytosine, thymine, uracil), a sugar, and usually more than one phosphate. DNA is negatively charged at PH 7. DNA is therefore water soluble, but the solubility decreases with decreasing PH. Following a mechanical (mixing) disintegration of the organic cell material supported by a detergent and often by an enzyme, the DNA is separated from the cell matrix by centrifugation or filtration. It is then often precipitated from a slightly acidic aqueous buffer medium by addition of ethanol or by extraction between an organic and an aqueous phase. More sophisticated methods use solid phase beads which are adsorbing DNA/RNA and are paramagnetic. To extract the target from solution, the beads are isolated, and the DNA extracted into aqueous buffered solution. Cleaning is commercially supported by kits which use quite straightforward sample preparation methods.

For amplification the two DNA strings must be separated. The chemical forces between the two strings are hydrogen bonds between two bases (e.g., cytosine and guanine). The helix structure formed between the strings generates additional strong structural forces, so-called stacking interactions, which define the stability of the double helix. Separation of the two strings at moderate temperature leaves the strings undamaged and able to reassemble. The temperature required for denaturation depends on the sequence of bases. A typical denaturation temperature is in the range 90 to 100 °C. The temperature can be lowered by addition of chemicals. Denaturation can as well be induced by radiation or high pressure.

The preparation is cooled down to slightly below the melting point of the primer and the primer is attached to the target DNA sequence.

Step 4 is running at a temperature which is optimal with respect to the activity of the enzyme.

The enzyme needed for amplification should be stable enough to endure step 2, the denaturation step, without being inactivated. Under these conditions the composition of the sample for amplification can be made up, placed into the PCR system, and run using an appropriate automated program (see below). The steps 3 to 6 are run now either with a pre-defined number of amplification steps or with quantitative control by real-time detection.

The PCR unit consists basically of a programmable thermal block. The program temperatures, heating as well as cooling, should be reached quickly and kept very accurately within fractions of a degree. The speed of heating and cooling depends on the amount of liquid to be processed, size and electrical power of the thermal block, and type of vessels used as sample containers. Statements about the heating and cooling rate may be based on maximum or average heating or cooling of the block or the average heating/cooling of the sample. The latter is the most significant information and should be used when time/temperature programs are listed. Although many successful methods have been published [147] to speed up PCR programs, the mostly widely used systems use standard vessels and, consequently, time-consuming programs. The average heating rates are in the range between 2 and 5 °C/s, the average cooling rates around 1 °C/s A typical temperature/time program of a thermocycler is listed in Table 3.1.

Step one is optional, (a preaheat of the instrument before the samples are added). Steps 3 to 5 are the amplification cycles. Assumed that 30 cycles are necessary to obtain the target DNA amount, the total program requires 11,480 s or roughly 3¼ h.

PCR temperature programs can be optimized for a specific section of the genome (single plex PCR) or compromise conditions can be used for amplification and detection of several genome sections (multiplex PCR) in one batch.

Table 3.1: PCR amplification; time temperature program.

Step	Temp	Action	Hold time	Ramp	Total
	°C		s	°/s	s
1	95	Block pre heat	120	4	140
2	95	Inital denaturation	180	None	180
3	95	Denaturation	30	4	36
4	62	Annealing	60	1	93
5	72	Elongation	60	4	63
		Repeat steps 2-4			372

Quantitation during amplification

The term "real time" is somewhat misleading as the measurement is not continuous but discontinuously repeated per cycle. The time of measurement depends on the function of the activity which generates radiation or quenches radiation. Usually, the samples are in standardized containers in the heating block. Detection must be possible through the polymer container. The detection itself is comparatively simple; the actions behind the light event are complex.

A straightforward method uses dyes which are binding specifically to DNA via intercalation. The fluorochrome is activated by light and releases the absorbed energy with characteristic fluorescence radiation. With increasing amount of target DNA, the intensity of the fluorescent dye is increasing as well. The fluorescent radiation is determined at the end of the elongation step (see above) cycle per cycle. The dye reacts with different PCR products and thus the specificity of the method is low. Multiplex measurements are not directly possible this way. The specificity can be increased, if a so-called melting curve analysis is run at the end of the amplification process. The temperature of the preparation is slowly increased from 60 to 95 ° C and the change in fluorescence intensity is monitored. Different DNA fragments have different melting temperatures. Reciprocal to the increase during amplification the fluorescence is decreasing with each denatured DNA fragment. Usually, the derivative of the curve is used to identify or distinguish the fragments from the target DNA. A quantitative determination of the peaks is hardly possible, however.

More specific and chemically more complex are methods which are based on energy transfer between fluorescence activated molecules and molecules ready to accept energy. This so-called Förster resonance energy transfer (FRET) is important for many biological processes, among them photosynthesis of plants. The FRET process is involved in the OLED technique as an example. In RT-PCR the effect is used to activate or disactivate fluorescence in molecules which are specific for the target

DNA. These "reporter" molecules are binding to a nucleotide involved in the PCR process. The enhancement of light or the quenching of light is used to quantitate the PCR processes. Multiplexing is generally feasible. Required are two molecules which are transferring energy if their distance is very small. The energy transfer is a near field effect. As soon as the distance becomes bigger, due to a progress in the PCR process, the energy transfer drops with the 6th power of the distance. The electromagnetic process at the core of FRET is similar to mechanisms in a modern cash card which couples to the receiving device only if the distance is very close to zero.

A FRET donor molecule and a FRET acceptor molecule, for example, are binding to the target DNA fragment such that the distance between the molecules is minimized. Energy transfer between the molecules becomes possible only under the influence of the generated DNA. The stimulated fluorescence of the donor and/or the acceptor can be quantitated at the end of the annealing phase. Like above, the method can be used as well to generate a melting curve under optical control.

Another example of the wide variety of optical probes is the Taqman probe. A nucleotide which is complementary to the target DNA contains a fluorescent molecule and a quenching molecule. The distance is short so that fluorescence is quenched. In the absence of the target DNA fluorescence is minimal. With increasing amount of target DNA more and more probe molecules encounter the polymerase and separate the fluorescent end from the quenching end of the nucleotide. Fluorescing radiation is generated which is proportional to the amount of target DNA produced. Determination is preformed at the end of the elongation cycle.

Illumination and detection

A typical PCR container holds 92 wells for plates or single cups. They must be equipped with light transparent caps and ought not generate any auto fluorescence. For real-time PCR all wells must be illuminated and read out. This may be realized with fiber-optic light guides which transport the light, required for activation, to the wells, and the fluorescent light back to the detector.

The activating light must be intense. Laser light or radiation of a flashlight source is focused on a so-called multiplexer unit, which distributes the light via the fibers to the wells. The same multiplexer collects the light from the fibers again and sends them to the detector. Each well is individually addressed.

The wavelength used for activation depends on the fluorescing molecule. In PCR the activating energy is close to 500 nm. The frequently used Ar laser emits, e.g., at 488 nm. White light sources are filtered respectively. The fluorescent light has lower energy/longer wavelength. The evaluated wavelength is between 500 and 650 nm. Filters or a dichroic mirror separate the activating light from the evaluated light. The latter is guided via a filter photometer or a simple monochromator to a detector, which may be based on APS or PMT technology.

All technical components which illuminate and read out the wells during or after the amplification process are not different from the techniques discussed earlier. One difference, however, is the separation of the illuminating light into several (e.g., 96) channels, and the selective read out of several channels with one detector. The device for temporal separation of the wells is an optical multiplexer (muxer) which separates the light to and from the well spatially and temporarily. The multiplexer technology is well known to separate parallel electromagnetic signals into time separated information. It is quasi a fast switch connecting input channels to output channels in both directions (mux/demux). Selection of the channels in data transmission is mostly wavelength selective; i.e., different wavelengths are separated in an electronically controlled multiplexing process. Multiplexing in plate reading is limited to a single wavelength for illumination and a single (different) wavelength for read out. The individual channel is selected for a limited amount of time; the other channels are blocked out. Read out becomes fast sequential. The illumination is brief compared to channel selection by the optics. The illumination takes usually less than 100 ms. An example of a real-time PCR optical system based on a tungsten halogen lamp as the photon source is sketched in Figure 3.10.

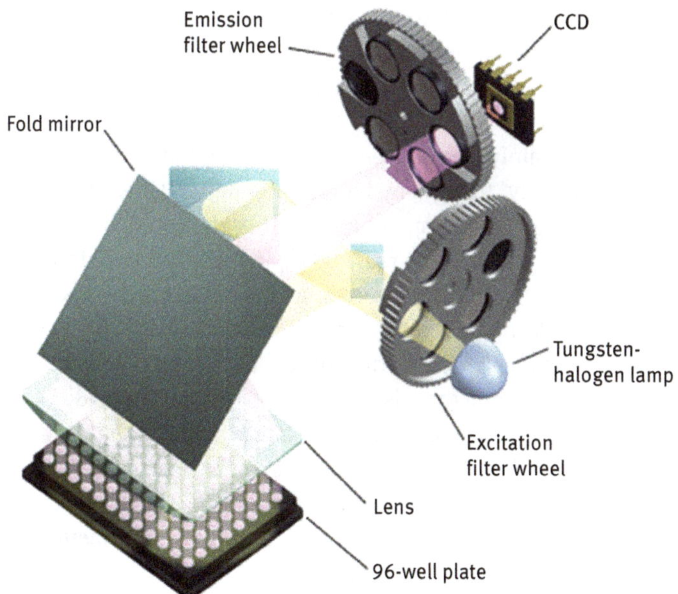

Figure 3.10: iQ™5 real-time PCR detection system optics. Courtesy of Biorad Hercules Claifornia, USA.

Read out and data evaluation of real-time PCR systems are usually less time critical. Read out is an integrated process of the amplification cycle which is the time limiting factor. If the amplification process is separated from quantitation, time of read out may be a quality criterion for sample throughput. Illumination and read out depend on the technology applied. Lasers usually allows higher intensity and better signal to noise than flash lamps. CCD read out is usually faster than PMT. Typical read out time for a 96 well plate may range from 1 to 10 min, depending on the technology applied and on the required FOM.

Treatment of data

The data of the individual channels are collected and treated by suitable algorithms. The fluorescence intensity is changing as result of the chemical reaction taking place during the generation of the target DNA. This change in intensity can, where appropriate, be compared in situ with a passive reference fluorescence dye. Data evaluation takes place via a quotient of measured fluorescence relative to the reference. Additionally, the background emission of the first few amplification cycles can be subtracted from the analyte intensity.

After several cycles, the analyte fluorescence becomes larger than the baseline. This value must take standard deviation of blank and baseline into consideration. It is named "threshold cycle" C_T. The threshold cycle relates to the number of copies at the starting point of the amplification. Suitable algorithms will indicate the number of starting copies. As the underlying process of amplification is a doubling of target DNA per cycle (2^n!) the C_T value becomes smaller by one if the number of copies in the beginning is doubled.

A PCR plate read contains mainly qualitative and semi-quantitative information. The data stem, quasi simultaneously, from many individual samples. Often the speed of availability of the information is of high importance, e.g., results of a Covid-19 test, including the different variations of the virus must be rapidly available. Data treatment, visualization of results, and reporting require sophisticated algorithms, intuitive operation of the program and capturing of the results and/or trends. Software is often the most important quality criterion between instruments of different manufacturers. A software package of one major manufacturer is explained exemplary in [148]. Most straightforward is the display of the fluorescence intensity as a function of the numbers of amplification cycles. An example is features in Figure 3.11.

The result may be displayed in corrected or non-corrected form, using logarithmic or linear intensity axes. Provided the reader is equipped with a spectrometer, the available intensity can be displayed as a function of the wavelength as well.

The direct comparison of the spectrally resolved fluorescence allows validation of the process. Differences between the relative spectral intensity of sample and standard indicate inconsistencies during sample preparation. Algorithms allow to calculate a best fit of the curves.

Chemical information from PCR analysis is very comprehensive. This information is obtained from the targeted preparation of the samples. Read out of the spectra during or after amplification provides the analytical answer. As always, calibration with

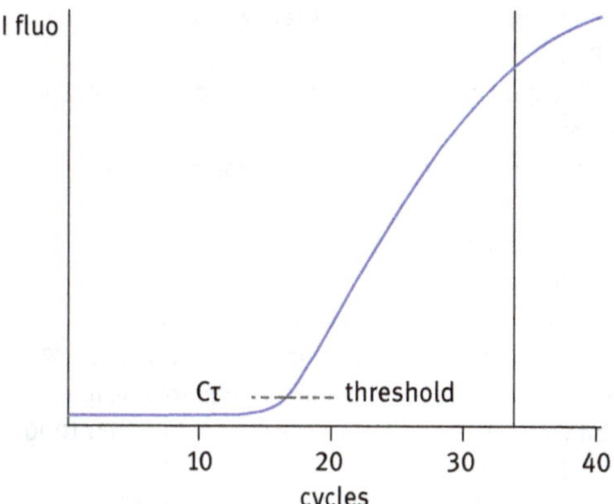

Figure 3.11: Fluorescence versus numbers of cycles for one selected well. The fluorescence intensity is below detection limit until it reaches the threshold value. From there it increases rapidly until it flattens at the end of the exponential phase (black vertical line). Beyond this line, the non-exponential plateau phase begins. Amplification is no longer by a fixed factor in each step.

suitable reference samples is required to evaluate and validate the results. The measured value (intensity, wavelength, and number of amplification cycles) is simple. The chemistry within the samples is highly complex. Numerous software features allow to infer subtle differences between the samples. Their description would exceed the scope of this paragraph. RT-PCR is an example of an analytical method where the spectroscopic process is just a resource to visualize the chemical processes taking place in the sample. The analytical information is caused by sophisticated biochemical reactions within the sample.

> *May, in spite of all distractions generated by technology, all of you succeed in turning information into knowledge, knowledge into understanding, and understanding into wisdom*
>
> *(Edsger W. Dijkstra, 1998).*

Literature

[125] Flores EMDM. (Ed) Microwave-Assisted Sample Preparation for Trace Element Determination, Elsevier B.V., Amsterdam, The Netherlands 2014, ISBN 978-0-444-59420-4.

[126] Hippler M, Bestimmung von Chrom und seinen Spezies im Boden, Dissertation, Dortmund University.

[127] van Deemter JJ, Zuiderweg FJ, Klinkenberg A. Longitudinal diffusion and resistance to mass transfer as causes of non-ideality in chromatography, Chem. Eng. Sci. 1956, 5, 271–289.

[128] Menzel O (2001) Charakterisierung makroporöser Materialien mit den Methoden der digitalen Bildverarbeitung und Bestimmung von effektiven Diffusionskoeffizienten durch Computersimulationen. Dissertation, Hannover university, 2001.

[129] Bard AJ, Inzelt G, Scholz F. Electrochemical Dictionary, Springer Science & Business Media, Luxemburg, 2008, ISBN 978-3-540-74598-3, 167.
[130] Horvath C, Melander W, Molnar I, Molnar P. Liquid chromatography of ionogenic substances with nonpolar stationary phases, Anal. Chem. 1977, 49, 2295.
[131] Horvath C, Melander W, Molnar I. Solvophobic interactions in lliquid chromatography with nonpolar stationary phases, J. Chrom. 1976, 125, 129.
[132] Kraak JC, Jonker KM, Huber JFK. Solvent-generated ion-exchange systems with anionic surfactants for rapid separations of amino acids, J. Chrom. 1977, 142, 671.
[133] Hoffmann NE, Liao JC. Reversed phase high performance liquid chromatographic separations of nucleotides in the presence of solvophobic ions, Anal. Chem. 1977, 49, 2231.
[134] Haddad P, Jackson Peter SM. Developments in suppressor technology for inorganic ion analysis by ion chromatography using conductivity detection, J. Chromatogr. A 2003, 1000, 725–742.
[135] Monaghan JM, Cook K, Gara D, Crowther D. Determination of nitrite and nitrate in human serum, J. Chrom. A 1997, 770, 143.
[136] Stratford MRL, Daennis MF, Cochrane R, Parkins CS, Everett SA. Measurement of nitrite and nitrate by high-performance ion chromatography, J. Chrom. A 1997, 770, 151.
[137] Roth M, Hampai H. Column chromatography of amino acids with fluoresence detection, J. Chrom. 1973, 83, 353.
[138] ISO 10304-1, International Organization for Standardization, Geneva, Switzerland, (2007) Water quality – Determination of dissolved anions by liquid chromatography of ions – Part 1: Determination of bromide, chloride, fluoride, nitrate, nitrite, phosphate and sulfate.
[139] Jensen D. Grundlagen der Ionenchromatographie; Modernste Trenntechnik, Thermo Fisher Scientific, ISBN 978-3-00-044477-7, 2013.
[140] Wuilloud JCA, Wuilloud RG, Vonderheide AP, Caruso JA. Gas chromatography/plasma spectrometry – An important analytical tool for elemental speciation studies, Spectrochim. Acta B 2004, 59, 755–792.
[141] Agilent Technologies, Inc.: Handbook of Hyphenated ICP-MS Applications, 2007, 2012, 2015.
[142] Brandes W, Wendelin S, Eder R. Bestimmung charakteristischer Aromastoffe in Barriqueweinen mittels HPLC und GC-MS, Mitteilungen Klosterneuburg 2002, 52, 97–105.
[143] Contado C. Field flow fractionation to explore the nano world, Anal. Bioanall. Chem. 2017, 409, 2501–2518.
[144] Rabel F, Sherma J. Review of the state of the art of preparative thin layer chromatography. J. Liq. Chromatogr. Relat. Technol. 2017, 40(4), 165–176, 10.1080/10826076.2017.1294081.
[145] Clegg DL. Paper chromatography. Anal. Chem. 1950, 22(1), 48–59.
[146] Sanger F, Nicklen S, Coulson AR. DNA sequencing with chain-terminating inhibitors. Proc. Natl. Acad. Sci. U.S.A. 1977, 74(12), 5463–5467, 10.1073/pnas.74.12.5463.
[147] Zhu H, et al.. PCR past, present and future. Biotechniques 2020, 69(4) 317–325.
[148] Schild TA. Einführung in die Real-Time TaqMan PCR-Technologie; Applied Biosystems GmbH. Weiterstadt. Rev. 2, 1

Index

https://doi.org/10.1515/9783110689662-004

www.ingramcontent.com/pod-product-compliance
Lightning Source LLC
Chambersburg PA
CBHW061339210326
41598CB00035B/5825

9783110689648